AF358480

Biomonitoring and Conservation of Freshwater & Marine Fishes

Biomonitoring and Conservation of Freshwater & Marine Fishes

Guest Editors

Robert L. Vadas, Jr.
Robert M. Hughes

Basel • Beijing • Wuhan • Barcelona • Belgrade • Novi Sad • Cluj • Manchester

Guest Editors

Robert L. Vadas, Jr.
Independent Researcher
Olympia
USA

Robert M. Hughes
Department of Fisheries, Wildlife,
and Conservation Sciences
Oregon State University
Corvallis
USA

Editorial Office
MDPI AG
Grosspeteranlage 5
4052 Basel, Switzerland

This is a reprint of the Special Issue, published open access by the journal *Fishes* (ISSN 2410-3888), freely accessible at: https://www.mdpi.com/journal/fishes/special_issues/1N7J6D5S01.

For citation purposes, cite each article independently as indicated on the article page online and as indicated below:

Lastname, A.A.; Lastname, B.B. Article Title. *Journal Name* **Year**, *Volume Number*, Page Range.

ISBN 978-3-7258-3293-4 (Hbk)
ISBN 978-3-7258-3294-1 (PDF)
https://doi.org/10.3390/books978-3-7258-3294-1

Contents

About the Editors

Robert L. Vadas, Jr.

Dr. Vadas has over 40 years of experience in sampling and analyzing data for watershed (aquatic/riparian) biotic assemblages in North America, with additional fin- and shellfish publications with collaborators worldwide. He worked for two consulting companies in Alberta and completed two postdoctoral projects on the Pacific Coast for fish-habitat and other issues, with a collective focus on riparian and instream-flow impacts to freshwater and estuarine ecosystems. He has since worked as a state-agency biologist, including past estuarine fish-habitat work with the Florida Marine Research Institute and present fresh/saltwater (especially salmonid) work with the Washington Department of Fish and Wildlife's Habitat Program. His focus has been on the hydrologic, habitat, and dietary needs of fishes, using field sampling, statistical analysis, and ecological synthesis. He is a Research Scientist for riparian/wetland, instream flow, dam, fish passage, overwater structure, and marine hydrokinetic topics. Over half of his field experience is with western North America, the rest being with the U.S. Eastern Seaboard. This includes sampling and modeling fish habitat across spatial scales, but also research at other trophic levels: wildlife, micro- and macroinvertebrates, trees, and phytoplankton. Bob has a strong background in biological and physicochemical sciences and statistics (including teaching of all 3 topics), with a strong publication record that includes theoretical–ecology and science–philosophy concerns. He has led three long-term, collaborative projects for fish biology and habitat research for his latest job. He co-authored (with Dr. Hughes) the Runner-up of the Best Paper Award 2023 in 'Water Biology and Security' for aquatic multimetric indices, which spawned their present guest-editorial roles for 'Fishes'. Vadas received the A.B. Massey (Outstanding Graduate Student in the Department of Fisheries and Wildlife Sciences) Award (Virginia Tech, 1992), Coelacanth Award for science/management excellence from the WA-BC Chapter (2023) of the American Fisheries Society (AFS); and Western Division (WD) President's Award of Special Recognition on Snake River issues (along with Dr. Hughes, 2023), while both have served on the Resource Policy and Environmental Concerns Committee of WDAFS.

Robert M. Hughes

Dr. Hughes is a Senior Scientist with the Amnis Opes Institute and a Courtesy Associate Professor at Oregon State University. Hughes has over 45 years of experience in sampling and analyzing data for fish and macroinvertebrate assemblages in Asia, Europe, North America, and South America. He has used his expertise to develop and evaluate indicators for the USEPA's Environmental Monitoring and Assessment Program (EMAP, now its National Aquatic Resource Survey), to evaluate ecoregions, and to generate biological criteria. He also has extensive experience in sampling and analyzing sedimentary diatom, zooplankton, and periphyton assemblage data. His experience includes sampling small streams and ponds, as well as the Laurentian Great Lakes and large navigable rivers. Hughes was a key member of the research team that developed and field-tested the ecoregion concept that led to the map of the ecoregions of the USA. He co-chaired the National Workshop on Instream Biological Monitoring and Criteria in 1987, co-authored EPA's Rapid Bioassessment Protocols for Fish and Benthic Macroinvertebrates in 1989 and its EMAP Field Protocols for Fish Assemblages and Fish Tissue in 2006, and provided technical expertise to the EPA's Steering Committee on Biological Criteria 1988-1990. Hughes has edited six books, authored over 290 peer-reviewed publications, and has been a guest speaker 85 times in Asia, Australia, Europe, and South America. He was the 2013–2014 President of the American Fisheries Society (AFS). He received 10 EPA awards for best scientific paper or technical contribution and co-authored the best papers in Transactions of the American Fisheries Society in 2008 and Lake and Reservoir Management in 2014. Hughes received the 2006 Environmental Stewardship Award from the North American Benthological Society, 2011 Fisheries Worker of the Year Award, and 2017 Lifetime Achievement Award from the AFS Oregon Chapter, 2013 AFS Distinguished Service Award, 3 Fulbright Scholarships (2007, 2010, 2022-23), and the 2022 AFS Award of Excellence. Dr. Hughes is a Fellow of the AFS and the Society for Freshwater Science.

Preface

Globally, native migratory and resident fishes are declining because of aquatic and terrestrial ecosystem degradation caused by physicochemical habitat alteration, migration barriers, overexploitation, excessive hatchery supplementation, non-native species introductions, and the climate crisis—all driven by human overpopulation and excessive energy and material consumption. Loss of diadromous fishes reduces marine-derived nutrients that are important for freshwater and floodplain biota, including riparian trees that protect freshwater ecosystems from land use. The depletion of marine and freshwater fisheries threatens natural-resource industries, human food supplies, and ecosystem processes. Healthy aquatic ecosystems have diverse habitats that house a diversity of fish species, including various trophic, habitat, reproductive, and life-history guilds. To better protect fish resources, which provide recreation and sustenance for millions of people, rigorous monitoring is important for assessing fish assemblage and population health and their limiting factors. Therefore, this Special Issue focuses on ecological analyses based on large sample sizes over relatively large areas.

Robert L. Vadas, Jr. and Robert M. Hughes
Guest Editors

 *fishes*

Editorial

Monitoring and Conservation of Freshwater and Marine Fishes: Synopsis

Robert L. Vadas, Jr. [1,*] and Robert M. Hughes [2,3,*]

1 Independent Researcher, 2909 Boulevard Rd. SE, Olympia, WA 98501, USA
2 Amnis Opes Institute, 2895 SE Glenn, Corvallis, OR 97333, USA
3 Department of Fisheries, Wildlife, and Conservation Sciences, Oregon State University, Box 104 Nash Hall, Corvallis, OR 97331, USA
* Correspondence: bobesan@comcast.net (R.L.V.J.); hughes.bob@amnisopes.com (R.M.H.)

1. Introduction

Globally, native migratory and resident fishes are declining from aquatic and terrestrial ecosystem degradation resulting from physicochemical habitat alteration, migration barriers, over-exploitation, hatchery supplementation, non-native species introductions, and the climate crisis [1]—all driven by human overpopulation and excessive energy and materials consumption [2]. Loss of diadromous fishes reduces marine-derived nutrients that are important for freshwater and floodplain biota [1,3], including riparian trees that protect freshwater ecosystems from land use [4]. The depletion of marine and freshwater fishes threaten natural-resource industries, human food supplies, and ecosystem processes [5,6]. Healthy aquatic ecosystems have diverse habitats that house a diversity of fish species, including various trophic, habitat, reproductive, and life-history guilds [1]. To better protect fish resources, which provide recreation and sustenance for millions of people, rigorous monitoring is important for assessing fish assemblage and population health and their limiting factors [1,7]. Therefore, this Special Issue focuses on ecological analyses based on large sample sizes over relatively large areas https://www.mdpi.com/journal/fishes/special_issues/1N7J6D5S01 (accessed on 15 July 2024).

Recently, we reviewed the use of multimetric indices (MMIs) for assessing the ecosystem condition of aquatic and riparian ecosystems [1,8]. The former paper was stimulated by prior research indicating natural, longitudinal shifts in food webs and positive relationships between sample size and fish species richness. We concluded that insufficient and inconsistent sampling confounded anthropogenic impact analyses when too few fish are collected at sites or too few sites are sampled [1]. We were also concerned that MMIs are subject to ad hoc modifications of metrics, thus requiring calibration across regions [8–11]. Those calibrations highlight the need for more general MMI metrics [1], which we hope to better achieve with this Special Issue. Others have indicated the critical importance of rigor in determining reference conditions for making biological-impact assessments [8,10]. Again, several papers in this Special Issue lend credence to those concerns.

We volunteered to edit this Special Issue because of our concern for better mechanistic understanding of fish-environmental relationships globally to improve fish-assemblage monitoring. So, we encouraged submitting authors to examine aquatic conservation at multiple biotic and spatiotemporal extents for more-effective management and monitoring [12,13]. Notably, fishes suffer cumulative impacts that often complicate their recovery efforts [12–14], especially in the face of the climate crisis [5,6,15–17] and non-native fish invasions [1,4,16–19].

Citation: Vadas, R.L., Jr.; Hughes, R.M. Monitoring and Conservation of Freshwater and Marine Fishes: Synopsis. *Fishes* **2024**, *9*, 470. https://doi.org/10.3390/fishes9120470

Received: 13 October 2024
Accepted: 11 November 2024
Published: 21 November 2024

2. Synopsis for Special Issue

2.1. Assessments of Drought or the Climate Crisis

Several papers dealt with drought or the climate crisis, which are increasing problems for stream fishes globally. Pompeu et al. [20] demonstrated the importance of ichthyoplankton surveys for identifying critical habitats for riverine fishes in the face of dams and other impacts. This Brazilian study was undertaken during a severe drought year and indicated the essential need to use eDNA and frequent sampling to accurately determine spawning rivers and periodicities [20]. We expect that eDNA will become increasingly important for monitoring fish assemblages [21–23], especially for those species that are rarely encountered with traditional techniques [1,20].

Hamilton et al. [24] was a microbiologic study. Although fish diseases have often been incorporated into MMIs to assess anthropogenic impacts [1], these relationships need better development, especially considering that disease spread is exacerbated by non-native fishes [25,26] and the climate crisis [27]. Hamilton et al.'s [24] salmonid-oriented paper, which admirably used Inuit First Nation help for the research, suggests that lake whitefish may be maladapted in northern Canada to deal with the climate crisis. Such climate-crisis disease sensitivity needs further elucidation in both temperate and tropical regions [27], as do the synergistic effects of eutrophication and the climate crisis [1,28].

Robinson et al. [26] examined a water-limited, agricultural river basin housing both native and non-native fish species. The natives often showed reduced ranges more recently but not obvious spatially extensive presence/absence trends over the last two decades, partly because of three different lifespan classes (<3 y, 3–6 y, >6 y) that require different study durations. Likewise, 12 years of data were needed to reveal a statistically significant ecohydrological trend in a Washington, USA, stream [7]. Power analyses to reduce Type-II statistical errors of not finding real impacts are critical for aquatic bioassessments in the face of continuing development pressures [1,5,29]. Robinson et al. [26] also addressed the concept of 'shifting baselines' [7] that plague biomonitoring and power analysis [1] to enhance the general usefulness of their paper for climate crisis research. Australian fish assemblages are adapted to "boom vs. bust" climatic fluctuations via rapid dispersal thereafter, but climate-mitigation management is nevertheless needed to prevent fish kills. This should include both fish-passage and smaller-extent rehabilitation efforts [26]. Wetland-specific sampling is also needed given wetland losses that affect fishes [1,3,30]. Robinson et al. [26] should prove to be an important paper for climate crisis, wetland, and instream-flow fields.

Bergström et al. [31] examined climate-based behavioral evolution. They found that a Swedish population of Wels catfish (*Silurus glanis*) in a mesotrophic lake showed different adult-foraging behavior than this catfish did farther south in Europe, based on a mark-recapture study and comparative literature. For the Swedish lake, summer/fall activity included nocturnal, pelagic feeding, but settling near the bottom for diurnal resting. This contradicts the general tenet that zooplankton and forage fishes are less vulnerable to nocturnal predation in lake epilimnia, although lower trophic levels went unstudied there. In not preferring warmer waters, including their display of late-winter activity under ice, this catfish's behavior differed from that of southern populations, which are dormant at such lower temperatures; prefer shallow, vegetated, sheltered bays; and differ in genetics, growth rates, and longevity. Moreover, habitat connectivity was important for successful lake feeding and creek reproduction (in late spring) in such northern, migrating populations, which have been long-isolated from their southern conspecifics [31]. Lake warming also was reported to decrease the length structure of northern pike (*Esox lucius*) in Lake Windermere (UK [32]). The short-duration, northern summers likely promoted such evolutionarily divergent seasonal behavior.

2.2. Assemblage Assessments Dealing with Non-Native or Stocked Species

In their spatially extensive mapping analysis to better achieve migratory-fish management, Kajee et al. [33] located native, endemic, threatened, and non-native species hotspots in South Africa in an important gap analysis study. Notably, non-native vs. threatened species records overlapped in over 50% of the area, as nearly half the threatened species records were outside protected areas and non-natives occurred in over a third of the protected areas [33]. Jelks et al. [34] also reported that habitat degradation and non-native fish were the major threats of at-risk North American fish species. Stream barriers were useful for excluding non-natives from the spawning habitats of native, migratory fishes in larger basins, which was also reviewed globally by Jones et al. [35].

Aparicio et al. [11] considered both native and non-native fishes in Spanish streams. Although non-native fishes typically degrade ecosystem processes and natural biodiversity [1,16], non-native fishes are often excluded from MMIs (but see [10,36]). Aparicio et al. [11] nicely included a separate metric for non-native pressure on native species and found that explicitly including non-native fish pressure provided a more comprehensive assessment of ecosystem health than did the European MMI without that metric.

Faro et al. [37] also addressed native and non-native fishes by holistically examining land use (especially agriculture), eutrophication, and hydromorphologic (e.g., dam-hydropeaking) criteria to classify Portuguese sites into four levels of human impacts. They emphasized the biophysical importance of intact riparian areas as native-fish habitats. Perhaps the most interesting is that they relied on just four fish-assemblage metrics (as percentages) in their MMI: native lithophils (for spawning), non-natives, migrants (via diadromy/potamodromy), and freshwater natives. The three metrics besides non-native fishes were associated with less-disturbed conditions, whereas non-natives were associated with more-stable flow regimes than native species, which preferred naturally varying flows [37]. Ruaro et al. [18] reported that fish MMI scores declined with increased abundance of non-native species in two Brazilian river basins. Faro et al. [37] found only partial support of better biotic conditions away from hydropeaking impacts; however, based on a large, Europe-wide database, Schinegger et al. [38] found that fishes were intolerant of hydrological stressors alone and when combined with morphological stressors. Dams are well-known for blocking migratory-fish access to upstream, downstream, and floodplain habitats, which also get degraded for landlocked fishes and their foods [1,3–5].

Wildhaber et al. [39] addressed dam impacts via a long-term database. They undertook a fish-habitat analysis for sicklefin chub (*Macrhybopsis meeki*) and sturgeon chub (*M. gelida*) and their piscine predators in the mainstem Missouri River, where hydropower production and channelization prevail. Some of the piscivores both predated on and competed with the chubs. The chubs were best caught by benthic trawling, but many netting and other methods were used for more-complete sampling intra- and interspecifically. The two cyprinids were subjected to habitat-occupancy modeling, including their use of both main- and off-channel habitats, and showed marked spawning-flow relationships, suggesting flow-regime naturalization as a needed management tool [39], which was also recommended elsewhere [40,41].

2.3. Assemblage Assessments with Macrohabitat Considerations

Monahan et al. [42] assessed 23 hand-picked, wadeable stream sites across the USA by electrofishing. The highest fish-assemblage alpha and beta diversities were found in warmer, lowland rivers in Atlantic basins, where fish body sizes tended to be smaller. This paper highlighted the species-depauperate nature of USA Pacific basins, where colder headwaters favored larger-bodied salmonids [42]. Based on a 2554-site database from a probability survey, Hughes et al. [29] reported the same alpha and beta diversity patterns.

Heppell et al. [43] examined estuarine fin- and shellfishes in an Oregon estuary, focusing on abundance (CPUE) and biodiversity parameters in trawl samples over 3.5 decades. They found a shift from (i) English sole (*Parophrys vetulus*) and other demersal fishes to (ii) Dungeness crab (*Cancer magister*) and other epibenthic crustaceans. Sculpins (Cottidae) had also become more prevalent. Hence, there has been a shift away from pelagic fishes [43], as has been noted for other altered estuaries [44,45]. Although Heppell et al. [43] did not examine causal mechanisms, it is likely that both estuarine development (e.g., shoreline armoring, channel dredging, and a public marina) and the climate crisis were responsible.

2.4. Eurasian-Minnow Genetics, Hybridization, and Speciation

Valić et al. [46] performed genetic analyses on Illyrian chub (*Squalius illyricus*) and Zrmanja chub (*Squalius zrmanjae*) from the Krka River in Croatia, comparing them with sequences in GenBank. They found that *S. zrmanjae* had a nuclear region resembling Dalmatian rudd (*Scardinius dergle*), suggesting the transfer of genetic information across genera [46]. Notably, fish hybridization has occasionally been incorporated into MMIs [1] because habitat damage, pollution, and the climate crisis may limit interspecific-niche separation. This is of particular concern for rare, threatened, or endangered fish species [47,48].

Laskar et al. [49] examined morphology, genetics, and ranges of *Osteobrama vigorsii* and *O. tikarpadaensis*. Their paper helped resolve a long-term quandary regarding unusual distributions of *Osteobrama* species in India that should improve fish conservation efforts. Such an integrated approach with morphologic and molecular data should enhance the robustness of species assessments, with usefulness for fish conservation beyond India. Further consideration of life-history divergences [31,50,51] could help define evolutionarily significant units or distinct population segments for fish species.

3. Conclusions

This Special Issue collectively addressed biotic scales from (i) salmonid skin microbiomes to (ii) cyprinid genetics and ecology to (iii) assessment of ichthyoplankton and older fishes across freshwater and estuarine habitats. The focus was typically guild- or assemblage-oriented to better assess anthropogenic impacts, but it also included a single-species study [31] that examined geographic variation in catfish ecology in the face of climate crisis pressures. The genetic, microbiologic, habitat, trophic, and hydrologic ecology of fishes that were discussed should help us to better assess anthropogenic impacts in other contexts [1]. We hope this Special Issue provides a springboard for other aquatic ecologists to formulate more holistic, ecosystem-health assessments—especially by spatiotemporal planning—to minimize impacts to rare and migratory species.

Increasingly, ecologists must consider applied ecology, which is why most papers in our issue had a Conclusions section with management recommendations. Scientists can no longer shy away from environmental advocacy in a rapidly changing world, which requires us to make scientifically backed diagnoses [52] like what medical doctors must do to protect people and public health [2,5]. Hence, long-term biomonitoring with a stronger focus on aquatic biodiversity protection across biotic and spatiotemporal scales is needed [1,3,53]. That biomonitoring is required for rehabilitation projects to (i) verify that such efforts succeed [54,55], (ii) document ecosystem dynamics [56,57], and (iii) improve MMI and other impact-assessment analyses [1,14,58]. Clearly, true adaptive management is needed, which presently receives more lip service than effective implementation [2,5,6,55].

Author Contributions: Both authors conceived this article and defined the Special Issue's scope. R.L.V.J. acted as main editor of all submitted articles and provided the first draft of this summary article, which both authors further revised. R.M.H. solicited several articles that he also helped edit, besides determining which of them qualified for page-charge reductions, given our global emphasis. All authors have read and agreed to the published version of the manuscript.

Acknowledgments: We thank the Special Issue authors for their patience with the careful revisions requested by the peer reviewers and guest editors. The regular Fishes editors thankfully helped put this summary into proper format.

Conflicts of Interest: The authors (including R.L.V.J. as an independent researcher) declare no conflict of interest.

References

1. Vadas, R.L., Jr.; Hughes, R.M.; Bae, Y.J.; Baek, M.J.; Gonzáles, O.C.; Callisto, M.; de Carvalho, D.R.; Chen, K.; Ferreira, M.T.; Fierro, P.; et al. Assemblage-based biomonitoring of aquatic ecosystem health via multimetric indices: A critical review and suggestions for improving their applicability. *Water Biol. Secur.* **2022**, *1*, 100054.
2. Hughes, R.M.; Vadas, R.L., Jr.; Michael, J.H., Jr.; Knutson, A.C., Jr.; DellaSala, D.A.; Burroughs, J.; Beecher, H. Why advocate—And how? In *Conservation Science and Advocacy for a Planet in Crisis: Speaking Truth to Power*; DellaSala, D., Ed.; Elsevier: Cambridge, MA, USA, 2021; pp. 177–197.
3. Sulliván, S.M.P.; Hughes, R.M.; Vadas, R.L., Jr.; Davies, G.T.; Shirey, P.D.; Colvin, S.A.R.; Infante, D.M.; Danehy, R.J.; Sanchez, N.K.; Keast, R.B. Waterbody connectivity: Linking science and policy for improved waterbody protection. *BioScience* **2024**, *in press*.
4. Hauer, F.R.; Locke, H.; Dreitz, V.J.; Hebblewhite, M.; Lowe, W.H.; Muhlfeld, C.C.; Nelson, C.R.; Proctor, M.F.; Rood, S.B. Gravel-bed river floodplains are the ecological nexus of glaciated mountain landscapes. *Sci. Adv.* **2016**, *2*, e1600026. [CrossRef] [PubMed]
5. Hughes, R.M.; Karr, J.R.; Vadas, R.L.; DellaSala, D.A.; Callisto, M.; Feio, M.J.; Ferreira, T.; Kleynhans, N.; Ruaro, R.; Yoder, C.O.; et al. Global concerns related to water biology and security: The need for language and policies that safeguard living resources versus those that dilute scientific knowledge. *Water Biol. Secur.* **2023**, *2*, 100191. [CrossRef]
6. Hughes, R.M.; Chambers, D.M.; DellaSala, D.A.; Karr, J.R.; Lubetkin, S.C.; O'Neal, S.; Vadas, R.L., Jr.; Woody, C.A. Environmental impact assessments should include rigorous scientific peer review. *Water Biol. Secur.* **2024**, *3*, 100269. [CrossRef]
7. Vadas, R.L., Jr.; Beecher, H.A.; Boessow, S.N.; Kohr, J.H. Coastal Cutthroat Trout redd counts impacted by natural water supply variations. *North Am. J. Fish. Manag.* **2016**, *36*, 900–912. [CrossRef]
8. Ruaro, R.; Gubiani, É.A.; Padial, A.A.; Karr, J.R.; Hughes, R.M.; Mormul, R.P. Responses of multimetric indices to disturbance are affected by index construction features. *Environ. Rev.* **2024**, *32*, 278–293. [CrossRef]
9. Resh, V.H.; McElravy, E.P. Contemporary quantitative approaches to biomonitoring using benthic macroinvertebrates. In *Freshwater Biomonitoring and Benthic Macroinvertebrates*; Rosenberg, D.M., Resh, V.H., Eds.; Chapman & Hall: New York, NY, USA, 1993; pp. 159–194.
10. Whittier, T.R.; Hughes, R.M.; Stoddard, J.L.; Lomnicky, G.A.; Peck, D.V.; Herlihy, A.T. A structured approach to developing indices of biotic integrity: Three examples from western USA streams and rivers. *Trans. Am. Fish. Soc.* **2007**, *136*, 718–735. [CrossRef]
11. Aparicio, E.; Alcaraz, C.; Rocaspana, R.; Pou-Rovira, Q.; García-Berthou, E. Adaptation of the European Fish Index (EFI+) to include the alien fish pressure. *Fishes* **2024**, *9*, 13. [CrossRef]
12. Allen, A.P.; Whittier, T.R.; Larsen, D.P.; Kaufmann, P.R.; O'Connor, R.J.; Hughes, R.M.; Stemberger, R.S.; Dixit, S.S.; Brinkhurst, R.O.; Herlihy, A.T.; et al. Concordance of taxonomic composition patterns across multiple lake assemblages: Effects of scale, body size, and land use. *Can. J. Fish. Aquat. Sci.* **1999**, *56*, 2029–2040. [CrossRef]
13. Feio, M.J.; Hughes, R.M.; Serra, S.R.; Nichols, S.J.; Kefford, B.J.; Lintermans, M.; Robinson, W.; Odume, O.N.; Callisto, M.; Macedo, D.R.; et al. Fish and macroinvertebrate assemblages reveal extensive degradation of the world's rivers. *Glob. Change Biol.* **2023**, *29*, 355–374. [CrossRef] [PubMed]
14. MacDonald, L.H. Evaluating and managing cumulative impacts: Processes and constraints. *Environ. Manag.* **2000**, *26*, 299–315. [CrossRef] [PubMed]
15. Pörtner, H.O.; Peck, M.A. Climate change effects of fishes and fisheries: Towards a cause-and-effect understanding. *J. Fish Biol.* **2010**, *77*, 1745–1779. [CrossRef] [PubMed]
16. Christensen, N.L.; Bartuska, A.M.; Brown, J.H.; Carpenter, S.; d'Antonio, C.; Francis, R.; Franklin, J.F.; MacMahon, J.A.; Noss, R.F.; Parsons, D.J.; et al. The report of the Ecological Society of America Committee on the Scientific Basis for Ecosystem Management. *Ecol. Appl.* **1996**, *6*, 665–691. [CrossRef]
17. Rahel, F.J. Homogenization of freshwater faunas. *Annu. Rev. Ecol. Syst.* **2002**, *33*, 291–315. [CrossRef]
18. Leprieur, F.; Beauchard, O.; Blanchet, S.; Oberdorff, T.; Brosse, S. Fish invasions in the world's river systems: When natural processes are blurred by human activities. *PLoS Biol.* **2008**, *6*, e28.
19. Ruaro, R.; Mormul, R.P.; Gubiani, É.A.; Piana, P.A.; Cunico, A.M.; da Graca, W.J. Non-native fish species are related to the loss of ecological integrity in neotropical streams: A multimetric approach. *Hydrobiologia* **2018**, *817*, 413–430. [CrossRef]
20. Pompeu, P.S.; Wouters, L.; Hilário, H.O.; Loures, R.C.; Peressin, A.; Prado, I.G.; Suzuki, F.M.; Carvalho, D.C. Inadequate sampling frequency and imprecise taxonomic identification mask results in studies of migratory freshwater fish ichthyoplankton. *Fishes* **2023**, *8*, 518. [CrossRef]
21. Pawlowski, J.; Bonin, A.; Boyer, F.; Cordier, T.; Taberlet, P. Environmental DNA for biomonitoring. *Mol. Ecol.* **2021**, *30*, 2931–2936. [CrossRef]
22. Takahashi, M.; Saccò, M.; Kestel, J.H.; Nester, G.; Campbell, M.A.; Van Der Heyde, M.; Heydenrych, M.J.; Juszkiewicz, D.J.; Nevill, P.; Dawkins, K.L.; et al. Aquatic environmental DNA: A review of the macro-organismal biomonitoring revolution. *Sci. Total Environ.* **2023**, *873*, 162322. [CrossRef]
23. von der Heyden, S. Environmental DNA surveys of African biodiversity: State of knowledge, challenges, and opportunities. *Environ. DNA* **2023**, *5*, 12–17. [CrossRef]
24. Hamilton, E.F.; Juurakko, C.L.; Engel, K.; Neufeld, J.D.; Casselman, J.M.; Greer, C.W.; Walker, V.K. Environmental impacts on skin microbiomes of sympatric high Arctic salmonids. *Fishes* **2023**, *8*, 214. [CrossRef]

25. Pinder, A.C.; Gozlan, R.E.; Britton, J.R. Dispersal of the invasive topmouth gudgeon, *Pseudorasbora parva* in the UK: A vector for an emergent infectious disease. *Fish. Manag. Ecol.* **2005**, *12*, 411–414. [CrossRef]

26. Robinson, W.; Koehn, J.; Lintermans, M. Contemporary trends in the spatial extent of common riverine fish species in Australia's Murray–Darling Basin. *Fishes* **2024**, *9*, 221. [CrossRef]

27. AFS (American Fisheries Society). Statement of world aquatic scientific societies on the need to take urgent action against human-caused climate change, based on scientific evidence. *Fisheries* **2021**, *46*, 413–422. Available online: https://climate.fisheries.org/world-climate-statement (accessed on 10 November 2024). [CrossRef]

28. Winfield, I.J.; Hateley, J.; Fletcher, J.M.; James, J.B.; Bean, C.W.; Clabburn, P. Population trends of Arctic Charr (*Salvelinus alpinus*) in the UK: Assessing the evidence for a widespread decline in response to climate change. *Hydrobiologia* **2010**, *650*, 55–65. [CrossRef]

29. Hughes, R.M.; Herlihy, A.T.; Comeleo, R.; Peck, D.V.; Mitchell, R.M.; Paulsen, S. Patterns in and predictors of stream and river macroinvertebrate genera and fish species richness across the conterminous USA. *Knowl. Manag. Aquat. Ecosyst.* **2023**, *424*, 19. [CrossRef]

30. TNCA. The nationally significant wetlands of Brindingabba. In *The Nature Conservancy-Australia, Land and Freshwater Stories (Victoria)*; TNCA: Melbourne, VIC, Australia, 2022; 1p. Available online: https://www.natureaustralia.org.au/what-we-do/our-priorities/land-and-freshwater/land-freshwater-stories/brindingabba (accessed on 10 November 2024).

31. Bergström, K.; Berggren, H.; Nordahl, O.; Koch-Schmidt, P.; Tibblin, P.; Larsson, P. Seasonal and daily movement patterns by Wels catfish (*Silurus glanis*) at the northern fringe of its distribution range. *Fishes* **2024**, *9*, 280. [CrossRef]

32. Vindenes, Y.; Edeline, E.; Ohlberger, J.; Langangen, Ø.; Winfield, I.J.; Stenseth, N.C.; Vøllestad, L.A. Effects of climate change on trait-based dynamics of a top predator in freshwater ecosystems. *Am. Nat.* **2014**, *183*, 243–256. [CrossRef]

33. Kajee, M.; Dallas, H.F.; Griffiths, C.L.; Kleynhans, C.J.; Shelton, J.M. The status of South Africa's freshwater fish fauna: A spatial analysis of diversity, threat, invasion, and protection. *Fishes* **2023**, *8*, 571. [CrossRef]

34. Jelks, H.L.; Walsh, S.J.; Burkhead, N.M.; Contreras-Balderas, S.; Diaz-Pardo, E.; Hendrickson, D.A.; Lyons, J.; Mandrak, N.E.; McCormick, F.; Nelson, J.S.; et al. Conservation status of imperiled North American freshwater and diadromous fishes. *Fisheries* **2008**, *33*, 372–407.

35. Jones, P.E.; Tummers, J.S.; Galib, S.M.; Woodford, D.J.; Hume, J.B.; Silva, L.G.; Braga, R.R.; Garcia de Leaniz, C.; Vitule, J.R.; Herder, J.E.; et al. The use of barriers to limit the spread of aquatic invasive animal species: A global review. *Front. Ecol. Evol.* **2021**, *9*, 611631.

36. de Carvalho, D.R.; Leal, C.G.; Junqueira, N.T.; de Castro, M.A.; Fagundes, D.C.; Alves, C.B.; Hughes, R.M.; Pompeu, P.S. A fish-based multimetric index for Brazilian savanna streams. *Ecol. Indic.* **2017**, *77*, 386–396.

37. Faro, A.T.; Ferreira, M.T.; Oliveira, J.M. A fish-based tool for the quality assessment of Portuguese large rivers. *Fishes* **2024**, *9*, 149. [CrossRef]

38. Schinegger, R.; Palt, M.; Segurado, P.; Schmutz, S. Untangling the effects of multiple human stressors and their impacts on fish assemblages in European running waters. *Sci. Total Environ.* **2016**, *573*, 1079–1088. [CrossRef]

39. Wildhaber, M.L.; West, B.M.; Bennett, K.R.; May, J.H.; Albers, J.L.; Green, N.S. Sicklefin chub (*Macrhybopsis meeki*) and sturgeon chub (*M. gelida*) temporal and spatial patterns from extant population monitoring and habitat data spanning 23 years. *Fishes* **2024**, *9*, 43. [CrossRef]

40. Hughes, R.M.; Rinne, J.N.; Calamusso, B. Historical changes in large river fish assemblages of the Americas: A synthesis. *Am. Fish. Soc. Symp.* **2005**, *45*, 603–612.

41. Schmutz, S.; Jurajda, P.; Kaufmann, S.; Lorenz, A.W.; Muhar, S.; Paillex, A.; Poppe, M.; Wolter, C. Response of fish assemblages to hydromorphological restoration in central and northern European rivers. *Hydrobiologia* **2016**, *769*, 67–78. [CrossRef]

42. Monahan, D.; Wesner, J.S.; Parker, S.M.; Schartel, H. Spatial patterns in fish assemblages across the National Ecological Observation Network (NEON): The first six years. *Fishes* **2023**, *8*, 55. [CrossRef]

43. Heppell, S.A.; Heppell, S.S.; Arbuckle, N.S.; Gallagher, M.B. A cross-decadal change in the fish and crustacean community of lower Yaquina Bay, Oregon, USA. *Fishes* **2024**, *9*, 125. [CrossRef]

44. Baptista, J.; Martinho, F.; Nyitrai, D.; Pardal, M.A. Long-term functional changes in an estuarine fish assemblage. *Mar. Pollut. Bull.* **2015**, *97*, 125–134. [CrossRef] [PubMed]

45. Lambert, M.; Ojala-Barbour, R.; Vadas, R., Jr.; McIntyre, A.; Quinn, T. Do small overwater structures impact marine habitats and biota? *Pac. Conserv. Biol.* **2024**, *30*, PC22037. [CrossRef]

46. Valić, D.; Mirković, M.K.; Besendorfer, V.; Teskeredžić, E. Molecular analysis of two endemic *Squalius* species: Evidence for intergeneric introgression among cyprinids and conservation issues. *Fishes* **2024**, *9*, 4. [CrossRef]

47. Muhlfeld, C.C.; Kovach, R.P.; Al-Chokhachy, R.; Amish, S.J.; Kershner, J.L.; Leary, R.F.; Lowe, W.H.; Luikart, G.; Matson, P.; Schmetterling, D.A.; et al. Legacy introductions and climatic variation explain spatiotemporal patterns of invasive hybridization in a native trout. *Glob. Change Biol.* **2017**, *23*, 4663–4674. [CrossRef]

48. Neville, H.M.; Dunham, J.B. Patterns of hybridization of nonnative cutthroat trout and hatchery rainbow trout with native redband trout in the Boise River, Idaho. *N. Am. J. Fish. Manag.* **2011**, *31*, 1163–1176. [CrossRef]

49. Laskar, B.A.; Banerjee, D.; Chung, S.; Kim, H.W.; Kim, A.R.; Kundu, S. Integrative taxonomy clarifies the historical flaws in the systematics and distributions of two *Osteobrama* fishes (Cypriniformes: Cyprinidae) in India. *Fishes* **2024**, *9*, 87. [CrossRef]

50. Winemiller, K.O.; Rose, K.A. Patterns of life-history diversification in North American fishes: Implications for population regulation. *Can. J. Fish. Aquat. Sci.* **1992**, *49*, 2196–2218. [CrossRef]

51. Schindler, D.E.; Armstrong, J.B.; Reed, T.E. The portfolio concept in ecology and evolution. *Front. Ecol. Evol.* **2015**, *13*, 257–263. [CrossRef]
52. Krebs, J.; Hassell, M.; Godfray, C. Lord Robert May of Oxford OM. 8 January 1936–28 April 2020. *Biogr. Mem. Fellows R. Soc.* **2021**, *71*, 375–398. [CrossRef]
53. Fausch, K.D.; Torgersen, C.E.; Baxter, C.V.; Li, H.W. Landscapes to riverscapes: Bridging the gap between research and conservation of stream fishes. *BioScience* **2002**, *52*, 483–498. [CrossRef]
54. Spence, B.C.; Lomnicky, G.A.; Hughes, R.M.; Novitzki, R.P. *An Ecosystem Approach to Salmonid Conservation*; TR-4501-96-6057; National Marine Fisheries Service: Portland, OR, USA, 1996; 356p.
55. Maas-Hebner, K.G.; Schreck, C.B.; Hughes, R.M.; Yeakley, J.A.; Molina, N. Scientifically defensible fish conservation and recovery plans: Addressing diffuse threats and developing rigorous adaptive management plans. *Fisheries* **2016**, *41*, 276–285. [CrossRef]
56. Karr, J.R.; Dionne, M. Designing surveys to assess biological integrity in lakes and reservoirs. In *Biological Criteria: Research and Regulation*; Office of Water, EPA-440/5-91-005; U.S. Environmental Protection Agency, Ed.; U.S. Environmental Protection Agency: Washington, DC, USA, 1991; pp. 62–72. Available online: https://nepis.epa.gov/Exe/ZyPURL.cgi?Dockey=00001OUI.TXT (accessed on 10 November 2024).
57. Vadas, R.L., Jr.; Vadas, R.L., Sr. Toward a unified ecology (book review). *Maine Nat.* **1995**, *2*, 55–57. [CrossRef]
58. Schlosser, I.J. Environmental variation, life history attributes, and community structure in stream fishes: Implications for environmental management and assessment. *Environ. Manag.* **1990**, *14*, 621–628. [CrossRef]

 *fishes*

Article

Environmental Impacts on Skin Microbiomes of Sympatric High Arctic Salmonids

Erin F. Hamilton [1,†], Collin L. Juurakko [1,†], Katja Engel [2], Josh D. Neufeld [2], John M. Casselman [1], Charles W. Greer [3,4] and Virginia K. Walker [1,5,*]

[1] Department of Biology, Queen's University, Kingston, ON K7L 3N6, Canada
[2] Department of Biology, University of Waterloo, Waterloo, ON N2L 3G1, Canada
[3] Department of Natural Resource Sciences, McGill University, Ste. Anne de Bellevue, QC H9X 3V9, Canada
[4] National Research Council Canada, Montreal, QC H4P 2R2, Canada
[5] School of Environmental Studies, Queen's University, Kingston, ON K7L 3N6, Canada
* Correspondence: walkervk@queensu.ca; Tel.: +1-(613)-533-6000
† These authors have contributed equally.

Abstract: In the region of King William Island, Nunavut, in the Canadian high Arctic, populations of salmonids including Arctic char (*Salvelinus alpinus*), cisco (*Coregonus autumnalis* and *C. sardinella*) as well as lake whitefish (*C. clupeaformis*) are diadromous, overwintering in freshwater and transitioning to saline waters following ice melt. Since these fish were sampled at the same time and from the same traditional fishing sites, comparison of their skin structures, as revealed by 16S rRNA gene sequencing, has allowed an assessment of influences on wild fish bacterial communities. Arctic char skin microbiota underwent turnover in different seasonal habitats, but these striking differences in dispersion and diversity metrics, as well as prominent taxa involving primarily Proteobacteria and Firmicutes, were less apparent in the sympatric salmonids. Not only do these results refute the hypothesis that skin communities, for the most part, reflect water microbiota, but they also indicate that differential recruitment of bacteria is influenced by the host genome and physiology. In comparison to the well-adapted Arctic char, lake whitefish at the northern edge of their range may be particularly vulnerable, and we suggest the use of skin microbiomes as a supplemental tool to monitor a sustainable Indigenous salmonid harvest during this period of change in the high Arctic.

Keywords: Arctic char; *Salvelinus alpinus*; *Coregonus* spp.; lake whitefish; cisco; microbiomes; Arctic; Nunavut; diadromy

Key Contribution: Skin-associated microbial communities of high Arctic salmonids are not simply dependent on water communities, reflecting host genome and physiology. Arctic char skin-associated microbial communities undergo striking changes in response to changing seasonal habitat and water salinity compared to lake whitefish, possibly suggesting lake whitefish maladaptation and vulnerability.

Citation: Hamilton, E.F.; Juurakko, C.L.; Engel, K.; Neufeld, J.D.; Casselman, J.M.; Greer, C.W.; Walker, V.K. Environmental Impacts on Skin Microbiomes of Sympatric High Arctic Salmonids. *Fishes* **2023**, *8*, 214. https://doi.org/10.3390/fishes8040214

Academic Editors: Robert L. Vadas Jr. and Robert M. Hughes

Received: 19 March 2023
Revised: 12 April 2023
Accepted: 12 April 2023
Published: 18 April 2023

1. Introduction

Arctic char (*Salvelinus alpinus*) have a circumpolar distribution and represent the northernmost fish species on Earth [1]. At high latitudes, populations can be diadromous with seasonal migration to escape sub-zero temperatures in the sea by overwintering in freshwater lakes and rivers, with a return to saline waters to feed in the spring. Another salmonid, the closely related lake whitefish (*Coregonus clupeaformis*), is commonly found in freshwater lakes and rivers all year. Nonetheless, members of the *Coregonus* species complex (CSC) including lake whitefish and cisco (Arctic cisco, *Coregonus autumnalis*, and sardine cisco, *Coregonus sardinella*), are sympatric with Arctic char in the high Arctic, on King William Island and at adjacent mainland fishing sites in Nunavut, Canada. This region includes the northern extent of the lake whitefish range [2]. Here, as well as in the James-Hudson

Bay area and the Yukon River, CSC can also be diadromous [3–5]. Indeed, traditional Indigenous knowledge shared by community members, or Inuit Qaujimajatuqangit (IQ), teaches that CSC in this region follow the annual migration of Arctic char and can be fished swimming upriver within days of the peak char autumn "run". Such migration demands that these fish species physiologically and behaviorally adjust to seasonal environmental changes, but less known are any changes to their skin-associated microbiota. Here, we track migration-associated changes to skin microbiota in these sympatric salmonids to determine if these communities are influenced by environmental or host-specific factors.

Skin shares microbial species with the surrounding water. Indeed, most of the microbiota differences between distinct populations of Atlantic salmon (*Salmo salar*) could be attributed to their host waters [6–8]. As well, analyses of Atlantic salmon and Arctic char populations revealed that variations in water salinity could impact skin structure [7,9–11]. However, other abiotic and biotic factors may also influence fish microbiomes [12–15]. Immune health could play a role, with mucosal-associated skin lymph tissue, a mucous complex of immunoglobulins, antimicrobial peptides, mucins, and commensal bacteria, being critical to innate immunity [16,17]. Indeed, teleosts appear to promote the association of symbiotic bacteria, likely to help maintain skin immune function stability, with any disruption possibly resulting in dysbiosis, or the loss of beneficial microbes and an increased pathogen abundance that could culminate in an inflammatory response [15,16,18–21]. It is likely important that salmonids maintain immune function homeostasis and symbiotic skin bacteria during changes due to seasonal migration, and we suggest that such turnover could be orchestrated by the host.

Despite its importance to fish health, little is known about the drivers that influence skin microbiomes. Experimental work presents conflicting results. For example, skin microbiota in Atlantic salmon and catfish, *Silurus glanis*, were not prominently shaped by the host [7,22]. However, species differences were reported to have the largest influence on skin microbiota among three factors investigated in six different Gulf of Mexico teleosts [23]. Host influence on skin consortia was also shown in hybrids produced by crosses between domestic and wild brook char, *Salvelinus fontinalis*, and indicated that certain bacterial genera were influenced by three quantitative trait loci [24]. In non-human land mammals, skin microbiota appears to be most influenced by the host species, with geographical habitat being less influential [25]. The latter findings argue that the ecology of the bacterial community is inexorably woven into the phylogenetic history of the host, a process dubbed phylosymbiosis [25–27].

As noted, in the region of the Arctic under study, Arctic char and CSC are migratory and are fished from the same traditional sites. Therefore, these salmonids present a unique opportunity to compare skin microbial communities in wild sympatric species sampled from freshwater and sea fishing sites. Such investigation should allow insight into abiotic influences on the bacterial communities, provide a baseline for microbial populations in the face of anthropogenic change, and also illuminate any even minor host-specific genotype influences in salmonids that diverged ~50 million years ago [28]. Such monitoring can positively contribute to fish population health surveillance and be useful for future management of sustainable Arctic fishery ventures, in addition to informing local Inuit of skin bacteria that could be of some concern when consuming raw fish.

2. Materials and Methods

2.1. Study Area, Fish, and Water Sampling

Fishing sites in the Kitikmeot region of Nunavut (NU) were located within 400 km of King William Island (KWI) and the community of Gjoa Haven, including the saline water bodies of Rasmussen Basin and Chantrey Inlet. Freshwater sites included six lakes and rivers, including a traditional stone weir river site (Figure 1). The subsistence fishing locations were chosen based on IQ sharing by local Inuit elders in association with the Hunters and Trappers Association of Gjoa Haven, NU. Licenses to fish for scientific purposes were obtained in accordance with section 52 of the general fishery regulations of the

Fisheries Act, Department of Fisheries and Oceans Canada (DFO), and water was sampled as permitted by the Nunavut Impact Review Board. Animal care permits were issued by the Freshwater Institute Animal Care Committee of DFO (S-18/19-1045-NU and FWI-ACC AUP-2018-63).

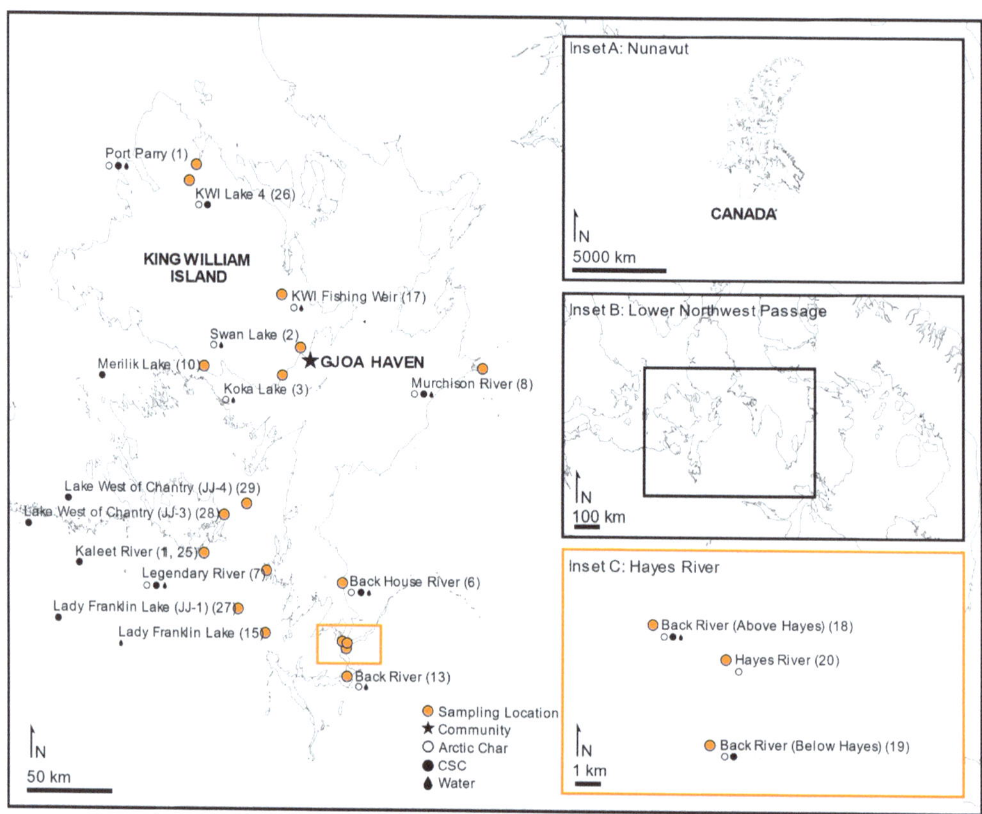

Figure 1. Study area of the lower Northwest Passage located in the Kitikmeot region within the Canadian territory of Nunavut. Sampling sites are shown indicating where fished Arctic char (open circles), members of the *Coregonus* species complex (CSC; closed circles) or water (dark drop) was collected. Map produced using ArcGIS Online and Affinity Designer.

Fish samples were aseptically collected from net and traditionally spear-harvested Arctic char and CSC. The majority of fish were humanely euthanized according to standard procedures with a blow to the head, while traditionally spear-harvested fish were killed according to traditional Inuit fishing practices. Each sampled fish was assigned a unique barcode [29]. Skin mucous samples were taken along the left lateral line of each fish using a sterile scalpel or cotton-tipped swab, stored in sterile barcode-labeled 5 mL tubes, frozen, packed into coolers with frozen freezer packs and shipped by plane [11]. Because of the distance to the laboratory, shipping coolers were kept in walk-in freezers during overnight layovers, and the skin samples were subsequently stored in −20 °C freezers. Once the aseptic skin samples were obtained at the fishing sites, the fish were weighed, measured for fork length (mm), and dissected to obtain otoliths, which were subsequently dried and used for age analysis as previously described [30,31].

Water samples were taken from as many fishing sites as was logistically possible (Figure 1), with up to 2 L of water filtered through sterile 0.22 μm filters (Pall) in triplicate. The filters were then frozen and transported in insulated containers at −20 °C, then stored at −80 °C. Additional water samples were collected in 50 mL plastic tubes, shipped with the skin samples but stored at −80 °C upon arrival at the laboratory. The water samples were thawed and assessed for specific conductivity using a conductivity meter (Traceable Fisherbrand, Fisher Scientific, Hampton, NH, USA).

2.2. Fish Condition and Growth Curve Calculations

Fulton's condition factor was calculated according to Barnham and Baxter as:

$$K = \frac{10^5 \times W}{L^3} \tag{1}$$

where mass (or weight, W) was measured in g and length (L) in mm [32]. Otolith age data were extrapolated for 12 Arctic char, four lake whitefish, and two ciscos, using a size-at-age key [33]. Growth curves were calculated as previously described by dividing the mean fork-length (FL), measured in mm, by age for fish aged 3–28 (Arctic char), 4–43 (lake whitefish) and 2–27 years (cisco species) and log10 transformed to construct plots [11]. A standard incremental annual growth curve was constructed by line of best fit, and deviations from these standards were calculated as percent relative differences where the mean growth standard is determined as the value of $FL \cdot Age^{-1}$, predicted by the calculated mean annual incremental growth curve at the specified age of the fish as previously described [11].

2.3. DNA Extractions and Sequencing

DNA was extracted from skin mucosal samples using the NucleoSpin Soil Extraction Kit (Machery-Nagel GmbH, Düren, Germany) with modifications including a final elution with double-distilled sterile water (ddH$_2$O) as previously described [11]. DNA extracts were diluted to ~50 ng μL^{-1}, and polymerase chain reaction (PCR) amplification was performed using primers 8F and 1406R to generate the V1–V9 region of the bacterial 16S ribosomal RNA (rRNA) gene and then subsequently re-amplified using the V4–V5 region using primers 515F-Y and 926R [34,35]. Skin-derived Illumina libraries were sequenced on a MiSeq instrument (Illumina Inc., San Diego, CA, USA). For water samples, the 16S rRNA gene V4–V5 region was amplified from each water sample as previously described and sequenced using a MiSeq instrument [36,37].

A total of 682 skin and 50 water samples, in addition to controls, were analyzed using Quantitative Insights Into Microbial Ecology 2 (QIIME2) (version 2020.6) managed by automated exploration of microbial diversity (AXIOME3) [38,39]. DADA2 (version 2020.6) was used to remove primer sequences and chimeras, dereplicate, and denoise reads [40]. Taxonomy was assigned to amplicon sequence variants (ASVs) using a naive Bayesian classifier pre-trained with the SILVA database (release 138) [41]. The prevalence method in Decontam was used to identify contaminants using a threshold of 0.5 as described previously [42,43]. Beta diversity was assessed using PCoA ordination with a Bray–Curtis dissimilarity matrix, and alpha diversity using Chao1 and Shannon index metrics. Diversity analysis was conducted using phyloseq (version 1.40.0) in Rstudio (version 2022.2.03) running R (version 4.2.0) [44]. Skin and water sequences obtained have been made available in the European Bioinformatics Institute (EBI) database under accession number PRJEB48811.

2.4. Statistical Analyses, Data Availability, and Efforts to Reduce Environmental Impact

Beta diversity between groups was tested through both PERMDISP and PERMANOVA using the adonis2 functions in the vegan R package (version 2.6.2) and the pairwise Adonis function from the corresponding package (version 0.4) using 10,000 permutations [45–48]. As noted, alpha diversity was determined using the Chao1 and Shannon index as a measure of taxonomic abundance and diversity, respectively. One-way ANOVAs with post-hoc Tukey's honest significant difference tests were performed to determine significant differences in means between factor groups with a 95% confidence threshold. Compact letter displays representing statistically significant groupings were generated using the R package multcompView (version 0.1.8). Bubble plots for taxonomic visualization were generated using ggplot2 (version 3.3.6) [49]. Similarity percentages (SIMPER) analyses were conducted using Bray–Curtis dissimilarity matrices in PAST (Paleontological Statistics) (version 4.08) [50]. Core microbiomes were determined using the microbiome package in R (version 1.18.0) with phyloseq and thresholds of 0.001 and prevalence of 50% [51].

Fish samples and otoliths have been archived for future access, and fish sample metadata is available in the Polar Data Catalogue (PDC) as open access (PDC#312992; NA profile of IOS 19115:2003, uploaded 5 February 2020, doi: 10.21963/12992). Measures were taken by the authors to reduce the environmental impact of research activities as well as to include Indigenous community members. These efforts included the hiring of local Inuit fishers, employing community youth to prepare samples, and making the fish available to the local "food bank" and other community programs. Furthermore, the coordination of multiple investigations encouraged southern visitors to volunteer for social science projects and enabled the bulk purchase of reagents to share among other research groups. Additionally, supplies and samples were shipped as personal baggage to reduce packaging and costs.

3. Results

3.1. Condition Factors and Annual Incremental Growth

The use of extrapolated age data for ~2.5% of the salmonids allowed the calculation of the mean annual incremental growth for 377 Arctic char, 188 lake whitefish, and 136 cisco, for which measurements were available. Annual incremental growth of individual fish was plotted against the standard curves (Figure S1), which allowed the calculation of relative differences from the standard growth curves and showed an average percent deviation of 3.7% (standard deviation, SD = 19) for Arctic char, 0.8% (SD = 10.5) for lake whitefish and 2.4% (SD = 15.5) for the two cisco species. There were no statistical differences in deviations for any of the salmonid groupings from different freshwater sites, indicating that there was no phenotype divergence that might suggest resource polymorphism within the taxa.

Arctic char condition factor (K) was significantly higher at the grouped saline fishing sites compared to all freshwater habitats ($p < 0.001$; one-way ANOVA; Figure S2). As previously reported, there was no significant difference in the condition factor for cisco between different seasonal habitats, and in lake whitefish the condition factor was significantly higher when caught in freshwater than in saline environments ($p < 0.001$, one-way ANOVA; Figure S2; [43]).

3.2. Arctic Char Skin Microbiome

A total of 441 Arctic char skin samples from fresh (n = 317) and saline (n = 124) waters were analyzed, including those obtained representing the change in seasonal habitats: autumn saline water (n = 124), autumn freshwater (n = 106), winter freshwater (n = 63), and spring freshwater (n = 148) (Figure 1). Across all seasonal habitats, Arctic char skin microbiomes were dominated by Proteobacteria, Cyanobacteria, Firmicutes, and Actinobacteriota (Figure S3).

Freshwater Arctic char skin samples had significantly higher ($p < 0.001$, one-way ANOVA) Shannon diversity than samples from saline waters (Figure 2A), with species richness likewise significantly higher ($p < 0.001$, one-way ANOVA) in freshwater-caught samples (Figure 2B). When considering seasonal habitats along with water conditions, autumn freshwater-caught Arctic char had significantly greater ($p < 0.05$, one-way ANOVA) Shannon diversity (Figure 2C) and species richness (Figure 2D) than all other seasonal habitats.

Bray–Curtis dissimilarity plots of Arctic char skin highlight the differences between fresh and saline waters, as well as autumn saline and other seasonal habitats (Figure 3) with significance verified using both PERMDISP ($p < 0.001$) and PERMANOVA ($p < 0.001$). Indeed, accounting for seasonality, autumn freshwater communities were significantly different from all others using both PERMDISP ($p < 0.05$) and PERMANOVA ($p = 0.001$).

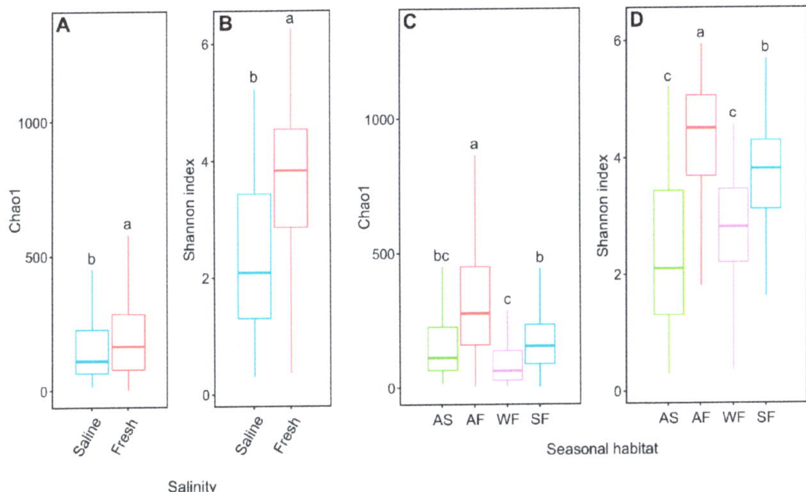

Figure 2. Alpha diversity metrics of Chao1 and Shannon entropy assessments of Arctic char skin community richness and diversity, respectively. Plots show differences between samples obtained from saline water (S; n = 124) and freshwater (F; n = 317) (**A**,**B**) and different seasonal habitats (**C**,**D**) including samples obtained from autumn saline water (AS; n = 124), autumn fresh water (AF; n = 106), winter fresh water (WF; n = 63), and spring fresh water (SF; n = 148). Different lowercase letters within the graphs display significantly different ($p < 0.001$) groupings as determined by one-way ANOVA.

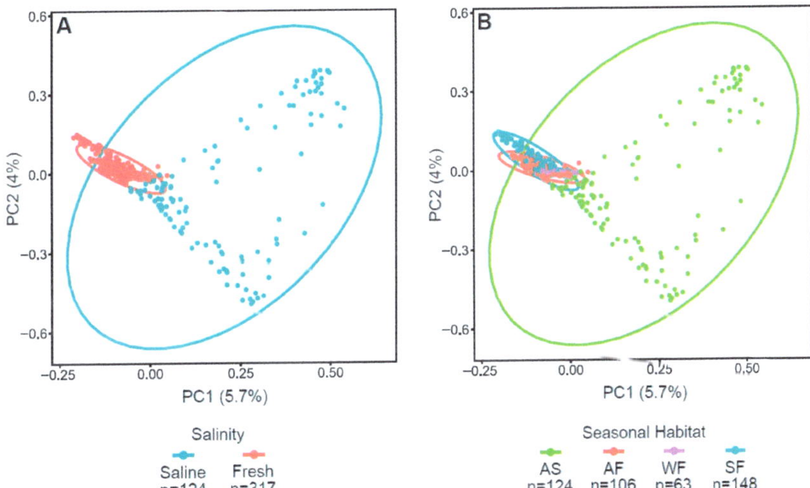

Figure 3. Principal coordinate analysis (PCoA) plots displaying dissimilarities in Arctic char skin samples using Bray–Curtis or beta diversity calculations. PCoA plots with (**A**) saline and freshwater environments and (**B**) seasonal habitats of autumn saline water (AS), autumn fresh water (AF), winter fresh water (WF), spring fresh water (SF). Number of fish samples per group (n) are indicated below the graphs.

Similarity percentages (SIMPER) analyses were performed to determine which ASVs were key to the differences between the saline and freshwater and seasonal habitat communities (Table S1). Proteobacteria, Cyanobacteria, and Firmicutes were the primary contributors to the distinctiveness of saline and freshwater communities. *Photobacterium*

(Proteobacteria) and *Tychonema* (Cyanobacteria) consistently contributed the most to microbiota dissimilarity, at 9.7% and 5% between saline and freshwater and between autumn saline and autumn fresh conditions, respectively. In contrast, the same genera had a much lower average relative abundance in freshwater, consisting of 0.05% and 0.04%, 0.008% and 0%, 0%, and 0.0003% in autumn fresh, winter fresh, and spring freshwater habitats, respectively (Table S1). More diverse taxa were responsible for dissimilarity between the different freshwater seasonal habitats, including Firmicutes, Proteobacteria, Cyanobacteria, Planctomycetota, Verrucomicrobiota, and Actinobacteriota. When comparing autumn to winter freshwater, *Staphylococcus* (Firmicutes) contributed the most to dissimilarity, at 3.7%, with *Escherichia-Shigella* (Proteobacteria) second, at 3.4%. *Staphylococcus*, at 3.8%, was again the greatest contributor to dissimilarity in the transition from winter to spring freshwater, and *Escherichia-Shigella*, at 3.5% dissimilarity, was noted when the autumn and spring freshwaters were compared.

ASVs present in ≥50% of skin samples and >0.1% relative abundance are defined as representing core taxa, but when considering every seasonal habitat, this criterion was not satisfied for a single ASV. However, when only skin samples from saline fishing sites were considered, there were six core taxa identified (Table S2). Four belong to Cyanobacteria (*Rivularia, Phormidesmis, Synechococcus* sp., *Tychonema*, and *Cyanobium*) and two to Gammaproteobacteria (*Psychrobacter* and *Photobacterium*). No core bacteria were noted from all freshwater-caught char, but if these were classified as to seasonal habitat, eight taxa were identified from autumn freshwater-caught fish, including two Gammaproteobacteria (*Polynucleobacter* and *Rhodoferax*), two Verrucomicrobiae, (*Chthoniobacter* and *Luteolibacter*), one Planctomycetota (a Gemmataceae), one Cyanobacterim (*Cyanobium*), and two Actinomycetota (a Sporichthyaceae and an Acidimicrobiia). With a prolonged stay in freshwater, the core skin microbiomes of winter and spring-caught fish were reduced to single genera, the kleptoplastic or photosynthetic-associated taxa Formanifera (*Planoglabratella opercularis*) and Gammaproteobacteria (*Rhodoferax*), respectively.

3.3. Influence of Surrounding Water on Arctic Char Skin Communities

Water samples were dominated by Proteobacteria, Actinobacteriota, Bacteroidota, and Cyanobacteria for both saline and freshwater sites (Figure S4), as previously reported from coastal waters [52,53]. Alpha diversity metrics (Figure S5) and PCoA plots of the water taxa from different seasonal habitats (Figure S6) showed no distinct groupings for saline and fresh waters. Two of the prominent water phyla, the Proteobacteria and Cyanobacteria (average relative abundances of 35% and 31% in fresh and saline, respectively, for Proteobacteria and 12% and 15% in fresh and saline, respectively, for Cyanobacteria), were also found on Arctic char skin (Figure S3). However, they had a different relative abundance on the skin (46% and 43% in fresh and saline skin samples, respectively, for Proteobacteria and 32% and 13% in fresh and saline, respectively, for Cyanobacteria), suggesting colonization bias. Notably, the relative abundance of Actinobacteriota decreased in spring freshwater compared to other water samples, but there was no significant change in the Arctic char skin community, reflecting that change in the fishing site waters. Likewise, the relative abundance of skin-associated Cyanobacteria decreased in spring freshwater habitats, but water microbiomes did not change with respect to this taxon. Bacteroidota also decreased in relative abundance on char caught in autumn saline and winter freshwater habitats, whereas water samples showed that phyla were reasonably consistent in these fished waters. Overall, individual taxa from the water communities are undoubtedly recruited from the water to Arctic char skin, but fish genomes and physiology appear to influence the relative bacterial abundance.

3.4. CSC Skin Microbiomes, Fishing Sites, and Water Microbiota

Similar to Arctic char, ordination based on Bray–Curtis dissimilarity showed distinct groupings between lake whitefish (n = 140) as well as cisco (n = 101) caught in fresh and saline waters (Figure 4). As in Arctic char, CSC skin communities did not simply reflect the surrounding water microbiota (Figures 5 and S4). For example, CSC skin showed a lower average relative abundance of Bacteroidota (3%) compared to water samples (29%) across all seasonal habitats. Firmicutes was relatively abundant in CSC skin, whereas this phylum represented less than 1% in water collected across all seasonal habitats. Cyanobacteria made up a higher average relative abundance on CSC skin (27%) compared to water samples (13%) with CSC skin having 46% relative abundance in the sea compared to 15% in saline water alone. Therefore, after environmental exposure to bacterial communities, there appears to be differential microbial recruitment on the CSC skin.

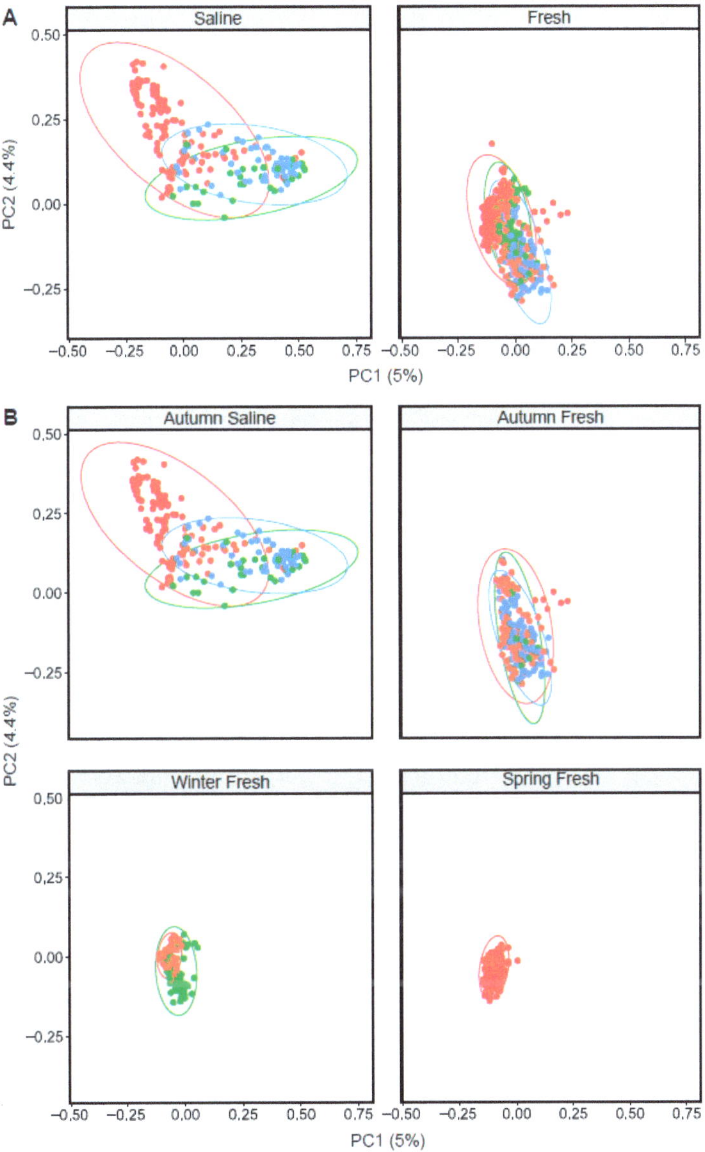

Figure 4. Principal coordinate analysis plots displaying dissimilarity using Bray–Curtis distances between Arctic char (red dots), cisco (bright green) and lake whitefish (blue) in (**A**) saline and freshwater, as well as (**B**) different seasonal habitats as indicated above each graph.

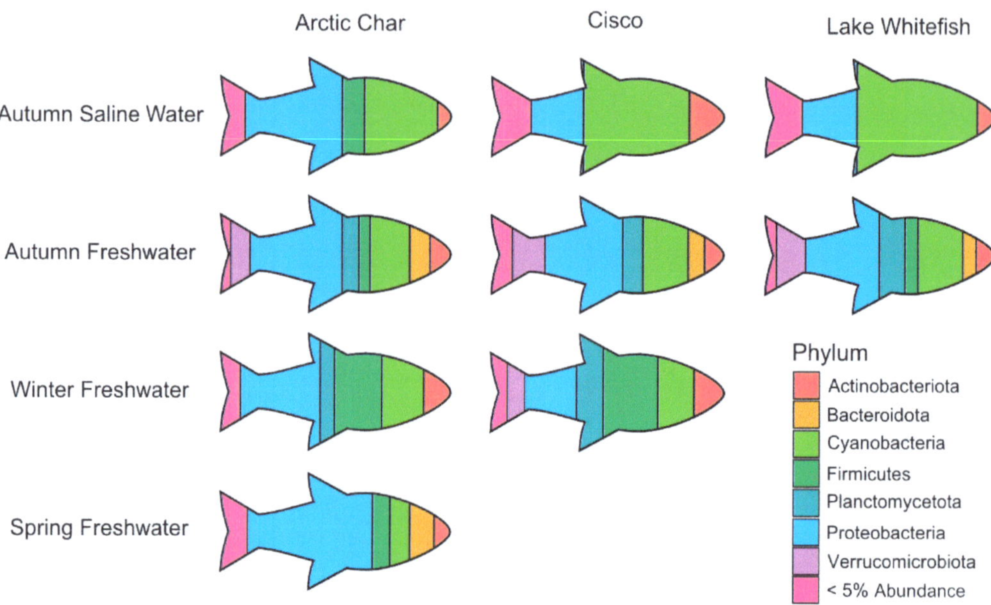

Figure 5. A pictograph showing abundant skin-associated microbiota in the high Arctic salmonids under study and across seasonal gradients. For simplicity, only the most prominent phyla are shown at their average relative abundance in Arctic char, cisco and lake whitefish. Phyla comprising < 5% average relative abundance are grouped together and only those seasonal habitats with sufficient number of caught fish are shown.

3.5. Impact of Fish Host on Skin Microbiomes

Across all seasonal habitats, skin from all three salmonids showed a large relative abundance of Proteobacteria. However, in autumn saline waters, cisco and lake whitefish skin microbiomes were dominated by Cyanobacteria, in contrast to a 50% lesser abundance of that taxa in Arctic char (Figure 5). As the different salmonids swam up freshwater rivers in the autumn, the skin communities became more similar, but diverged again during the winter when cisco contained relatively more Verrucomicrobiota and Planctomycetota, compared to Arctic char with average relative abundance of Planctomycetota and Verrucomicrobiota at least doubling (6% to 12% and 3% to 8%, respectively). By the time spring arrived, Arctic char skin microbiota showed the highest relative abundance of Proteobacteria among all characterized skin-associated microbiomes.

When these high Arctic sympatric salmonids were compared, skin alpha diversity, Chao1, in ocean-caught Arctic char was significantly (Chao1: $p < 0.001$, one-way ANOVA) lower than in both cisco and lake whitefish (Figure 6A). Community diversity in these chars were also lower than in freshwater samples ($p < 0.001$, one-way ANOVA). Likewise, Arctic char skin community diversity was significantly lower ($p < 0.001$, one-way ANOVA) than in lake whitefish and cisco under saline conditions and also significantly lower than in lake whitefish in freshwater samples. Overall, skin communities from Arctic char showed significantly decreased ($p < 0.001$, one-way ANOVA) community abundance and diversity in autumn saline and winter fresh habitats. Notably, taxa changes along migratory routes, as seen by the shifts between seasonal habitats, were more pronounced in Arctic char compared to cisco and lake whitefish.

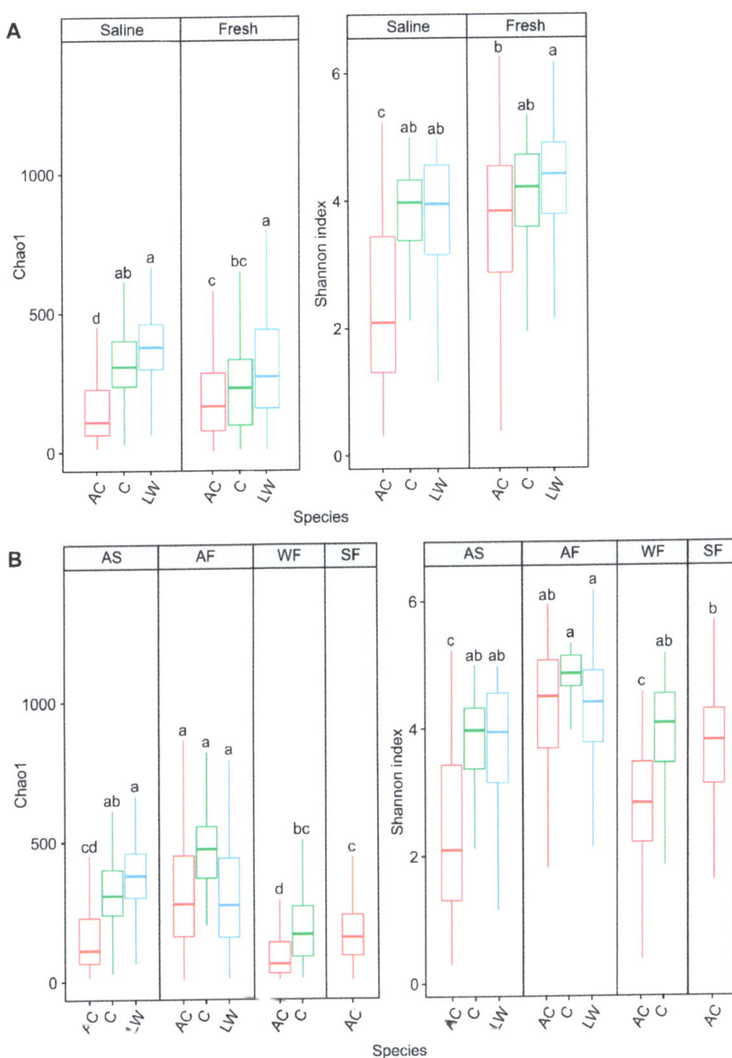

Figure 6. Alpha diversity metrics, Chao1 and Shannon diversity, measuring skin microbial community abundance and diversity, respectively, of Arctic char (AC) and *Coregonus* species complex cisco (C) as well as lake whitefish (LW) in (**A**) saline and freshwater and (**B**) in different seasonal habitats (abbreviated as described in Figure 3 and situated above each graph). Lower case letters within each individual graph display significantly different ($p < 0.001$) groupings as determined by one-way ANOVA. It should be noted that data from lake whitefish and cisco (Hamilton et al., 2023 [43]) are presented here for comparison with the Arctic char calculated diversity.

Bray–Curtis dissimilarity matrices comparing Arctic char and CSC skin communities in fish caught in waters of different salinities and seasonal habitat showed significant centroid differences ($p < 0.001$, PERMDISP and $p < 0.005$, PERMANOVA; Figure 4). Considering only seasonal habitat, Arctic char was consistently significantly different in diversity metrics than either cisco or lake whitefish. For example, in autumn saline and freshwater habitats as well as in winter freshwater habitats, Arctic char were significantly different (PERMDISP; $p < 0.001$, $p < 0.05$ and $p < 0.005$, respectively, and PERMANOVA; $p < 0.005$, $p < 0.01$, and $p < 0.005$, respectively).

4. Discussion

4.1. Skin Microbiota in Related Migrating Salmonids

Salmonoid skin epithelia are protected by a layer of mucus, which presumably acts to defend against pathogens, osmotic and mechanical stress, environmental perturbations as well as to conserve energy by reducing drag [12,54]. These secreted glycosylated mucins presumably foster colonization, particularly by biofilm-formers that are recruited from the surrounding waters, with fish benefits including protection against freeze-thaw, the sequestering of metals in oligotrophic environments and the production of antimicrobial metabolites to reduce pathogen colonization [7,55–59]. Although there are some differences in pathogen adhesion to different mucins [54], we are unaware of any differences in mucin chemistry in closely related salmonids. Thus, we hypothesized that related and sympatric Arctic char and CSC, fished from the same waters, would have the same or very similar communities. Indeed, the surrounding water microbiota has been argued to have the biggest influence on the teleost microbiome [7,58,60–63]. However, our hypothesis was not correct; even though water microbiota appeared to be relatively similar and overlapping, independent of salinity and seasonal change, the skin microbiota of the different salmonids changed (Figures S3 and S4). Only a portion of the water taxa were recovered on the skin, and depending on the seasonal habitat, there were some distinct fish skin communities (Figure 5). This strongly argues that wild salmonid hosts exert an important role on skin microbiome establishment.

As indicated, water samples had generally similar microbiota, but there was a decrease in Shannon diversity and species richness in winter- and spring-collected waters. Significantly, despite a similar diversity in autumn-sampled waters, independent of salinity, diversity of Arctic char skin communities in different autumn environmental habitats was not the same. For example, six taxa were classified as core community members in ocean-caught char, but none of these could be consistently identified at freshwater sites. Arctic char skin diversity increased after entry to freshwater, both as assessed by Shannon metrics, species richness and by PCoA plots, and this shift was also apparent when comparing the two autumn seasonal habitats (Figures 2–4). In comparison, few core taxa were identified in CSC samples. In addition, there were no significant differences in the diversity of lake whitefish skin communities obtained from ocean and freshwater fishing sites (Figure 4; [43]). Indeed, our initial expectation that core microbiota would be shared among the different salmonids and between environments was not corroborated; distinct differences between Arctic char and CSC, cisco and most notably, lake whitefish, were observed.

Both Arctic char and cisco skin communities were relatively abundant in ASVs representing Actinobacteria, which are likely psychrophilic and commensal, with previous reports in rainbow trout gut microbiota [64–66]. Core taxa in ocean Arctic char skin communities also included Proteobacteria, represented by the low-temperature-tolerant *Psychrobacter* and *Photobacterium*. Both of these are known to form biofilms, and *Photobacterium* was one of the drivers of overall seasonal dissimilarity, with an average relative abundance ~400–1500-fold greater in ocean-caught fish than in freshwater samples. All three groups of ocean-caught salmonids were colonized by Cyanobacteria, which is not surprising considering that these primary producers, many of which are tolerant to low temperatures, synthesize bio-reactive compounds and can form multi-taxa biofilms [57,59,67,68]. Together, this phylum made up ~32% and ~25% of the ASVs in Arctic char and CSC, respectively. Nevertheless, of the five cyanobacteria found as part of the core in Arctic char sampled from saline fishing sites, only one, *Cyanobium*, was regularly found in cisco. Lake whitefish carried both *Cyanobium* and *Planktothrix*, with the latter genus not a core taxon in autumn saline-caught Arctic char or cisco. This may be cause for some concern, since some of these species produce microcystins that are associated with whitefish toxicity [43,69].

According to IQ, lake whitefish and cisco follow Arctic char upriver and indeed, nets and spears pulled from these waters frequently contained char and CSC. This autumn migration was associated with a partial turnover of skin taxa in these salmonids. As noted, diversity and species richness increased in Arctic char during this run but not in CSC. In

Arctic char, *Cyanobium* was retained during the swim upriver, but turnover resulted in replacement by autumn freshwater taxa, including *Luteolibacter* and planktonic *Polynucleobacter* and *Rhodoferax* that all associate with biofilms [70–73]. A few taxa overlapped in the sympatric salmonids, and although not consistent enough to be part of a core, *Luteolibacter* and *Chthoniobacter*, the latter a genus previously reported in ice-covered lakes, were the most frequent ASVs in communities from autumn freshwater-caught lake whitefish and in cisco, a ASV corresponding to *Candidatus Bacilloplasma*, which was previously identified as part of the Atlantic salmon microbiome, predominated [74,75]. Fewer fish were obtained in the winter and spring from lines set under the ice, but in overwintering Arctic char there was a loss of Shannon diversity and species richness associated with these seasonal habitats, which in this case, was also seen in the water communities. The number of relatively consistent distinct core skin bacteria declined so much that the Arctic char skin was represented by a single genus in each of these under ice habitats.

4.2. Adaptation to Environmental Conditions, a Changing Climate, and Fisheries Management

The three salmonids shared a common ancestor ~50 million years ago in the ice-free Eocene Arctic Ocean. Arctic char retained their circumpolar distribution after the last glaciation, but ancestral CSC colonized North American lakes and are currently extending their northern range [1,2,28,76]. Thus, Arctic char should be well adapted to high Arctic conditions, and cisco may also be so, since they frequently seasonally migrate throughout their range. In contrast, diadromy is not frequent in lake whitefish worldwide. However, at their northern limits, this behaviour may be mandated by the low resources in high Arctic lakes [1,3,77–79]. Migration to and from oligotrophic lakes and the sea requires that their skin, a major component of their immune system [80,81], is safeguarded with commensal bacteria that form biofilms. Certain microbes from the water column appear to join as a core part of the consortium depending on water salinity, with the salmonid skin facilitating the proliferation of specific microbiota from the water column as shown here and supported by previous findings [24]. As Arctic char swim upriver in the autumn there is a transient increase in microbiota diversity, which is associated with a turnover of members of the skin community, presumably increasing fitness by recruiting beneficial taxa from the water. For example, *Photobacterium* represented more than 19% of the taxa in Arctic char caught in saline waters, and since species within this bacterial genus produce antibacterial compounds, these could inhibit the growth of competing bacteria [82]. Wild Arctic char appear to have developed a symbiotic relationship with this taxon, as suggested by its prevalence in both skin- and intestine-associated microbiomes in the autumn saline seasonal habitat (Figure S1; [83]). *Cyanobium* sp. with antibacterial and antiviral properties likely also contributes to this role, and in this case is part of the autumn saline core bacteria in all three salmonids.

As indicated, diadromous Arctic char are well adapted to their environment. These salmonids showed a higher condition (K) at the start of their autumn migration before swimming upriver, while cisco showed no condition differences between these seasonal habitats. In contrast, lake whitefish may not be as well adapted, since they had a significantly lower average condition upon their return from summer feeding compared to migrating freshwater-caught fish. Certain lake whitefish year classes were absent in the otolith data set, suggesting that there may have been lower recruitment in certain year classes, and in contrast to an increase in diversity that accompanied the autumn habitat transition seen in Arctic char, mean community richness did not increase in lake whitefish. It has been suggested that a shift in microbiota can assist fish to cope with hypotonic stress [84], and thus this difference may also reflect the fitness cost to lake whitefish at the edge of their range. In addition, the presence of possibly microcystin-producing *Planktothrix* as a core taxon only in lake whitefish may be cause for concern. Taken together, it is migratory lake whitefish, and not Arctic char or cisco, which may be less able to cope with climate change. We therefore recommend that these populations be targeted for future monitoring using this baseline data.

Diadromous Arctic char present with cisco and lake whitefish at traditional fishing sites have been harvested by Indigenous peoples throughout their oral history and, we hope, well into the future. As noted, the additional stress associated with climate change may be particularly challenging for migratory lake whitefish with biotic and abiotic stresses telegraphed to the skin consortium, offering a new tool to monitor the health of fish populations, in addition to enumeration and condition calculations. It is not known if any members of the skin consortium pose a risk to humans, but since Inuit frequently consume raw fish, knowledge of the timing of pathogen risk could help inform fishers to mitigate potential human health risks. For example, since skin microbiomes may be more stable in the autumn saline environment and therefore less susceptible to disease, community members might consider choosing to fish the autumn "runs" closer to the river outflow rather than upriver, where the microbiota is more likely to be in a transition state. To recapitulate, we have shown that distinct differences between sympatric salmonid skin-associated communities reflect salmonid genomic differences that help drive differential colonization. Such analysis, it is hoped, will contribute to future sustainable management of Arctic fisheries, particularly under increasing intergovernmental claims and commercial interests in the region, whilst maintaining Indigenous fishing rights and the interests and health of local communities.

Supplementary Materials: The following supporting information can be downloaded at: https://www.mdpi.com/article/10.3390/fishes8040214/s1, All supplemental information including Table S1: SIMPER analyses; Table S2: core microbiomes; Figure S1: growth curves; Figure S2: condition factors; Figure S3: all skin microbiota phyla; Figure S4: all water microbiota phyla; Figure S5: alpha diversity of water microbiomes; Figure S6: beta diversity of water microbiomes.

Author Contributions: Sample processing was done by E.F.H., sequencing was performed by C.W.G., J.D.N. and K.E., data analysis by J.D.N., C.W.G., E.F.H., K.E., J.M.C. and C.L.J., figures made by C.L.J., study design by V.K.W., initial draft written by V.K.W. and C.L.J. All authors have read and agreed to the published version of the manuscript.

Funding: This work was funded by the "Towards a Sustainable Fishery for Nunavummiut" project, a large-scale Genome Canada project funded by the Government of Canada through Genome Canada and the Ontario Genomics Institute (OGI-096), as well as associated and in-kind support from the Ontario Ministry of Research and Innovation, CanNor, Nunavut Arctic College, and the Government of Nunavut. Important additional funding was provided by the Northern Scientific Training Program (Polar Knowledge Canada) and Queen's University to EH, and a Discovery Grant from the Natural Sciences and Engineering Research Council (Canada) to VKW.

Institutional Review Board Statement: The study was approved by the Freshwater Institute Animal Care Committee of the Department of Fisheries and Oceans Canada (S-18/19-1045-NU and FWI-ACC AUP-2018-63).

Data Availability Statement: The data presented in this study is openly available in the European Bioinformatics Institute (EBI) database under accession number PRJEB48811 at https://www.ebi.ac.uk/ena/browser/view/PRJEB48811 (uploaded on 1 September 2022).

Acknowledgments: We thank the residents of Gjoa Haven, Nunavut, the Gjoa Haven Hunters and Trappers Association, community fishers, and youth for identifying fishing sites, for their invaluable IQ, as well as sampling expertise and preparation. Geraint Element and Peter van Coverden de Groot facilitated the collection and shipping of many samples to the laboratory, and GIS services consultant, Eric Daechsel, made the initial map. Kristy Moniz is thanked for general laboratory assistance, with age and growth interpretation assistance from Bronte McPhedran, Jordan Balson and Kate Brouwer. We are grateful to Stephan Schott and his social science team as well as W. Leggett, L. Harris, and K. Hedges for supporting our efforts to investigate Arctic fish.

Conflicts of Interest: The authors declare no conflict of interest.

References

1. Klemetsen, A.; Amundsen, P.A.; Dempson, J.B.; Jonsson, B.; Jonsson, N.; O'Connell, M.F.; Mortensen, E. Atlantic Salmon *Salmo salar* L., Brown Trout *Salmo trutta* L. and Arctic Charr *Salvelinus alpinus* (L.): A Review of Aspects of Their Life Histories. *Ecol. Freshw. Fish* **2003**, *12*, 1–59. [CrossRef]
2. Scott, W.B.; Crossman, E.J. Freshwater Fishes of Canada. *Fish. Res. Board Can. Bull.* **1973**, *184*, 966.
3. Morin, R.; Dodson, J.J.; Power, G. Life History Variations of Anadromous Cisco (*Coregonus artedii*), Lake Whitefish (*C. clupeaformis*), and Round Whitefish (*Prosopium cylindraceum*) Populations of Eastern James–Hudson Bay. *Can. J. Fish. Aquat. Sci.* **1982**, *39*, 958–967. [CrossRef]
4. Brown, R.J.; Bickford, N.; Severin, K. Otolith Trace Element Chemistry as an Indicator of Anadromy in Yukon River Drainage Coregonine Fishes. *Trans. Am. Fish. Soc.* **2007**, *136*, 678–690. [CrossRef]
5. Koch, I.; Das, P.; McPhedran, B.E.; Casselman, J.M.; Moniz, K.L.; van Coeverden de Groot, P.; Chen, C.Y.; Walker, V.K. Correlation of Mercury Occurrence with Age, Elemental Composition, and Life History in Sea-Run Food Fish from the Canadian Arctic Archipelago's Lower Northwest Passage. *Foods* **2021**, *10*, 2621. [CrossRef] [PubMed]
6. Wang, C.; Sun, G.; Li, S.; Li, X.; Liu, Y. Intestinal Microbiota of Healthy and Unhealthy Atlantic Salmon *Salmo salar* L. in a Recirculating Aquaculture System. *J. Ocean Limnol.* **2018**, *36*, 414–426. [CrossRef]
7. Webster, T.M.U.; Consuegra, S.; Hitchings, M.; de Leaniz, C.G. Interpopulation Variation in the Atlantic Salmon Microbiome Reflects Environmental and Genetic Diversity. *Appl. Environ. Microbiol.* **2018**, *84*, e00691-18. [CrossRef] [PubMed]
8. Sylvain, F.É.; Holland, A.; Bouslama, S.; Audet-Gilbert, É.; Lavoie, C.; Val, A.L.; Derome, N. Fish Skin and Gut Microbiomes Show Contrasting Signatures of Host Species and Habitat. *Appl. Environ. Microbiol.* **2020**, *86*, e00789-20. [CrossRef]
9. Lokesh, J.; Kiron, V. Transition from Freshwater to Seawater Reshapes the Skin-Associated Microbiota of Atlantic Salmon. *Sci. Rep.* **2016**, *6*, 19707. [CrossRef]
10. Dehler, C.E.; Secombes, C.J.; Martin, S.A. Seawater Transfer Alters the Intestinal Microbiota Profiles of Atlantic Salmon (*Salmo salar* L.). *Sci. Rep.* **2017**, *7*, 13877. [CrossRef]
11. Hamilton, E.F.; Element, G.; van Coeverden de Groot, P.; Engel, K.; Neufeld, J.D.; Shah, V.; Walker, V.K. Anadromous Arctic Char Microbiomes: Bioprospecting in the High Arctic. *Front. Bioeng. Biotechnol.* **2019**, *7*, 32. [CrossRef]
12. Wilson, B.; Danilowicz, B.S.; Meijer, W.G. The Diversity of Bacterial Communities Associated with Atlantic Cod *Gadus morhua*. *Microb. Ecol.* **2008**, *55*, 425–434. [CrossRef] [PubMed]
13. Kueneman, J.G.; Parfrey, L.W.; Woodhams, D.C.; Archer, H.M.; Knight, R.; McKenzie, V.J. The Amphibian Skin-Associated Microbiome across Species, Space and Life History Stages. *Mol. Ecol.* **2014**, *23*, 1238–1250. [CrossRef] [PubMed]
14. Apprill, A. Marine Animal Microbiomes: Toward Understanding Host–Microbiome Interactions in a Changing Ocean. *Front. Mar. Sci.* **2017**, *4*, 222. [CrossRef]
15. Ghosh, S.K.; Wong, M.K.-S.; Hyodo, S.; Goto, S.; Hamasaki, K. Temperature Modulation Alters the Gut and Skin Microbial Profiles of Chum Salmon (*Oncorhynchus keta*). *Front. Mar. Sci.* **2022**, *9*, 1027621. [CrossRef]
16. Xu, Z.; Parra, D.; Gomez, D.; Salinas, I.; Zhang, Y.A.; Von Gersdorff Jorgensen, L.; LaPatra, S.E.; Sunyer, J.O. Teleost Skin, an Ancient Mucosal Surface That Elicits Gut-Like Immune Responses. *Proc. Natl. Acad. Sci. USA* **2013**, *110*, 13097–13102. [CrossRef]
17. Esteban, M.Á.; Cerezuela, R. Fish Mucosal Immunity: Skin. In *Mucosal Health in Aquaculture*; Beck, B.H., Peatman, E., Eds.; Academic Press: Cambridge, MA, USA, 2015; pp. 67–92. [CrossRef]
18. Scharschmidt, T.C.; Fischbach, M.A. What Lives on Our Skin: Ecology, Genomics and Therapeutic Opportunities of the Skin Microbiome. *Drug Discov. Today Dis. Mech.* **2013**, *10*, e83–e89. [CrossRef] [PubMed]
19. Klenerman, P.; Ogg, G. Killer T Cells Show Their Kinder Side. *Nature* **2018**, *555*, 594–595. [CrossRef]
20. Linehan, J.L.; Harrison, O.J.; Han, S.J.; Byrd, A.L.; Vujkovic-Cvijin, I.; Villarino, A.V.; Sen, S.K.; Shaik, J.; Smelkinson, M.; Tamoutounour, S.; et al. Non-classical Immunity Controls Microbiota Impact on Skin Immunity and Tissue Repair. *Cell* **2018**, *172*, 784–796. [CrossRef]
21. Sanford, J.A.; Zhang, L.J.; Williams, M.R.; Gangoiti, J.A.; Huang, C.M.; Gallo, R.L. Inhibition of HDAC8 and HDAC9 by Microbial Short-Chain Fatty Acids Breaks Immune Tolerance of the Epidermis to TLR Ligands. *Sci. Immunol.* **2016**, *1*, eaah4609. [CrossRef]
22. Chiarello, M.; Paz-Vinas, I.; Veyssière, C.; Santoul, F.; Loot, G.; Ferriol, J.; Boulêtreau, S. Environmental Conditions and Neutral Processes Shape the Skin Microbiome of European Catfish (*Silurus glanis*) Populations of Southwestern France. *Environ. Microbiol. Rep.* **2019**, *11*, 605–614. [CrossRef] [PubMed]
23. Larsen, A.M.; Tao, Z.; Bullard, S.A.; Arias, C.R. Diversity of the Skin Microbiota of Fishes: Evidence for Host Species Specificity. *FEMS Microbiol. Ecol.* **2013**, *85*, 483–494. [CrossRef] [PubMed]
24. Boutin, S.; Sauvage, C.; Bernatchez, L.; Audet, C.; Derome, N. Interindividual Variations of the Fish Skin Microbiota: Host Genetics Basis of Mutualism? *PLoS ONE* **2014**, *9*, e102649. [CrossRef] [PubMed]
25. Ross, A.A.; Müller, K.M.; Weese, J.S.; Neufeld, J.D. Comprehensive Skin Microbiome Analysis Reveals the Uniqueness of Human Skin and Evidence for Phylosymbiosis within the Class Mammalia. *Proc. Natl. Acad. Sci. USA* **2018**, *115*, E5786–E5795. [CrossRef]
26. Brooks, A.W.; Kohl, K.D.; Brucker, R.M.; van Opstal, E.J.; Bordenstein, S.R. Phylosymbiosis: Relationships and Functional Effects of Microbial Communities across Host Evolutionary History. *PLoS Biol.* **2016**, *14*, e2000225. [CrossRef] [PubMed]
27. Kelly, C.; Salinas, I. Under Pressure: Interactions between Commensal Microbiota and the Teleost Immune System. *Front. Immunol.* **2017**, *8*, 559. [CrossRef] [PubMed]

28. Crête-Lafrenière, A.; Weir, L.K.; Bernatchez, L. Framing the Salmonidae family phylogenetic portrait: A more complete picture from increased taxon sampling. *PLoS ONE* **2012**, *7*, e46662. [CrossRef]
29. Wu, Y.; Lougheed, D.R.; Lougheed, S.C.; Moniz, K.; Walker, V.K.; Colautti, R.I. baRcodeR: An Open-Source R Package for Sample Labelling. *Methods Ecol. Evol.* **2020**, *11*, 980–985. [CrossRef]
30. Campana, S.E.; Casselman, J.M.; Jones, C.M. Bomb Radiocarbon Chronologies in the Arctic, with Implications for the Age Validation of Lake Trout (*Salvelinus namaycush*) and Other Arctic Species. *Can. J. Fish. Aquat. Sci.* **2008**, *65*, 733–743. [CrossRef]
31. Casselman, J.M.; Jones, C.M.; Campana, S.E. Bomb Radiocarbon Age Validation for the Long-Lived, Unexploited Arctic Fish Species *Coregonus clupeaformis*. *Mar. Freshw. Res.* **2019**, *70*, 1781–1788. [CrossRef]
32. Barnham, C.; Baxter, A. Condition Factor, K, for Salmonid Fish. Fisheries Notes. *Nat. Resour. Environ.* **1998**, *5*, 1–3.
33. Isermann, D.A.; Knight, C.T. A Computer Program for Age-Length Keys Incorporating Age Assignment to Individual Fish. *N. Am. J. Fish. Manag.* **2005**, *25*, 1153–1160. [CrossRef]
34. Parada, A.E.; Needham, D.M.; Fuhrman, J.A. Every Base Matters: Assessing Small Subunit rRNA Primers for Marine Microbiomes with Mock Communities, Time Series and Global Field Samples. *Environ. Microbiol.* **2016**, *18*, 1403–1414. [CrossRef]
35. Quince, C.; Lanzen, A.; Davenport, R.J.; Turnbaugh, P.J. Removing Noise from Pyrosequenced Amplicons. *BMC Bioinform.* **2011**, *12*, 38. [CrossRef] [PubMed]
36. Tremblay, J.; Yergeau, E.; Fortin, N.; Cobanli, S.; Elias, M.; King, T.L.; Greer, C.W. Chemical Dispersants Enhance the Activity of Oil-and Gas Condensate-Degrading Marine Bacteria. *ISME J.* **2017**, *11*, 2793–2808. [CrossRef] [PubMed]
37. Cobanli, S.E.; Wohlgeschaffen, G.; Ryther, C.; MacDonald, J.; Gladwell, A.; Watts, T.; Greer, C.W.; Elias, M.; Wasserscheid, J.; Robinson, B.; et al. Microbial Community Response to Simulated Diluted Bitumen Spills in Coastal Seawater and Implications for Oil Spill Response. *FEMS Microbiol. Ecol.* **2022**, *98*, fiac033. [CrossRef]
38. Bolyen, E.; Rideout, J.R.; Dillon, M.R.; Bokulich, N.A.; Abnet, C.C.; Al-Ghalith, G.A.; Alexander, H.; Alm, E.J.; Arumugam, M.; Asnicar, F.; et al. Reproducible, Interactive, Scalable and Extensible Microbiome Data Science Using QIIME 2. *Nat. Biotechnol.* **2019**, *37*, 852–857. [CrossRef]
39. Min, D.; Doxey, A.C.; Neufeld, J.D. AXIOME3: Automation, Extension, and Integration of Microbial Ecology. *GigaScience* **2021**, *10*, giab006. [CrossRef]
40. Callahan, B.J.; McMurdie, P.J.; Rosen, M.J.; Han, A.W.; Johnson, A.J.A.; Holmes, S.P. DADA2: High-Resolution Sample Inference from Illumina Amplicon Data. *Nat. Methods* **2016**, *13*, 581–583. [CrossRef]
41. Pruesse, E.; Quast, C.; Knittel, K.; Fuchs, B.M.; Ludwig, W.; Peplies, J.; Glöckner, F.O. SILVA: A Comprehensive Online Resource for Quality Checked and Aligned Ribosomal RNA Sequence Data Compatible with ARB. *Nucleic Acids Res.* **2007**, *35*, 7188–7196. [CrossRef]
42. Davis, N.M.; Proctor, D.M.; Holmes, S.P.; Relman, D.A.; Callahan, B.J. Simple Statistical Identification and Removal of Contaminant Sequences in Marker-Gene and Metagenomics Data. *Microbiome* **2018**, *6*, 226. [CrossRef] [PubMed]
43. Hamilton, E.F.; Juurakko, C.L.; Engel, K.; de Groot, P.J.C.; Casselman, J.M.; Greer, C.W.; Neufeld, J.D.; Walker, V.K. Characterization of Skin- and Intestine Microbial Communities in Migrating High Arctic Lake Whitefish and Cisco. *bioRxiv.* 2023. Available online: https://www.biorxiv.org/content/10.1101/2023.03.08.531621v1 (accessed on 12 April 2023). [CrossRef]
44. McMurdie, P.J.; Holmes, S. phyloseq: An R Package for Reproducible Interactive Analysis and Graphics of Microbiome Census Data. *PLoS ONE* **2013**, *8*, e61217. [CrossRef] [PubMed]
45. Anderson, M.J. A New Method for Non-Parametric Multivariate Analysis of Variance. *Austral Ecol.* **2001**, *26*, 32–46. [CrossRef]
46. Anderson, M.J. Distance-Based Tests for Homogeneity of Multivariate Dispersion. *Biometrics* **2006**, *62*, 245–253. [CrossRef]
47. McArdle, B.H.; Anderson, M.J. Fitting Multivariate Models to Community Data: A Comment on Distance-Based Redundancy Analysis. *Ecology* **2001**, *82*, 290–297. [CrossRef]
48. Arbizu, P.M. PairwiseAdonis: Pairwise Multilevel Comparison Using Adonis, R Package Version 0.4. 2020. Available online: https://github.com/pmartinezarbizu/pairwiseAdonis (accessed on 12 April 2023).
49. Wickham, H. *Elegant Graphics for Data Analysis (ggplot2)*; Springer: New York, NY, USA, 2009.
50. Hammer, Ø.; Harper, D.A.T.; Ryan, P.D. PAST: Paleontological Statistics Software Package for Education and Data Analysis. *Palaeontol. Electron.* **2001**, *4*, 9.
51. Lahti, L.; Shetty, S. Microbiome R Package. *Bioconductor* **2017**. Available online: https://bioconductor.org/packages/release/bioc/html/microbiome.html (accessed on 12 April 2023). [CrossRef]
52. Liu, S.; Wawrik, B.; Liu, Z. Different Bacterial Communities Involved in Peptide Decomposition between Normoxic and Hypoxic Coastal Waters. *Front. Microbiol.* **2017**, *8*, 353. [CrossRef]
53. Jiang, X.; Zhu, Z.; Wu, J.; Lian, E.; Liu, D.; Yang, S.; Zhang, R. Bacterial and Protistan Community Variation across the Changjiang Estuary to the Ocean with Multiple Environmental Gradients. *Microorganisms* **2022**, *10*, 991. [CrossRef]
54. Padra, J.T.; Murugan, A.V.; Sundell, K.; Sundh, H.; Benktander, J.; Lindén, S.K. Fish Pathogen Binding to Mucins from Atlantic Salmon and Arctic Char Differs in Avidity and Specificity and Is Modulated by Fluid Velocity. *PLoS ONE* **2019**, *14*, e0215583. [CrossRef]
55. Wu, Z.; Kan, F.W.K.; She, Y.M.; Walker, V.K. Biofilm, Ice Recrystallization Inhibition and Freeze-Thaw Protection in an Epiphyte Community. *Appl. Biochem. Microbiol.* **2012**, *48*, 363–370. [CrossRef]
56. Cai, W.; De La Fuente, L.; Arias, C.R. Biofilm Formation by the Fish Pathogen Flavobacterium columnare: Development and Parameters Affecting Surface Attachment. *Appl. Environ. Microbiol.* **2013**, *79*, 5633–5642. [CrossRef]

57. Christmas, N.A.; Barker, G.; Anesio, A.M.; Sánchez-Baracaldo, P. Genomic Mechanisms for Cold Tolerance and Production of Exopolysaccharides in the Arctic Cyanobacterium *Phormidesmis Priestleyi* BC1401. *BMC Genom.* **2016**, *17*, 533. [CrossRef]
58. Webster, T.M.U.; Rodriguez-Barreto, D.; Castaldo, G.; Gough, P.; Consuegra, S.; de Leaniz, C.G. Environmental Plasticity and Colonisation History in the Atlantic Salmon Microbiome: A Translocation Experiment. *Mol. Ecol.* **2020**, *29*, 886–898. [CrossRef] [PubMed]
59. Pagliara, P.; De Benedetto, G.E.; Francavilla, M.; Barca, A.; Caroppo, C. Bioactive Potential of Two Marine Picocyanobacteria Belonging to *Cyanobium* and *Synechococcus* Genera. *Microorganisms* **2021**, *9*, 2048. [CrossRef]
60. Boutin, S.; Bernatchez, L.; Audet, C.; Derôme, N. Network Analysis Highlights Complex Interactions between Pathogen, Host and Commensal Microbiota. *PLoS ONE* **2013**, *8*, e84772. [CrossRef] [PubMed]
61. Chiarello, M.; Villeger, S.; Bouvier, C.; Bettarel, Y.; Bouvier, T. High Diversity of Skin-Associated Bacterial Communities of Marine Fishes Is Promoted by Their High Variability among Body Parts, Individuals and Species. *FEMS Microbiol. Ecol.* **2015**, *91*, fiv061. [CrossRef] [PubMed]
62. Llewellyn, M.S.; Boutin, S.; Hoseinifar, S.H.; Derome, N. Teleost Microbiomes: The State of the Art in Their Characterization, Manipulation and Importance in Aquaculture and Fisheries. *Front. Microbiol.* **2014**, *5*, 207. [CrossRef]
63. Øygarden, E.T. Influence of Genetic Background and Environmental Factors on the Skin and Gut Microbiota of Atlantic Salmon (*Salmo salar*) Fry. Master's Thesis, Norwegian University of Science and Technology, Trondheim, Norway, 2017. Available online: https://ntnuopen.ntnu.no/ntnu-xmlui/handle/11250/2454377 (accessed on 1 May 2021).
64. Guardabassi, L.; Dalsgaard, A.; Olsen, J.E. Phenotypic Characterization and Antibiotic Resistance of Acinetobacter spp. Isolated from Aquatic Sources. *J. Appl. Microbiol.* **1999**, *87*, 659–667. [CrossRef]
65. Junge, K.; Christner, B.; Staley, J. Diversity of Psychrophilic Bacteria from Sea Ice and Glacial Ice Communities. In *Extremophiles Handbook*; Horikoshi, K., Ed.; Springer: Tokyo, Japan, 2011; pp. 794–815. [CrossRef]
66. Zhan, M.; Huang, Z.; Cheng, G.; Yu, Y.; Su, J.; Xu, Z. Alterations of the Mucosal Immune Response and Microbial Community of the Skin upon Viral Infection in Rainbow Trout (*Oncorhynchus mykiss*). *Int. J. Mol. Sci.* **2022**, *23*, 14037. [CrossRef]
67. Ernst, B.; Hoeger, S.J.; O'Brien, E.; Dietrich, D.R. Oral Toxicity of the Microcystin-Containing Cyanobacterium *Planktothrix rubescens* in European Whitefish (*Coregonus lavaretus*). *Aquat. Toxicol.* **2006**, *79*, 31–40. [CrossRef] [PubMed]
68. Faria, S.I.; Teixeira-Santos, R.; Romeu, M.J.; Morais, J.; Jong, E.D.; Sjollema, J.; Vasconcelos, V.; Mergulhão, F.J. Unveiling the Antifouling Performance of Different Marine Surfaces and Their Effect on the Development and Structure of Cyanobacterial Biofilms. *Microorganisms* **2021**, *9*, 1102. [CrossRef] [PubMed]
69. Pancrace, C.; Barny, M.A.; Ueoka, R.; Calteau, A.; Scalvenzi, T.; Pédron, J.; Humbert, J.-F.; Gugger, M. Insights into the *Planktothrix* Genus: Genomic and Metabolic Comparison of Benthic and Planktic Strains. *Sci. Rep.* **2017**, *7*, 41181. [CrossRef]
70. Hahn, M.W.; Jezberová, J.; Koll, U.; Saueressig-Beckm, T.; Schmidt, J. Complete Ecological Isolation and Cryptic Diversity in Polynucleobacter Bacteria Not Resolved by 16S rRNA Gene Sequences. *ISME J.* **2016**, *10*, 1642–1655. [CrossRef] [PubMed]
71. Cai, W.; Li, Y.; Niu, L.; Zhang, W.; Wang, C.; Wang, P.; Meng, F. New Insights into the Spatial Variability of Biofilm Communities and Potentially Negative Bacterial Groups in Hydraulic Concrete Structures. *Water Res.* **2017**, *123*, 495–504. [CrossRef] [PubMed]
72. Romero, F.; Acuña, V.; Sabater, S. Multiple Stressors Determine Community Structure and Estimated Function of River Biofilm Bacteria. *Appl. Environ. Microbiol.* **2020**, *86*, E00291-20. [CrossRef]
73. Li, P.; Wang, C.; Liu, G.; Luo, X.; Rauan, A.; Zhang, C.; Li, T.; Yu, H.; Dong, S.; Gao, Q. A Hydroponic Plants and Biofilm Combined Treatment System Efficiently Purified Wastewater from Cold Flowing Water Aquaculture. *Sci. Total Environ.* **2022**, *821*, 153534. [CrossRef]
74. Fournier, I.B.; Lovejoy, C.; Vincent, W.F. Changes in the Community Structure of Under-Ice and Open-Water Microbiomes in Urban Lakes Exposed to Road Salts. *Front. Microbiol.* **2021**, *12*, 660719. [CrossRef]
75. Webster, T.M.U.; Consuegra, S.; de Leaniz, C.G. Early Life Stress Causes Persistent Impacts on the Microbiome of Atlantic Salmon. *Comp. Biochem. Physiol. Part D Genom. Proteom.* **2021**, *40*, 100888. [CrossRef]
76. Eberle, J.J.; Greenwood, D.R. Life at the Top of the Greenhouse Eocene World—A Review of the Eocene Flora and Vertebrate Fauna from Canada's High Arctic. *Bulletin* **2012**, *124*, 3–23. [CrossRef]
77. Swanson, H.K.; Kidd, K.A.; Reist, J.D. Effects of Partially Anadromous Arctic Charr (*Salvelinus alpinus*) Populations on Ecology of Coastal Arctic Lakes. *Ecosystems* **2010**, *13*, 261–274. [CrossRef]
78. Corush, J.B. Evolutionary Patterns of Diadromy in Fishes: More than a Transitional State between Marine and Freshwater. *BMC Evol. Biol.* **2019**, *19*, 1–13. [CrossRef] [PubMed]
79. Laske, S.M.; Amundsen, P.A.; Christoffersen, K.S.; Erkinaro, J.; Guðbergsson, G.; Hayden, B.; Kahilainen, K.K.; Klemetsen, A.; Knudsen, R.; L'Abée-Lund, J.H.; et al. Circumpolar Patterns of Arctic Freshwater Fish Biodiversity: A Baseline for Monitoring. *Freshw. Biol.* **2022**, *67*, 176–193. [CrossRef]
80. Wang, S.; Wang, Y.; Ma, J.; Ding, Y.; Zhang, S. Phosvitin Plays a Critical Role in the Immunity of Zebrafish Embryos via Acting as a Pattern Recognition Receptor and an Antimicrobial Effector. *J. Biol. Chem.* **2011**, *286*, 22653–22664. [CrossRef]
81. Pietrzak, E.; Mazurkiewicz, J.; Slawinska, A. Innate Immune Responses of Skin Mucosa in Common Carp (*Cyprinus carpio*) Fed a Diet Supplemented with Galactooligosaccharides. *Animals* **2020**, *10*, 438. [CrossRef]
82. Mansson, M.; Nielsen, A.; Kjærulff, L.; Gotfredsen, C.H.; Wietz, M.; Ingmer, H.; Gram, L.; Larsen, T.O. Inhibition of Virulence Gene Expression in Staphylococcus aureus by Novel Depsipeptides from a Marine Photobacterium. *Mar. Drugs* **2011**, *9*, 2537–2552. [CrossRef]

83. Element, G.; Engel, K.; Neufeld, J.D.; Casselman, J.M.; van Coeverden de Groot, P.; Greer, C.W.; Walker, V.K. Seasonal Habitat Drives Intestinal Microbiome Composition in Anadromous Arctic Char (*Salvelinus alpinus*). *Environ. Microbiol. Rep.* **2020**, *22*, 3112–3125. [CrossRef]
84. Lai, K.P.; Lin, X.; Tam, N.; Ho, J.C.H.; Wong, M.K.-S.; Gu, J.; Chan, T.F.; Tse, W.K.F. Osmotic Stress Induces Gut Microbiota Community Shift in Fish. *Environ. Microbiol.* **2020**, *22*, 3784–3802. [CrossRef]

 *fishes*

Article

Inadequate Sampling Frequency and Imprecise Taxonomic Identification Mask Results in Studies of Migratory Freshwater Fish Ichthyoplankton

Paulo Santos Pompeu [1,*], Lídia Wouters [1], Heron Oliveira Hilário [2], Raquel Coelho Loures [3], Alexandre Peressin [4], Ivo Gavião Prado [5], Fábio Mineo Suzuki [5] and Daniel Cardoso Carvalho [2]

[1] Graduate Program in Applied Ecology, Departament of Ecology and Conservation, Federal University of Lavras, Lavras 37203-202, Brazil; lidia_wouters@hotmail.com

[2] Laboratório de Genética da Conservação, Graduate Program in Vertebrate Biology, Pontifícia Universidade Católica de Minas Gerais, Belo Horizonte 30535-610, Brazil; heronoh@gmail.com (H.O.H.); carvalho.lgc@gmail.com (D.C.C.)

[3] Cemig Geração e Transmissão SA, Rua Barbacena 1200, Santo Agostinho, Belo Horizonte 30190-131, Brazil; raquel.fontes@cemig.com.br

[4] Centro de Ciências da Natureza, Universidade Federal de São Carlos (UFSCar)—Campus Lagoa do Sino, Buri 75775-000, Brazil; alexandre.peressin@gmail.com

[5] Pisces—Consultoria e Serviços Ambientais, Lavras 37206-662, Brazil; ivogaviaoprado@gmail.com (I.G.P.); suzuki.fms@gmail.com (F.M.S.)

* Correspondence: pompeu@ufla.br

Abstract: In South America, knowledge of major spawning sites is crucial for maintaining migratory fish populations. In this study, we aimed to understand the spatio-temporal distribution of fish eggs in the upper São Francisco River using high sampling frequency and DNA metabarcoding identification. We evaluated the possible effects of the non-molecular identification of eggs and decreased sampling frequency on the determination of spawning sites and major breeding periods. Collections were carried out every three days from November 2019 to February 2020. We found that, if we had assumed that all of the free and non-adhesive sampled eggs belonged to migratory species, as is usual in the literature, this assumption would have been wrong for both the spawning sites and the breeding periods. Moreover, any decrease in the frequency of sampling could dramatically affect the determination of the major spawning rivers, and the spawning events of some of the migratory species may not have been detected. Therefore, without the proper identification and adequate sampling frequency of eggs, important spawning sites may be overlooked, leading to ineffective or inappropriate conservation measures.

Keywords: DNA metabarcoding; neotropics; South America; São Francisco River; spawning sites

Key Contribution: Studies with ichthyoplankton without molecular identification of eggs and with low sampling frequency may not detect important spawning events. This can lead to the incorrect identification of spawning sites and inadequate conservation strategies.

Citation: Pompeu, P.S.; Wouters, L.; Hilário, H.O.; Loures, R.C.; Peressin, A.; Prado, I.G.; Suzuki, F.M.; Carvalho, D.C. Inadequate Sampling Frequency and Imprecise Taxonomic Identification Mask Results in Studies of Migratory Freshwater Fish Ichthyoplankton. *Fishes* **2023**, *8*, 518. https://doi.org/10.3390/fishes8100518

Academic Editors: Robert L. Vadas, Jr. and Robert M. Hughes

Received: 7 October 2023
Accepted: 17 October 2023
Published: 19 October 2023

1. Introduction

In South America, large migratory fish species are important for commercial and sport fisheries because of their size and abundance [1]. These species are also impaired by river fragmentation by damming because they exhibit complex upstream and downstream movements and require different habitats in order to complete their reproductive cycles [2–4]. Moreover, a lack of connectivity hinders access to spawning sites, and the formation of large reservoirs directly interferes with egg and larvae drift [5], which would normally be carried to floodplains during floods [6].

Considering the high degree of fragmentation in South American basins and the many new hydroelectric projects planned [7], knowledge of the major spawning sites in each basin is crucial for the establishment of appropriate measures for maintaining migratory fish populations [8]. Thus, studies of the early life stages of fish have become increasingly common [9], especially those focusing on the ichthyoplankton of migratory species. These studies have investigated spawning periods and sites, spawning intensity e.g., [10,11], factors related to reproductive success, and the species-specific characteristics of those processes [12].

One of the major methodological challenges in these studies is the correct taxonomic identification of ichthyoplankton. In the case of larvae, despite the high diversity and high rates of endemism [13], taxonomic keys are available for some neotropical river basins [8,9,14], enabling the identification of family or genus. The identification of eggs, on the other hand, has been more limited. Because having free and non-adhesive eggs is a trait associated with migratory species [15], some authors have considered that, in conventional ichthyoplankton sampling, these eggs would predominantly originate from species that exhibit migratory behavior, e.g., [16,17]. Other studies have distinguished between migratory and non-migratory species by using information on the size of the perivitelline space [6,8,9], which is wider in the former. Nonetheless, these traits have been shown to have low accuracy in discriminating between those two species groups [18], which is particularly concerning, given that the presence of eggs is the most accurate indicator of the spawning location.

New genetic tools have shown promise as a means of enabling the improved identification of ichthyoplankton, especially by enhancing the taxonomic resolution of sampled eggs. High-throughput sequencing (HTS) platforms associated with DNA barcoding (i.e., DNA metabarcoding) allow for the identification of multiple species from environmental bulk samples. This methodology has tremendous potential for monitoring and assessing environmental quality [19] has been referred to as DNA-based next-generation biomonitoring, or "Biomonitoring 2.0" [20]. As such, non-invasive methods ensure a rapid and cost-efficient biodiversity assessment of many difficult-to-identify organisms, such as ichthyoplankton samples, which is particularly important when dealing with the megadiverse neotropical ichthyofauna [21].

Another key aspect of ichthyoplankton studies is related to the periodicity of sampling, because neotropical migratory species usually have a single spawning event during the breeding season [22], and the number of spawning events for each species is unknown for most of them. Many neotropical ichthyoplankton studies have been conducted either bi-weekly [23,24] or monthly [25,26]. Studies at a greater sampling frequency are less common but indicate that some spawning events occur over only a few days [4,16].

Therefore, we sought to understand the spatio-temporal distribution of fish eggs in the upper São Francisco River by using a high sampling frequency and DNA metabarcoding identification. We evaluated the possible effects of the non-molecular identification of eggs on the determination of potential spawning sites and on the main breeding periods of migratory species. We also determined how a decreased sampling frequency affected the determination of the major spawning sites in the basin and the identification of the spawning sites of each migratory species.

2. Materials and Methods

2.1. Study Area

Our study comprised the São Francisco River basin upstream of Três Marias Reservoir, Brazil. There, the São Francisco River mainstem has a lotic segment of approximately 400 km and a draining an area of 26,680 km^2. Some local tributaries, such as the Pará River, are recognized feeding sites, whereas the Samburá and Bambuí Rivers and the São Francisco River headwaters are important spawning sites for migratory species [4]. The average annual precipitation in the São Francisco Basin is 1036 mm, and the Köppen climate is classified as Cwa, which is characterized by an October-to-March rainy season [27]. Fish spawning is concentrated between November and February [4,28].

We selected six sampling sites for ichthyoplankton sampling. Once the ichthyoplankton was transported over a long stretch of river, each point represented its respective upstream river. One site was located in the Samburá River (SAM), one site in the Bambuí River (BAM), and one site in the Pará River (PAR), always near their mouths. Three sites were located along the São Francisco River mainstem (upstream of the Samburá River mouth—SFS; upstream of the Bambuí River mouth—SFB; and upstream of the Pará River mouth—SFP), (Figure 1). The distribution of the sites aimed to encompass all major tributaries of the upper basin, as well as intermediate stretches of the São Francisco River.

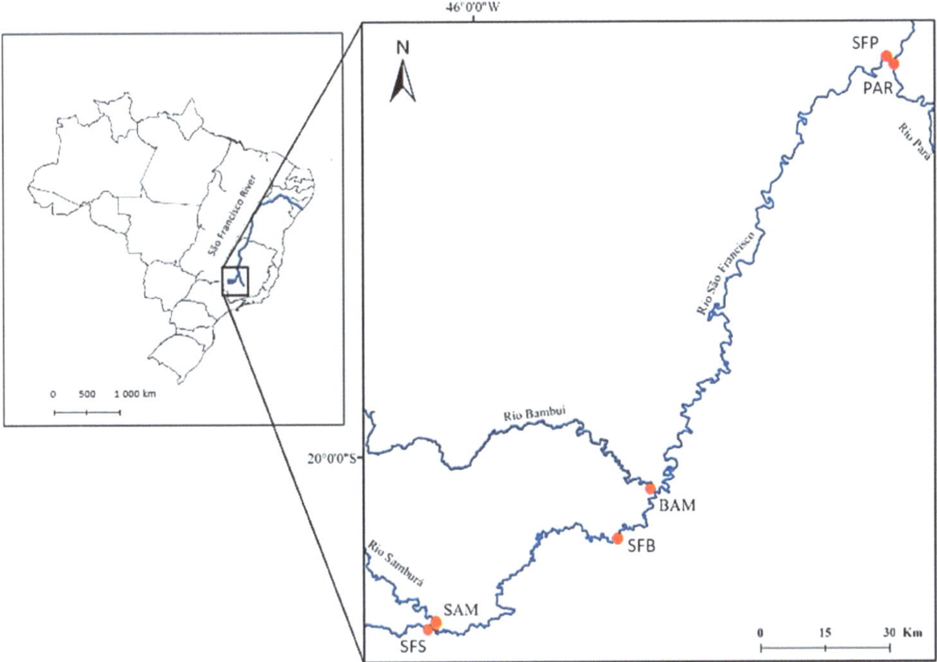

Figure 1. Map of the study area, showing the locations of the sampling sites in the upper São Francisco River, Minas Gerais, Brazil.

2.2. Methods

Migratory fish in the São Francisco and Paraná River basins do not spawn in floodplains. Instead, spawning often occurs in shallower and smaller-sized tributaries, as in the case of our study region [4]. The ichthyoplankton is then passively transported to the floodplains, where they arrive as larvae at the end of yolk sac absorption. Because spawning occurs in such shallow and turbulent rivers, no differences in their distribution in the water column are expected. In fact, even in deep South American rivers, no differences in ichthyoplankton abundance have been observed between samples taken at the surface and those at the river bottom [29]. In the same study, the authors reported differences in ichthyoplankton abundance between sampling hours; however, those also varied with distance from the collection site to the spawning region. Therefore, defining the ideal time is not straightforward, but has been resolved below.

Because the focus of this work was to assess the risks of not considering the genetic identification of eggs for defining spawning sites and times, as well as the effects of reducing the sampling frequency, we took care to replicate the sampling design commonly used in studies of this nature. Therefore, we maximized the sampling frequency (every three days at all points), a collection effort that is rare in Brazil, but that prevented sampling the same point multiple times per day. That was a choice that had to be made because of our available budget for both field collections and genetic analyses—a budget already much higher than the majority of ichthyoplankton studies in Brazil. Therefore, samples were taken only near the surface and at sunset, because no differences in egg abundance are expected at different depths [29], especially in shallow rivers, and because spawning occurs in the late afternoon and evening, near sunset [30].

The collections were carried out from 1 November 2019 to 29 February 2020, every three days, in the late afternoon, at all six points, resulting in 41 samples per point (246 samples in total). Because the hatching time of migratory fish in the São Francisco basin is always <24 h [31], the 3-day collection interval ensured that each sample represented a potentially independent spawning event. A conical net (40-cm diameter) with a flow meter attached was positioned at the location with the greatest current velocity for 10 min at a depth of 0.5 m. The collected material was placed in 600-mL plastic jars with absolute ethanol and taken to the laboratory for screening via a Bogorov plate and a stereomicroscope.

All samples with fish eggs were analyzed using the DNA metabarcoding method, which allows the determination of multiple species from a single sample through high-performance DNA sequencing (Illumina). DNA was extracted using the salting-out method adapted from [32]. The DNA was quantified on a NanoDrop 2000 spectrophotometer (Thermo Scientific, Cleveland, OH, USA), and then the samples were normalized to 100 ng/μL. Because our focus was to locate spawning rivers, larvae were not considered.

A 655 bp fragment of the 5′ end of the mitochondrial COI gene was amplified via PCR using a combination of different primers (F1 and FishR1), modified from sequences already published in the literature. The original sequences received a tail (Illumina pre-adapter) complementary to the adapter used in a subsequent second PCR. This reaction was performed with a water sample to monitor possible contamination (negative control), as well as a positive control. The PCR products were then purified with magnetic beads (Agencourt AMPure XP®—Beckman Coulter).

After purification, a PCR was performed with Nextera Index kit® adapters (Illumina Inc., San Diego, CA, USA) to amplify the amplicon set from the previous step. In this reaction, adapters compatible with the Illumina next-generation sequencing system (P5 and P7) were used as primers. A single index combination (specific sequences associated with the Illumina adapter) was used for the subsequent identification of each sample, because all points will form a single sequencing set. The amplification product was evaluated on a 1.5% agarose gel.

The samples were successfully amplified in the PCR because they showed the expected band pattern for the COI fragment (655 bp COI + 60 bp adapter + 64 bp index = 780 bp). No amplification was observed for the negative control, indicating no contamination in the reactions. The PCR products were then, again, purified with magnetic beads (Agencourt AMPure XP®—Beckman Coulter), quantified on a Nanodrop, and normalized to 20 ng/μL.

All samples were pooled together, and this pool was purified using a Zymoclean™ Large Fragment DNA Recovery kit (Zymo Research) to remove spurious fragments from the desired-sized fragment (655 bp COI + 60 bp adapter + 64 bp index = 779 bp). Using real-time PCR, performed with a KAPA Biosystems Quantification Kit (Illumina, São Paulo, Brazil) reagent, the pool was quantified, diluted to a concentration of 2 nM, and again quantified to confirm the final concentration. The final solution was denatured and loaded onto the MiSeq® equipment (Illumina) using a Miseq v3 300-cycle sequencing kit (2 × 150 bp), with a final concentration of 16 pM.

The bioinformatic analysis was performed using a custom pipeline written in R (R Core Team, 2022) and DADA2 [33] and Phyloseq [34] packages, as well as the cutadapt program [35]. Briefly, the demultiplexed sequencing reads were downloaded from Basespace (Illumina). An initial quality control step was carried out where reads with undetermined bases (Ns), or a Q-score of <20, were removed. Then, primer sequences were detected and removed. The remaining reads were submitted to dereplication determination of ASVs (amplicon sequencing variants) and chimera removal, using DADA2 core functions. The R1 and R2 reads were analyzed as complementary datasets and used independently of ASV determination, as the amplicon span hinders read merge by overlap. These ASVs were submitted to a first round of taxonomic classification, performed with a Bayesian classifier integrated into the DADA2 package, using a custom database built from 114,425 vertebrate sequences, available on BOLD (https://www.boldsystems.org/ accessed on 10 August 2022). A second round of taxonomic classification was performed using a similarity search in the NCBInt (nucleotide database) with a local implementation of the BLASTn [36], with thresholds of sequence coverage > 80% and sequence identity > 80%. The taxonomy identification obtained from these complementary classifications was manually curated for each ASV and sample, considering the BLASTn similarity and coverage values and the distribution in the basin of the identified species or close taxa.

Previous work using the DNA metabarcoding approach on ichthyoplankton bulk samples was able to successfully detect alpha diversity at the species level and has provided good estimates of the relative abundance of the larvae [37]. Thus, we used the relative read abundance (RRA) to estimate the species abundance for each bulk sample. The RRA of each ASV on each sample was obtained by dividing the ASV absolute abundance by the total absolute abundance of the sample.

The total density of eggs in each sample was calculated by standardizing the abundance per 10 m^3 of filtered water [6]. The density of eggs from migratory species was also estimated per sample based on their relative read abundance (RRA) in each sample. For both estimates (total and migratory species), periods with abundance peaks were visually compared, and their congruence was tested by Pearson correlation. Species were classified as migratory according to Sato and Godinho [38].

To evaluate the beta diversity patterns of sampling rivers and dates, we performed a non-metric multidimensional scaling (NMDS) analysis. The input data consisted of the taxa abundance of each ichthyoplankton pool as input variables and estimated distances were calculated with Bray–Curtis dissimilarity using the function metaMDS of the R-package vegan [39].

The importance of each river as a spawning ground was evaluated by comparing densities by ANOVA or Kruskal–Wallis, depending on their distribution, considering the total number of eggs, and considering only the estimated number of eggs from migratory species. This evaluation was performed for the entire dataset and for simulations of six- and fifteen-day sampling intervals. In the case of the six-day interval, these were produced considering a start date of November 1st or 4th. For the 15-day interval, scenarios were simulated with collections starting on November 1st, 4th, 7th, 10th, and 13th.

To evaluate the effect of sampling frequency, for the three sampling scenarios (three-, six-, and fifteen-day intervals), the number of migratory species with recorded spawning in each river and the number of spawning rivers for each migratory species were compared by ANOVA.

3. Results

A total of 48,465 fish eggs were collected from the upper São Francisco basin and 71% of the samples contained fish eggs. After DNA metabarcoding analysis, a total of 14,817,732 raw paired DNA reads were obtained for the 149 egg pools, with an average of 49,724 reads per pool. After quality control, error correction, dereplication, and chimera removal, 11,995,008 reads remained, with an average of ~80,500 DNA sequences per pool, corresponding to a total of 507 unique ASVs of 230 bp on average. This enabled the identification of 35 fish taxa, including 7 migratory species (Appendix A). Seven taxa were identified to genus level and three to family.

The majority of the identified ASVs (63.5%) were associated with a single species, mandi *Pimelodus pohli*. The migratory species, *Prochilodus argenteus*, *P. costatus*, *Megaleporinus obtusidens*, and *Leporinus taeniatus*, accounted for 2.92%, 2.88, 2.26%, and 0.9% of the detected DNA sequences, respectively. The other migratory species found (*Pseudoplatystoma corruscans*, *Brycon* sp., and *Megaleporinus reinhardti*) represented <0.01% each. Despite the dominance of a single species, considerable variation in the ichthyoplankton composition was observed between the sampling points and among the different samples from each river (Figure 2).

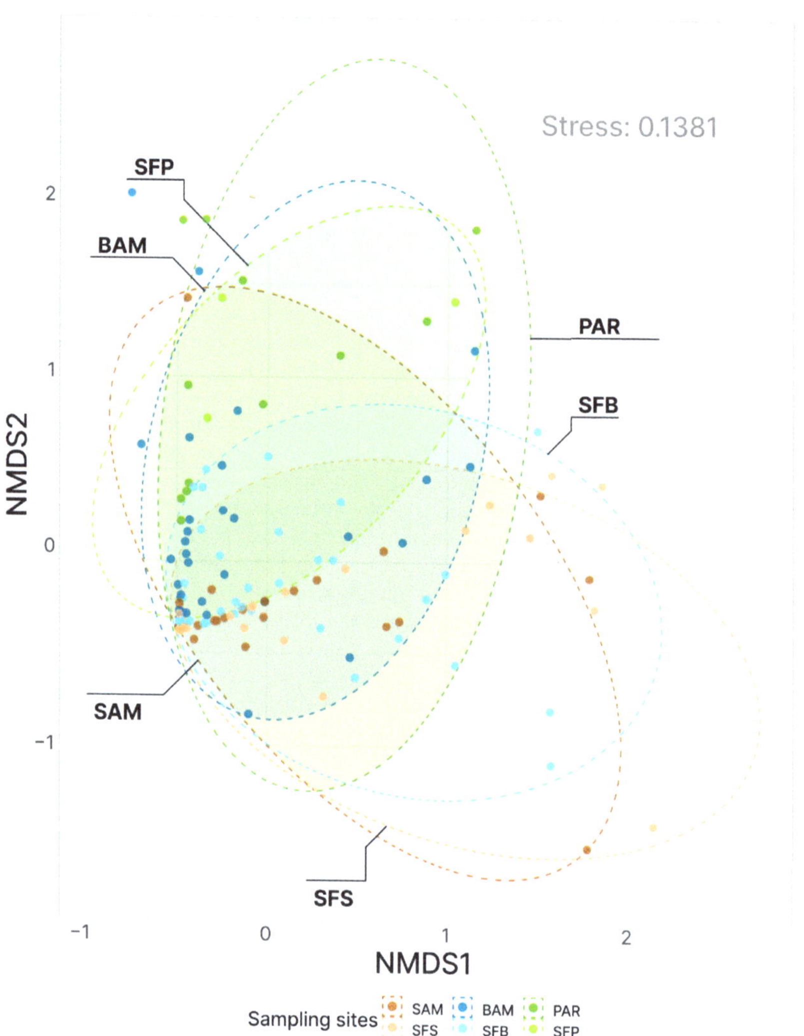

Figure 2. Non−metric multidimensional scaling (NMDS) plot showing ordination based on DNA metabarcoding of taxa abundance, with similarity estimated using Bray–Curtis. Ellipses encompass all samples from the same site. São Francisco River sites are represented by SFB, SFP, and SFS; and tributaries by PAR, BAM, and SAM.

When comparing the total egg densities, the SFB site stood out, having both the greatest averages and the greatest peaks of egg densities (Figure 3A) (KW = 2.47; $p = 0.01$). However, when considering only migratory-fish-egg abundances, the Bambuí (BAM) and Samburá (SAM) River sites indicated the greatest reproductive activity (Figure 3B) (F = 4.76; $p < 0.001$).

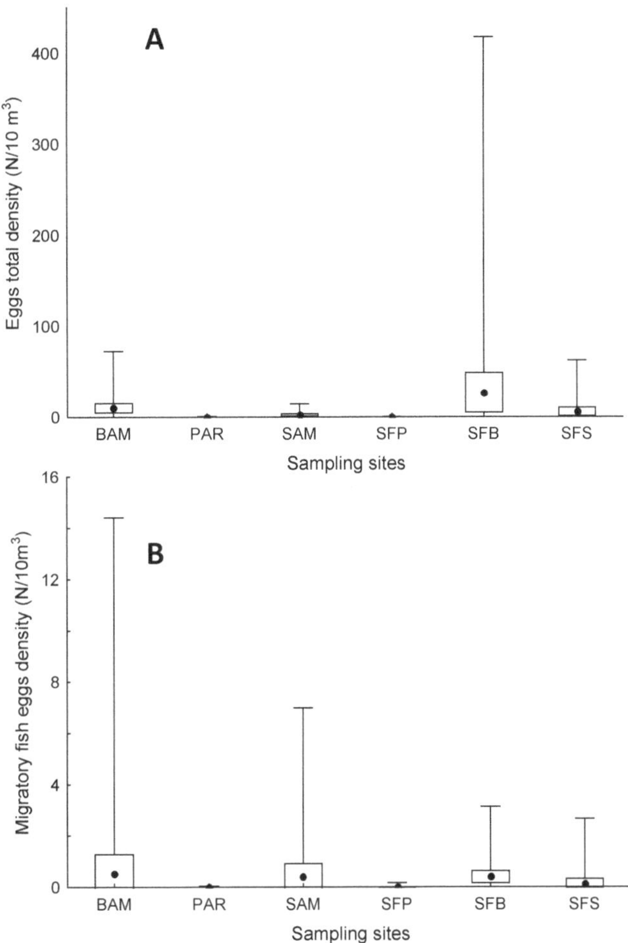

Figure 3. Estimated egg density of all species (**A**) and migratory species (**B**) per sampling point in the upper São Francisco River from samplings every three days (dot = mean; box = standard error; whisker = range).

When considering the total egg densities, the largest spawning events occurred between the second half of November and the first week of January. However, when considering only the abundance of the migratory fish eggs, the largest reproductive events occurred in the first half of November and from the second half of January to the first half of February (Figure 4), and both of these evaluations were not congruent (r = 0.05; $p = 0.47$).

For the total egg density, both of the six-day sampling simulations pointed to significant differences among the sites (F = 4.74; $p < 0.001$) but produced different results regarding the importance of each river for fish spawning. Whereas, in one simulation, the SFB site remained the most important breeding site, in the other, the BAM and SFS sites had similar migratory-fish-egg densities (Figure 5A). The two six-day interval simulations for estimating migratory-egg densities were not able to capture differences among the sites (F = 1.34; $p = 0.24$). In addition, in one simulation, the Bambuí River (BAM) stood out, whereas, in the other, Samburá River (SAM) was the most important breeding location (Figure 5B).

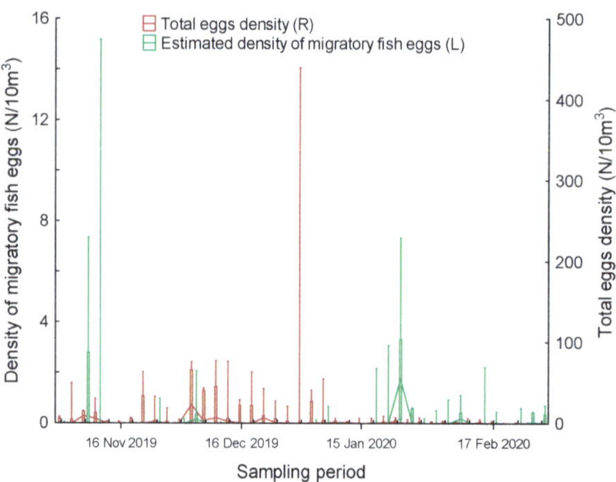

Figure 4. Total egg density (red) and estimated density of migratory species eggs (green) during the sampling period in the upper São Francisco River from sampling every three days (dot = mean; box = standard error; whisker = range).

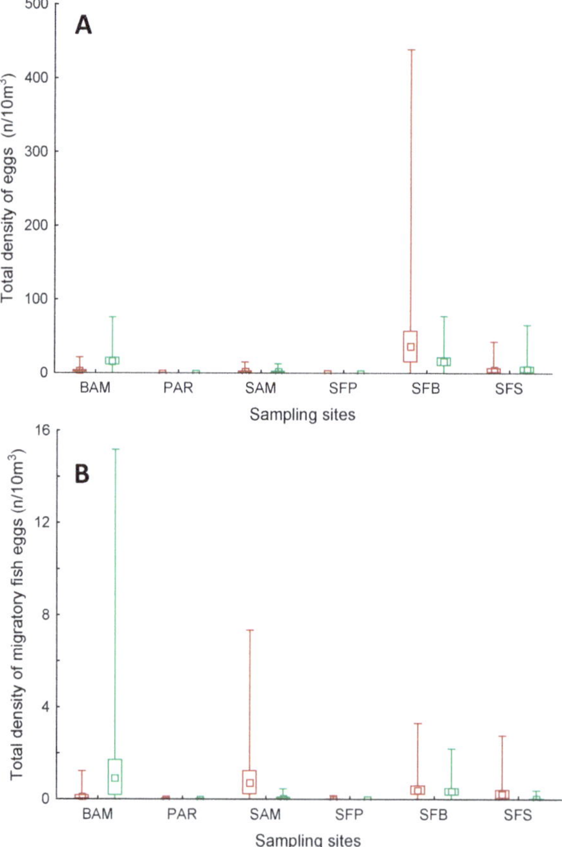

Figure 5. Total egg density (**A**) and estimated egg density of migratory species (**B**) by collection point in the upper São Francisco River, considering samplings every six days. Different colors represent each of the two simulations (dot = mean; box = standard error; whisker = range).

Even more divergent results were observed for the 15-day sampling intervals. For the total egg density, in only two simulations did the SFB site remain the most important, whereas, in the other three, the densities were more similar between the Bambuí (BAM), Samburá (SAM), and upper São Francisco River (SFS) sites (Figure 6A). For migratory fish, the importance of the Bambuí River (BAM) was shown in only one simulation, and in only two out of five simulations for the Samburá River (SAM) (Figure 6B). In all of these cases, the differences between the points were not statistically significant.

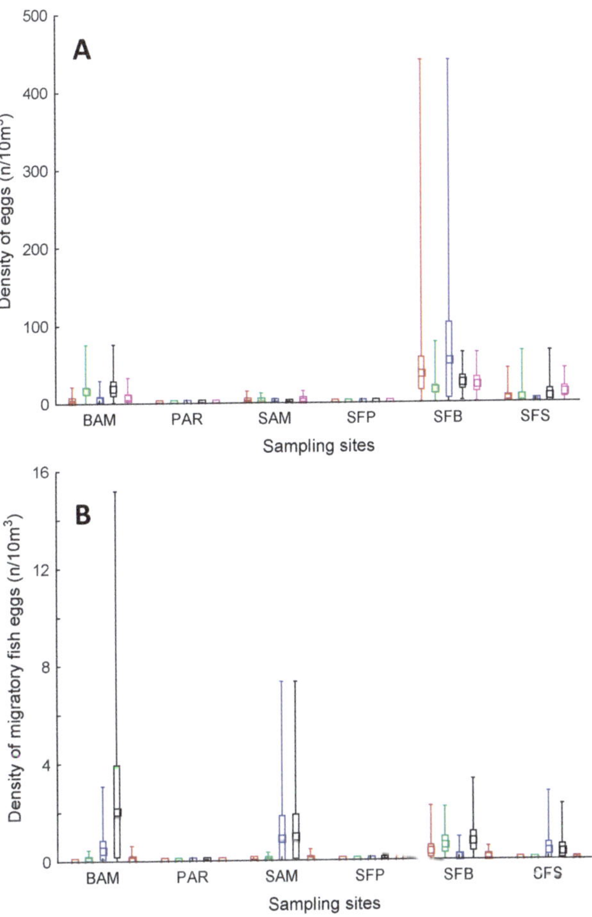

Figure 6. Total egg density (**A**) and estimated density of eggs from migratory species (**B**) per sampling site in the upper São Francisco River, considering sampling every 15 days. Different colors represent each of the five simulations (dot = mean; box = standard error; whisker = range).

Regarding the spawning grounds for migratory species, an increase in the sampling interval directly interfered with the quality of the information about the importance of the sampled sites (F = 8.88; $p < 0.001$). For all of the sites, increasing the sampling interval from three to six days prevented the identification of some species, and, for four sites, a fifteen-day interval resulted in a total non-recording of migratory species (Figure 7). Similarly, a progressive increase in the sampling interval decreased the number of spawning rivers identified for each migratory species (F = 14.26; $p < 0.001$), and four of them (*Brycon* sp., *P. corruscans*, *P. costatus*, and *M. reinhardti*) may not have their reproductive event recorded in the basin if sampled only every 15 days (Figure 8).

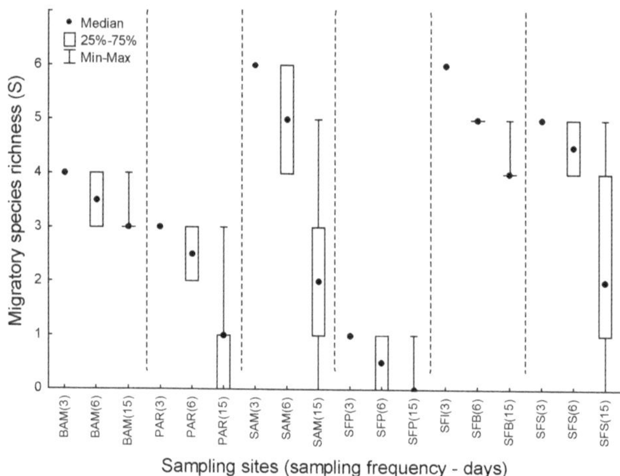

Figure 7. Number of migratory species with spawning recorded for each sampling point, considering intervals of 3, 6, or 15 days between collections.

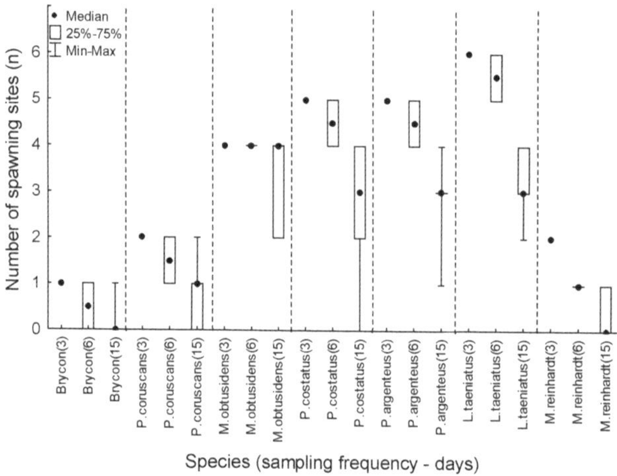

Figure 8. Number of spawning sites inferred for each migratory species, considering sampling intervals of 3, 6, or 15 days.

4. Discussion

We determined the spatio-temporal distribution of fish eggs in the upper São Francisco River using high frequency sampling and DNA metabarcoding identification. The molecular identification of eggs and high frequency sampling was necessary to obtain reliable data. For instance, our results showed that, if we had assumed that all of the sampled eggs belonged to migratory species, this assumption would have been wrong for both the spawning sites and the breeding periods. Moreover, the six- and fifteen-day sampling intervals dramatically affected the determination of the major spawning rivers in the upper basin, and the spawning events of some migratory species were not detected.

The DNA metabarcoding identification was able to determine the presence of eggs from seven migratory fish species. Of those known to occur in the basin [40], only the eggs of dourado (*Salminus franciscanus*) were undetected. Most of the eggs were identified to species level; however, the absence of complete genetic databases limited some identifications to genus or family levels only [21]. As expected, eggs from species with parental care, such as Loricariidae and Cichliformes [41], or with internal fertilization, such as Auchenipteridae [42] and Poeciliidae [43], were not recorded. However, eggs from rheophilic species that do not undertake long migrations predominated in the samples, especially *Pimelodus pohli*. This species performs fractional spawning and reproduces throughout the year, with spawning peaks occurring between November and February [44].

The identification of ichthyoplankton through morphological characteristics observation is highly complex, especially during the embryonic phase, when there is great morphological similarity among species [6,45]. Even considering the differences in the perivitelline space of the eggs, which tends to be larger in migratory species than in sedentary species [6], the distinction between the groups is imprecise [18]. Given this challenge, it is common for eggs to only be quantified and evaluated collectively. Although these analyses can contribute to a general understanding of reproductive dynamics, the use of such methods to evaluate the reproductive patterns of a specific group is inadequate. In this study, higher egg densities from migratory species were recorded in the Samburá and Bambuí Rivers, two recognized spawning rivers based on previous telemetry studies [4]. The importance of these two São Francisco River tributaries for migratory species would be greatly underestimated if all of the sampled eggs were presumed to be from migratory species. Similarly, non-molecular identification would point to spawning periods that would lead to different interpretations regarding the main environmental factors that trigger the reproductive process.

In addition to the method of identification, sampling frequency is another key factor that directly affects the detection of reproductive patterns. This factor was particularly evident in the significant temporal and spatial variation observed in ichthyoplankton composition. Ichthyoplankton studies conducted in neotropical rivers have commonly adopted biweekly or longer sampling intervals, e.g., [24,25,46–48]. Incomplete identification becomes even more critical when we consider the fish monitoring programs developed by the hydroelectric sector. In most cases, these studies are conducted with quarterly or biannual frequency [49], resulting in sporadic collections that may or may not be outside of the reproductive period of migratory species. The minimum ichthyoplankton sampling protocol [50], a reference used by environmental agencies to guide the environmental regulation of hydropower plants, recommends monthly collections over a period of one year for surveys and for monthly collection to be carried out for at least four months in the reproductive season for monitoring. Our results indicate that the greater the sampling interval, the lower the chances of recording important reproductive events. Migratory species exhibit spawning that is highly synchronized with environmental variables [4,11,24], and they exhibit reproductive homing [51], with fidelity to a single spawning site. Therefore, an inadequate sample frequency leads to the erroneous conclusion that a river or region is not important as a spawning site for some migratory species. An error such as this would result in inadequate management and conservation measures, because having quality data is essential for us to understand the impact of existing or new dams on migratory fish populations [8]. In the upper São Francisco River, the spawning of four out of the seven migratory species would have gone undetected if the sampling had been conducted biweekly or monthly. Among those species is Surubim *Pseudoplatystoma corruscans*, the largest-sized species in the basin, which is currently listed as endangered [52], and *Prochilodus argenteus*, which comprises half of the fisheries in some stretches of the São Francisco River basin [38,53].

5. Conclusions

Most South American river basins, which harbor an enormous biodiversity of fish, are already highly fragmented [54]. In these systems, free-flowing river segments are refuges for migratory species, some of which are rare and/or threatened [11,24,47,48,55]. In this context, the identification of critical habitats, especially spawning sites, is essential for taking appropriate conservation measures. We have found that, without the proper identification and adequate sampling frequency of eggs, important spawning sites will be overlooked, leading to ineffective or inappropriate conservation measures. Therefore, it is necessary to improve the ichthyoplankton surveys and monitoring programs conducted in environmental impact studies related to aquatic ecosystem projects. Given the importance of molecular techniques to improve species identification accuracy, it is crucial to provide incentives and support in order to make them more financially accessible.

Author Contributions: P.S.P., R.C.L., A.P. and L.W. conceived and designed the investigation. L.W., A.P., I.G.P. and F.M.S. performed the fieldwork. P.S.P., I.G.P., A.P., F.M.S. and L.W. analyzed the data. D.C.C. and H.O.H. performed the genetic analyses of eggs. P.S.P. wrote the paper. All co-authors revised the paper. All authors have read and agreed to the published version of the manuscript.

Funding: This research was funded by Companhia Energética de Minas Gerais (CEMIG) P&D Agência Nacional de Energia Elétrica (ANEEL) GT612. P.S.P. and D.C.C were supported by the CNPq productivity grants (302328/2022-0; 312102/2022-4, respectively).

Institutional Review Board Statement: All proceedings were carried out under capture permission (SISBIO 10327) and the approval of the ethics committee of the Universidade Federal de Lavras (CEUA UFLA 003/2019).

Informed Consent Statement: Not applicable.

Data Availability Statement: The data that support the findings of this study are available upon request from the corresponding author (P.S.P.).

Acknowledgments: We thank the local residents who volunteered to collect ichthyoplankton samples, as well as the collectors from the Fish Ecology Laboratory who assisted with sample screening. We also thank the anonymous reviewers and special-issue editors for their comments on the manuscript.

Conflicts of Interest: The authors declare no conflict of interest.

Appendix A

Table A1. Relative abundance (ASV %) of fish species identified by DNA metabarcoding in ichthyoplankton samples from the upper São Francisco River. Migratory species and parental care are also indicated [22,31,36,56,57].

Fish Species	ASV %	Migratory (Y/N)	Parental Care (Y/N)
ORDER CLUPEIFORMES			
Family Engraulidae			
Anchoviella vaillanti (Steindachner, 1908)	0.06	N	N
ORDER CHARACIFORMES			
Family Crenuchidae			
Characidium fasciatum Reinhardt, 1867	<0.01	N	N
Characidium zebra Eigenmann, 1909	0.07	N	N
Family Parodontidae			
Parodon hilarii Reinhardt, 1867	<0.01	N	N
Family Serrasalmidae			
Myleus micans (Lütken, 1875)	0.01	N	N
Myleus sp.	<0.01	N	N
Myloplus sp.	<0.01	N	N
Serrasalmus sp.	<0.01	N	Y
Family Anostomidae			
Leporellus vittatus (Valenciennes, 1850)	0.86	N	N
Leporinus piau Fowler, 1941	0.04	N	N
Leporinus taeniatus Lütken, 1875	2.26	Y	N
Megaleporinus obtusidens (Valenciennes, 1837)	0.90	Y	N
Megaleporinus reinhardti (Lütken, 1875)	<0.01	Y	N
Schizodon knerii (Steindachner, 1875)	1.94	N	N

Table A1. *Cont.*

Fish Species	ASV %	Migratory (Y/N)	Parental Care (Y/N)
Family Curimatidae			
Curimatidae sp.	<0.01	N	N
Family Prochilodontidae			
Prochilodus argenteus Spix and Agassiz, 1829+	2.92	Y	N
Prochilodus costatus Valenciennes, 1850 +	2.88	Y	N
Family Bryconidae			
Brycon sp.	<0.01	?	N
Family Characidae			
Astyanax scabripinnis (Jenyns, 1842)	0.02	N	N
Bryconamericus sp.	0.26	N	N
Piabina sp.	0.11	N	N
Psalidodon fasciatus (Cuvier, 1819)	0.08	N	N
Tetragonopterus chalceus Spix and Agassiz, 1829	<0.01	N	N
ORDER SILURIFORMES			
Family Cetopsidae			
Cetopsidae sp.	0.05	N	N
Family Doradidae			
Doradidae sp.	<0.01	N	N
Family Heptapteridae			
Cetopsorhamdia iheringi Schubart and Gomes, 1959	0.57	N	N
Imparfinis sp.	<0.01	N	N
Family Pimelodidae			
Bergiaria westermanni (Lütken, 1874)	4.84	N	N
Pimelodus fur (Lütken, 1874)	1.29	N	N
Pimelodus maculatus Lacepède, 1803	11.54	N	N
Pimelodus pohli Ribeiro and Lucena, 2006	63.54	N	N
Pseudoplatystoma corruscans (Spix and Agassiz, 1829)	0.01	Y	N
ORDER GYMNOTIFORMES			
Family Sternopygidae			
Eigenmannia sp.	<0.01	N	N
ORDER PERCIFORMES			
Family Sciaenidae			
Pachyurus sp.	3.89	N	N
Plagioscion squamosissimus (Heckel, 1840)	<0.01	N	N

References

1. Agostinho, A.A.; Pelicice, F.M.; Gomes, L.C Dams and the fish fauna of the Neotropical region: Impacts and management related to diversity and fisheries. *Braz. J. Biol.* **2008**, *68*, 1119–1132. [CrossRef] [PubMed]
2. Agostinho, A.A.; Thomaz, S.M.; Gomes, L.C. Conservação da biodiversidade em águas continentais do Brasil. *Megadiversidade* **2005**, *1*, 70–78.
3. Godinho, A.L.; Pompeu, P.D.S. A importância dos ribeirões para os peixes de piracema. In *Águas, Peixes e Pescadores do São Francisco da Minas Gerais*; PUC Minas: Belo Horizonte, Brazil, 2003; pp. 361–372.
4. Lopes, J.D.M.; Pompeu, P.S.; Alves, C.B.M.; Peressin, A.; Prado, I.G.; Suzuki, F.M.; Facchin, S.; Kalapothakis, E. The critical importance of an undammed river segment to the reproductive cycle of a migratory Neotropical fish. *Ecol. Freshw. Fish* **2019**, *28*, 302–316. [CrossRef]
5. Pelicice, F.M.; Pompeu, P.S.; Agostinho, A.A. Large reservoirs as ecological barriers to downstream movements of Neotropical migratory fish. *Fish Fishes* **2015**, *16*, 697–715. [CrossRef]
6. Nakatani, K.; Agostinho, A.A.; Baumgartner, G.; Bialetzki, A.; Sanches, P.V.; Makrakis, M.C.; Pavanelli, C.S. *Ovos e Larvas de Peixes de Água Doce: Desenvolvimento e Manual de Identificação*; EDUEM: Maringá, Brazil, 2001.
7. Brasil, Ministério de Minas e Energia, Empresa de Pesquisa Energética. *Plano Nacional de Energia 2050/Ministério de Minas e Energia*; Empresa de Pesquisa Energética, MME/EPE: Brasília, Brazil, 2020.
8. Pompeu, P.S.; Agostinho, A.A.; Pelicice, F.M. Existing and future challenges: The concept of successful fish passage in South America. *River Res. Appl.* **2012**, *28*, 504–512. [CrossRef]
9. Orsi, M.L.; Almeida, F.S.; Swarça, A.C.; Claro-García, A.; Vianna, N.C.; Garcia, D.A.Z.; Bialetzki, A. *Ovos, Larvas e Juvenis dos Peixes da Bacia do Rio Paranapanema uma Avaliação para a Conservação*; Triunfal Gráfica e Editora: Assis, Brazil, 2016; 136p.

10. Reynalte-Tataje, D.A.; Nakatani, K.; Fernandes, R.; Agostinho, A.A.; Bialetzki, A. Temporal distribution of ichthyoplankton in the Ivinhema River (Mato Grosso do Sul State/Brazil): Influence of environmental variables. *Neotrop. Ichthyol.* **2011**, *9*, 427–436. [CrossRef]
11. Suzuki, F.M.; Pompeu, P.S. Influence of abiotic factors on ichthyoplankton occurrence in stretches with and without dams in the upper Grande River basin, south-eastern Brazil. *Fish. Manag. Ecol.* **2016**, *23*, 99–108. [CrossRef]
12. Baumgartner, G.; Nakatani, K.; Gomes, L.C.; Bialetzki, A.; Sanches, P.V.; Makrakis, M.C. Fish larvae from the upper Paraná River: Do abiotic factors affect larval density? *Neotrop. Ichthyol.* **2008**, *6*, 551–558. [CrossRef]
13. Reis, R.E.; Albert, J.S.; Di Dario, F.; Mincarone, M.M.; Petry, P.; Rocha, L.A. Fish biodiversity and conservation in South America. *J. Fish Biol.* **2016**, *89*, 12–47. [CrossRef]
14. Nascimento, F.L.; Lima, C.A.R.M.A. *Descrição das Larvas das Principais Espécies de Peixes Utilizadas pela Pesca, no Pantanal*; Embrapa: Brasília, Brazil, 2000; 25p.
15. Rizzo, E.; Sato, Y.; Barreto, B.P.; Godinho, H.P. Adhesiveness and surface patterns of eggs in neotropical freshwater teleosts. *J. Fish Biol.* **2002**, *61*, 615–632. [CrossRef]
16. Pompeu, P.S.; Nogueira, L.B.; Godinho, H.P.; Martinez, C.B. Downstream passage of fish larvae and eggs through a small-sized reservoir, Mucuri River, Brazil. *Zoology* **2011**, *28*, 739–746. [CrossRef]
17. Reynalte-Tataje, D.A.; Agostinho, A.A.; Bialetzki, A.; Hermes-Silva, S.; Fernandes, R.; Zaniboni-Filho, E. Spatial and temporal variation of the ichthyoplankton in a subtropical river in Brazil. *Environ. Biol. Fishes* **2012**, *94*, 403–419. [CrossRef]
18. Becker, R.A.; Sales, N.G.; Santos, G.M.; Santos, G.B.; Carvalho, D.C. DNA barcoding and morphological identification of neotropical ichthyoplankton from the Upper Paraná and São Francisco. *J. Fish Biol.* **2015**, *87*, 159–168. [CrossRef]
19. Leese, F.; Bouchez, A.; Abarenkov, K.; Altermatt, F.; Borja, Á.; Bruce, K.; Ekrem, T.; Čiampor, F., Jr.; Čiamporová-Zaťovičová, Z.; Costa, F.O.; et al. Why we need sustainable networks bridging countries, disciplines, cultures and generations for aquatic biomonitoring 2.0: A perspective derived from the DNAqua-Net COST action. *Adv. Ecol. Res.* **2018**, *58*, 63–99.
20. Baird, D.J.; Hajibabaei, M. Biomonitoring 2.0: A new paradigm in ecosystem assessment made possible by next-generation DNA sequencing. *Mol. Ecol.* **2012**, *21*, 2039–2044. [CrossRef]
21. Carvalho, D.C. Ichthyoplankton DNA metabarcoding: Challenges and perspectives. *Mol. Ecol.* **2022**, *31*, 1612–1614. [CrossRef]
22. Godinho, A.L.; Lamas, I.R.; Godinho, H.P. Reproductive ecology of Brazilian freshwater fishes. *Environ. Biol. Fishes* **2010**, *87*, 143–162. [CrossRef]
23. Cheshire, K.J.M.; Ye, Q.; Gillanders, B.M.; King, A. Annual variation in larval fish assemblages in a heavily regulated river during differing hydrological conditions. *River Res. Appl.* **2015**, *32*, 1207–1219. [CrossRef]
24. Rosa, G.R.; Salvador, G.N.; Bialetzki, A.; Santos, G.B. Spatial and temporal distribution of ichthyoplankton during an unusual period of low flow in a tributary of the São Francisco River, Brazil. *River Res. Appl.* **2018**, *34*, 69–82. [CrossRef]
25. Soares, M.d.L.; Massaro, M.V.; Hartmann, P.B.; Siveris, S.E.; Pelicice, F.M.; Reynalte-Tataje, D.A. The main channel and river confluences as spawning sites for migratory fishes in the middle Uruguay River. *Neotrop. Ichthyol.* **2022**, *20*, e210094. [CrossRef]
26. Melo-Silva, M.; da Silva, J.C.; Bialetzki, A. Community structure of fish larvae in different biotopes of a neotropical river. *Community Ecol.* **2022**, *23*, 1–12. [CrossRef]
27. CBHSF (Comitê da Bacia Hidrográfica do Rio São Francisco). *Diagnóstico da Dimensão Técnica e Institucional Volume 2: Caracterização da Bacia*; CBHSF: Belo Horizonte, Brazil, 2016; 292p.
28. Bazzoli, N. Parâmetros reprodutivos de peixes de interesse comercial na região de Pirapora. In *Águas, Peixes e Pescadores do São Francisco das Minas Gerais*; PUC Minas: Belo Horizonte, Brazil, 2003; pp. 291–306.
29. Hermes-Silva, S.; Reynalte-Tataje, D.; Zaniboni-Filho, E. Spatial and temporal distribution of ichthyoplankton in the upper Uruguay River, Brazil. *Braz. Arch. Biol. Tech.* **2009**, *52*, 933–944. [CrossRef]
30. Baumgartner, G.; Nakatani, K.; Gomes, L.; Bialetzki, A.; Sanches, P. Identification of spawning sites and natural nurseries of fishes in the upper Paraná River, Brazil. *Environ. Biol. Fishes* **2004**, *71*, 115–125. [CrossRef]
31. Sato, Y.; Fenerich-Verani, N.; Nuñer, A.P.O.; Godinho, H.P.; Verani, J.R. Padrões reprodutivos de peixes da bacia do São Francisco. In *Águas, Peixes e Pescadores do São Francisco das Minas Gerais*; PUC Minas: Belo Horizonte, Brazil, 2003; pp. 229–274.
32. Aljanabi, S.M.; Martinez, I. Universal and rapid salt-extraction of high quality genomic DNA for PCR-based techniques. *Nucleic Acids Res.* **1997**, *25*, 4692–4693. [CrossRef]
33. Callahan, B.J.; McMurdie, P.J.; Rosen, M.J.; Han, A.W.; Johnson, A.J.A.; Holmes, S.P. DADA2: High-resolution sample inference from Illumina amplicon data. *Nat. Methods* **2016**, *13*, 581–583. [CrossRef]
34. McMurdie, P.J.; Holmes, S. phyloseq: An R package for reproducible interactive analysis and graphics of microbiome census data. *PLoS ONE* **2013**, *8*, e61217. [CrossRef]
35. Martin, M. Cutadapt removes adapter sequences from high-throughput sequencing reads. *EMBnet J.* **2011**, *17*, 10–12. [CrossRef]
36. Altschul, S.F.; Gish, W.; Miller, W.; Myers, E.W.; Lipman, D.J. Basic local alignment search tool. *J. Mol. Biol.* **1990**, *215*, 403–410. [CrossRef]
37. Nobile, A.B.; Freitas-Souza, D.; Ruiz-Ruano, F.J.; Nobile, M.L.M.; Costa, G.O.; De Lima, F.P.; Camacho, J.P.M.; Foresti, F.; Oliveira, C. DNA metabarcoding of Neotropical ichthyoplankton: Enabling high accuracy with lower cost. *MBMG* **2019**, *3*, e35060. [CrossRef]
38. Sato, Y.; Godinho, H.P. Migratory Fishes of the São Francisco River. In *Migratory Fishes of South America: Biology, Fisheries and Conservation Status*; Carolsfeld, J., Harvey, B., Ross, C., Baer, A., Eds.; IDRC-World Bank: Ottawa, ON, Canada, 2003; pp. 195–232.

39. Oksanen, J.; Blanchet, F.G.; Kindt, R.; Legendre, P.; Minchin, P.R.; O'Hara, R.B.; Simpson, G.; Solymos, P.; Stevens, M.; Wagner, H. Vegan: Community Ecology Package. R Package, Version 2.5.6.1-29; 2015. Available online: https://CRAN.R-project.org/package=vegan (accessed on 1 August 2022).

40. Sato, Y.; Cardoso, E.L.; Amorim, J.C.C. *Peixes das Lagoas Marginais do Rio São Francisco a Montante da Represa de Três Marias (Minas Gerais)*, 3rd ed.; CODEVASF: Brasília, Brazil, 1988.

41. Blumer, L.S. A bibliography and categorization of bony fishes exhibiting parental care. *Zool. J. Linn.* **1982**, *75*, 1–22. [CrossRef]

42. Calegari, B.B.; Vari, R.P.; Reis, R.E. Phylogenetic systematics of the driftwood catfishes (Siluriformes: Auchenipteridae): A combined morphological and molecular analysis. *Zool. J. Linn.* **2019**, *187*, 661–773. [CrossRef]

43. Thibault, R.E.; Schultz, R.J. Reproductive adaptations among viviparous fishes (Cyprinodontiformes: Poeciliidae). *Evolution* **1978**, *32*, 320–333. [CrossRef]

44. Veloso-Junior, V.C.; Arantes, F.P.; dos Santos, J.E. Short communication Population structure and reproduction of the Neotropical catfish *Pimelodus pohli*, Ribeiro and Lucena, 2006 (Teleostei: Pimelodidae). *J. Appl. Ichthyol.* **2014**, *30*, 399–402. [CrossRef]

45. Reynalte-Tataje, D.A.; Lopes, C.A.; Massaro, M.V.; Hartmann, P.B.; Sulzbacher, R.; Santos, J.A.; Bialetzki, A. State of the art of identification of eggs and larvae of freshwater fish in Brazil. *Acta Limnol. Bras.* **2020**, *32*, e6. [CrossRef]

46. Suzuki, F.M.; Pires, L.V.; Pompeu, P.S. Passage of fish larvae and eggs through the Funil, Itutinga and Camargos Reservoirs on the upper Rio Grande (Minas Gerais, Brazil). *Neotrop. Ichthyol.* **2011**, *9*, 617–622. [CrossRef]

47. Lopes, C.A.; Zaniboni-Filho, E. Mosaic environments shape the distribution of Neotropical freshwater ichthyoplankton. *Ecol. Freshw. Fish* **2019**, *28*, 544–553. [CrossRef]

48. Azevedo-Santos, V.M.; Daga, V.S.; Pelicice, F.M.; Henry, R. Drifting in a free-flowing river: Distribution of fish eggs and larvae in a small tributary of a Neotropical reservoir. *Biota Neotrop.* **2021**, *21*, e20211227. [CrossRef]

49. Loures, R.C. Effectiveness of Fish Monitoring Programs in Hydropower Plant Reservoirs, Lavras. Ph.D. Thesis, Federal University of Lavras, Lavras, Brazil, 2019.

50. Bialetzki, A.; Tataje, D.; Oliveira, E.; Zaniboni-Filho, E.; Baumgartner, G.; Makrakis, M.; Leite, R. Protocolo mínimo de amostragem do ictioplâncton de água doce para estudos de levantamento, inventário e monitoramento ambiental para implantação de empreendimentos hidrelétricos. *Bol. Soc. Bras. Ictiol.* **2015**, *113*, 32–34.

51. Godinho, A.L.; Kynard, B. Migration and spawning of radio-tagged zulega *Prochilodus argenteus* in a dammed Brazilian river. *Trans. Am. Fish. Soc.* **2006**, *135*, 811–824. [CrossRef]

52. MMA (Ministério meio Ambiente). *Portaria no 148 de 7 de Junho de*; Diario Oficial da União: Brasília, Brazil, 2022.

53. Sato, Y.; Cardoso, E.L.; Godinho, A.L.; Godinho, H.P. Hypophysation parameters of the fish *Prochilodus marggravii* obtained in routine hatchery station conditions. *Rev. Bras. Biol.* **1996**, *56*, 59–64.

54. Nilsson, C.; Reidy, C.A.; Dynesius, M.; Revenga, C. Fragmentation and flow regulation of the world's large river systems. *Science* **2005**, *308*, 405–408. [CrossRef]

55. Silva, J.C.; Rosa, R.R.; Galdioli, E.M.; Soares, C.M.; Domingues, W.M.; Veríssimo, S.; Bialetzki, A. Importance of dam-free stretches for fish reproduction: The last remnant in the upper Paraná River. *Acta Limnol. Bras.* **2017**, *29*, e106. [CrossRef]

56. Graça, W.J.; Pavanelli, G.C. *Peixes da Planície de Inundação do Alto Rio Paraná e Áreas Adjacentes*; Eduem: Maringá, Brazil, 2007; p. 241.

57. Suzuki, H.I.; Pelicice, F.M.; Luiz, E.A.; Latini, J.D.; Agostinho, A.A. Estratégias reprodutivas da assembléia de peixes da planície de inundação do alto rio Paraná. Pesquisas Ecológicas de Longa Duração–PELD.(Org.). In *A Planície Alagável do Rio Paraná: Estrutura e Processos Ambientais*; PELD: Maringá, Brazil, 2002.

Article

Spatial Patterns in Fish Assemblages across the National Ecological Observation Network (NEON): The First Six Years

Dylan Monahan [1,2,*], Jeff S. Wesner [3], Stephanie M. Parker [2] and Hannah Schartel [2]

[1] River Rat Watershed Solutions, Boulder, CO 80026, USA
[2] National Ecological Observatory Network, Battelle, Boulder, CO 80301, USA
[3] Biology, University of South Dakota, Vermillion, SD 57069, USA; jeff.wesner@usd.edu
*   Correspondence: river.rat.h20shed@gmail.com

Abstract: The National Ecological Observation Network (NEON) is a thirty-year, open-source, continental-scale ecological observation platform. The objective of the NEON project is to provide data to facilitate the understanding and forecasting of the ecological impacts of anthropogenic change at a continental scale. Fish are sentinel taxa in freshwater systems, and the NEON program has been sampling and collecting fish assemblage data at wadable stream sites for six years. One to two NEON wadable stream sites are located in sixteen domains from Alaska to Puerto Rico. The goal of site selection was that sites represent local conditions but with the intention that site data be analyzed at a continental observatory level. Site selection did not include fish assemblage criteria. Without using fish assemblage criteria, anomalies in fish assemblages at the site level may skew the expected spatial patterns of North American stream fish assemblages, thereby hindering change detection in subsequent years. However, if NEON stream sites are representative of the current spatial distributions of North American stream fish assemblages, we could expect to find the most diverse sites in Atlantic drainages and the most depauperate sites in Pacific drainages. Therefore, we calculated the alpha and regional (beta) diversities of wadable stream sites to highlight spatial patterns. As expected, NEON sites followed predictable spatial diversity patterns, which could facilitate future change detection and attribution to changes in environmental drivers, if any.

Keywords: NEON; fish; diversity; assemblages

Key Contribution: NEON sites have collected fish assemblage data for six years. Currently, NEON sites follow a predictable spatial pattern of fish assemblage diversity.

Citation: Monahan, D.; Wesner, J.S.; Parker, S.M.; Schartel, H. Spatial Patterns in Fish Assemblages across the National Ecological Observation Network (NEON): The First Six Years. *Fishes* **2023**, *8*, 552. https://doi.org/10.3390/fishes8110552

Academic Editors: Robert L. Vadas, Jr. and Robert M. Hughes

Received: 11 August 2023
Revised: 2 November 2023
Accepted: 3 November 2023
Published: 16 November 2023
Corrected: 8 February 2024

1. Introduction

Fish assemblage data can quantify the variety of fish species across any area and can provide important information about land use, water pollution, habitat degradation, and invasive species. Using fish assemblages to assess human-caused stream degradation requires an understanding of the expected fish assemblage for that site or region [1–4].

The National Ecological Observation Network (NEON) is a thirty-year, open-source, continental-scale ecological observation platform. The objective of the NEON project is to provide researchers and the public with data to facilitate an understanding and forecasting of the ecological impacts of climate, land use changes, and invasive species at a continental scale. NEON open-source data allow researchers to access continental data collected using uniform protocols. NEON provides infrastructure and consistent methodologies for the collection and analysis of these data [5]. Consistent observations at a continental extent can help users compare smaller-extent watershed studies to broader extents [6].

The NEON freshwater program is a vital part of NEON's goal of detecting and quantifying the drivers of ecological change by sampling community composition, measuring surface and groundwater chemistry, deploying micrometeorology and in situ water quality instrumentation in and around water bodies, and tracking habitat structures [5].

Fish assemblages are an important component of freshwater ecosystem data. Fish are considered sentinel taxa in freshwater systems because they are often mobile and play essential roles in energy and nutrient transfer. Therefore, quantitative fish data are an important component in detecting aquatic ecosystem patterns and changes (4). NEON fish sampling methods provide fish assemblage data, which are a vital tool for researchers now and in the future.

A foundational principle of the NEON program is that fish assemblages are a useful indicator of anthropogenic influences. NEON stream sites are collocated with other environmental data, allowing users to better understand the drivers of changes to fish assemblage data [7–9]. A challenge to using NEON fish assemblage data is that fish assemblages need to be understood at both the site and biogeographic levels to assess the effects of potential anthropogenic degradation. NEON's wadable stream sites were selected to answer a broad range of ecological questions at varying scales, but were not specifically selected to represent the full range of regional fish assemblages.

Nonetheless, a key question is as follows: do NEON sites represent expected continental-scale fish assemblage patterns? Fish assemblages in wadable stream sites are determined by several site-specific features: these include where the site is located, habitat conditions, and the historical pattern of fish colonization at the site [10,11]. The macroecological context within which a site sits is often an important fish assemblage predictor. In the United States, primarily due to glacial history and climate change, we would expect to see the highest alpha diversities in Atlantic drainage sites, particularly in sites found in warmer lowland river drainages [12]. Sites in Pacific drainages with colder winters, drier summers, and less stable river drainages would be expected to have the lowest biodiversity scores [13]. We would also expect beta diversity to show an effect on the drainage location when comparing site dissimilarity.

Here, we use alpha and beta diversity metrics and size composition data from NEON wadable stream sites to describe spatial patterns in NEON fish data. In describing these spatial patterns, we seek to confirm whether NEON sites in the first years of sampling (2017–2022) meet expected spatial fish assemblage patterns. We also check size composition to contextualize the ecological relationship of assemblages to their sites in a manner comparable to a continental scale. We use species occurrence data to inform potential differences in seasonal and temporal sampling.

2. Materials and Methods

2.1. NEON Site Selection

The NEON observatory is divided into 20 ecoclimatic domains based on statistical geographic clustering [14]. Fish were collected at 23 wadable stream sites in 16 domains, (Figure 1; Table 1; see neonscience.org for additional site information, accessed between 1 January 2017 and 31 December 2022). NEON does not sample fish at large river sites because the effort needed to conduct quantitative fish sampling in rivers exceeds NEON's resources.

2.2. Biological Sampling Windows

Fish sampling at NEON sites occurs twice per year, annually, in the spring and fall. Because of the wide seasonal range of sites spread out from Alaska to Puerto Rico, spring and fall cover a range of months depending on the site (Table 2). Spring sampling dates are intended to coincide with the start of warming degree days and the start of peak greenness, and fall sampling dates are determined to coincide with a decrease in light levels and temperature at the site. These criteria were chosen as the fish sampling windows, as they are an important biogeochemical catalyst and allow fish data to be associated with those collocated biogeochemical parameters [5,14].

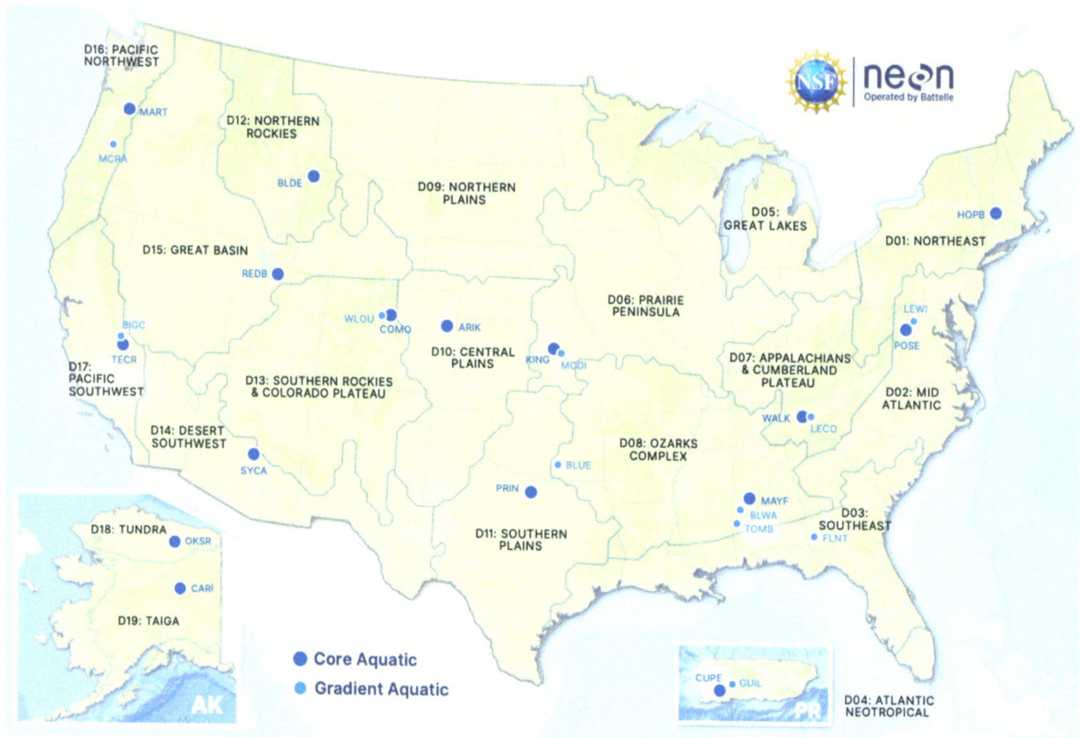

Figure 1. NEON wadable stream sites in 20 ecoclimatic domains. Core sites are wilderness sites, and gradient sites are sites with known anthropogenic stressors [15].

Table 1. NEON sites, drainages, domain numbers, and domain names.

NEON Site Name	Drainage	Domain Number	Domain Name
HOPB	Atlantic	01	Northeast
LEWI	Atlantic	02	Mid-Atlantic
POSE	Atlantic	02	Mid-Atlantic
CUPE	Atlantic	04	Atlantic Neotropical
GUIL	Atlantic	04	Atlantic Neotropical
MCDI	Atlantic	06	Prairie Peninsula
KING	Atlantic	06	Prairie Peninsula
LECO	Atlantic	07	Appalachian
WALK	Atlantic	07	Appalachian
MAYF	Atlantic	08	Ozark Complex
MAYF	Atlantic	08	Ozark Complex
ARIK	Atlantic	10	Central Plains
BLUE	Atlantic	11	Southern Plains
PRIN	Atlantic	11	Southern Plains
BLDE	Atlantic	12	Northern Rockies
WLOU	Pacific	13	Southern Rockies

Table 1. *Cont.*

NEON Site Name	Drainage	Domain Number	Domain Name
SYCA	Pacific	14	Desert Southwest
REDB	Pacific	15	Great Basin
MART	Pacific	16	Pacific Northwest
MCRA	Pacific	16	Pacific Northwest
BIGC	Pacific	17	Pacific Southwest
TECR	Pacific	17	Pacific Southwest
OKSR	Pacific	18	Tundra
CARI	Pacific	19	Taiga

Table 2. Domain site-sampling windows.

Domain	Site	Spring Sampling Window	Fall Sampling Window
1	HOPB	11 Apr–9 May	3 Oct–31 Oct
2	POSE	19 Mar–16 Apr	18 Oct–15 Nov
2	LEWI	19 Mar–16 Apr	18 Oct–15 Nov
4	CUPE	24 Jan–21 Feb	10 Nov–8 Dec
4	GUIL	26 Jan–23 Feb	9 Nov–7 Dec
6	KING	23 Mar–20 Apr	3 Oct–31 Oct
6	MCDI	20 Mar–17 Apr	27 Sep–25 Oct
7	LECO	15 Mar–12 Apr	12 Oct–9 Nov
7	WALK	09 Mar–06 Apr	19 Oct–16 Nov
8	MAYF	05 Mar–02 Apr	24 Oct–28 Nov
10	ARIK	21 Mar–18 Apr	20 Sep–18 Oct
11	PRIN	17 Feb–17 Mar	23 Oct–20 Nov
11	BLUE	07 Mar–04 Apr	12 Oct–9 Nov
12	BLDE	10 Jun–08 Jul	30 Aug–27 Sep
13	COMO	05 Jul–02 Aug	5 Sep–3 Oct
13	WLOU	02 Jul–30 Jul	3 Sep–1 Oct
14	SYCA	12 Jan–11 Feb	3 Jun–3 Jul
15	REDB	29 Mar–26 Apr	29 Sep–27 Oct
16	MCRA	10 Apr–08 May	23 Sep–21 Oct
16	MART	06 Apr–04 May	22 Sep–20 Oct
17	TECR	06 May–17 Jun	17 Sep–15 Oct
17	BIGC	02 Apr–30 Apr	28 Sep–26 Oct
18	OKSR	21 May–18 Jun	7 Aug–4 Sep
19	CARI	02 May–30 May	18 Aug–15 Sep

Sampling windows are 28 days long [14] and based on historic, publicly available air temperatures from the National Oceanographic and Atmospheric Administration (NOAA) and riparian phenology Moderate Resolution Imaging Spectroradiometer (MODIS) data. Contingent decisions include allowing fish sampling for up to 30 days after the end of the sampling window to accommodate staffing concerns, weather delays, and high or low water, as documented for the data users [14]. As more years of consecutive NEON data become available, data used to define the sampling windows are replaced with NEON sensor and stream discharge data, allowing the sampling to be flexible with changing site or climate conditions over the lifetime of the NEON project.

2.3. NEON Fish Data

NEON stream sites are 1 km long and divided into 10 (80–100 m) reaches, except for MCDI, where the sampling permit restricts the site to 500 m. Six (80–100 m) reaches are scheduled for DC backpack electrofishing at each site (except MCDI, with three reaches scheduled per bout), using the NEON wadable stream fish sampling protocol at every site [16]. No major protocol changes occurred over the six years of this study. Three of the six scheduled reaches were fixed reaches sampled every visit (Figure 2) by employing three-pass depletion sampling. The other three reaches included random reaches that came from a panel of seven random reaches sampled on a rotating schedule. Each random reach was randomly selected before the first year of sampling and scheduled for sampling so that each random reach was scheduled to be sampled at least once every three years, and random reaches were sampled on a single pass. At MCDI, where land ownership and permitting the restricted site length to 500 m occurred, there were five designated reaches; therefore, each year, only two fixed and one random reach were sampled per bout per year.

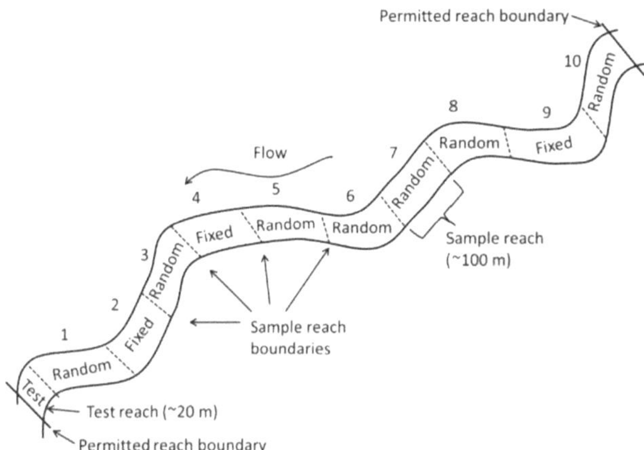

Figure 2. Schematic of a 1 km NEON stream site delineated into ten 100 m reaches: 3 fixed and 7 random sampling reaches. The three fixed reaches are sampled every visit; three random reaches are chosen each year for sampling [16].

All reaches were closed-sampled with fixed block nets set at the top and bottom of the reach. Not all the reaches scheduled are always sampled each bout because of weather, equipment, and logistic issues. One fixed reach per bout was the minimum effort required for fish data to be available to the public.

Captured fish were identified to the lowest taxonomic level possible based on [17] and the Integrated Taxonomic Information System (ITIS) online database (http://www.itis.gov accessed on 31 July 2017). When field scientists responsible for identification were uncertain, they used a morphospecies or identification qualifier. Some fish were identified only to their family or genus (followed by an SP. or SPP.). Federally listed species are obscured when published so that they appear identified at the family level; this protects listed species and is part of the NEON agreement with the U.S. Fish and Wildlife Service.

The first fifty individuals captured from the same taxonomic identification group per reach were wet-weighed (g) and measured to the total length (mm). After fifty individuals from the same taxonomic identification group were measured and weighed, all fish captured in that reach from that taxonomic group were bulk-counted and not measured. This process started again at the start of each reach sample.

2.4. Downloading and Compiling NEON Fish Data

NEON electrofishing data were downloaded on 16 April 2023 from the NEON data portal [18]. Taxonomic data were counted per bout from measured fish and bulk fish data. Only first-pass data were used unless a new species was collected in a 2nd or 3rd pass from a 3-pass depletion reach, and it was also caught at one of the single-pass reaches. The catch per unit effort was calculated and normalized to the hour for all taxonomies captured during a bout. Taxonomic data were counted per bout from the measured fish and bulk fish data.

2.5. Alpha Diversity

To describe the spatial distribution of fish taxonomy at NEON stream sites, the vegan R Package was used to calculate species richness and Shannon and Simpson metrics for each site on a per-bout basis [19]. All first-pass, electrofishing data of both fixed and random reaches from 2017 to 2022 were analyzed.

2.6. Beta Diversity

To test the diversity between drainages, beta diversity was mapped for all bouts from 2017 to 2022 for each of the 23 wadable stream sites, using the betadiver and betadisper command in Vegan [19]. CPUE data were used to calculate the dissimilarities between species observed from data via Atlantic and Pacific drainage and a principal coordinates analysis (PCoA) distribution using a beta z distribution.

2.7. Size Composition

NEON measures individual fish sizes (total lengths and field measured wet weight) for the first fifty individuals of each species on each electrofishing pass. The total length is measured to the nearest 0.1 mm, and wet weight is measured to the nearest 0.1 g, with a lower limit of 0.3 mg. The resulting dataset contained 52,882 individual measures of fish length and wet weight from 2015 through 2022. Lengths and wet weights were compiled and sorted by species, site, and years.

3. Results

3.1. Alpha Diversity Metrics

Since 2017, NEON has collected 112 species of fish at wadable stream sites. The greatest number of fish species sampled were from domains 01, 02, 04, 06, 07, 08, and 10, including 42 species sampled at NEON's MAYF site in domain 08 and 33 at NEON's BLUE site in domain 11. All of these domains were Atlantic draining. The lowest scores were recorded at domains 12, 13, 14, 15, 16, 17, 18, 19, and 20 (all Arctic and Mountain West sites). Except for SYCA in the Desert Southwest (domain 15), species at these lower-scoring sites were dominated by Salmonidae (Table A1).

Similarly, Shannon diversity scores were highest at NEON's Southeastern, Southern, Central Plains, Atlantic, and Caribbean sites and lowest at NEON's West Coast, Arctic, and Mountain West sites (Figure 3). Since 2017, ten sites produced fish species during spring sampling but not in fall sampling, and nine sites yielded occurrences of fish species during fall sampling but not in the spring sampling (Table A2). This indicates the usefulness of seasonal sampling, as well as its hindrance to assessing annual trends.

3.2. Beta Diversity Metrics

PCoA mapping shows that some site visits in Pacific drainages were similar to some of those in Atlantic drainages (Figure 4). However, Atlantic drainage sites have a greater range in the fish species found at those sites, indicating a greater level of beta diversity in Atlantic drainage sites.

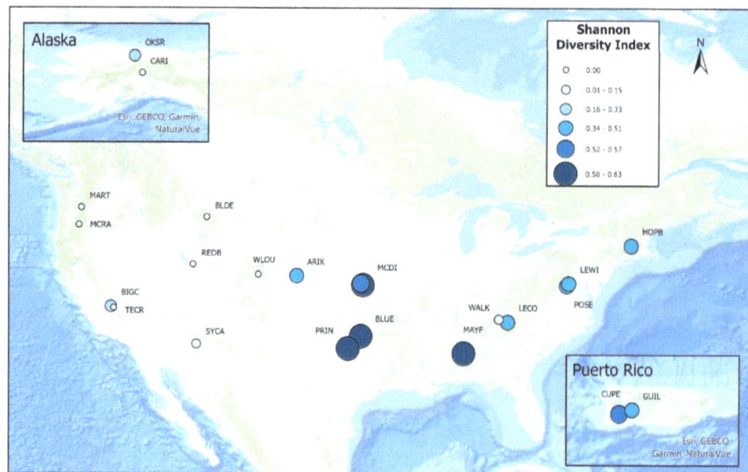

Figure 3. NEON stream—fish Shannon diversities.

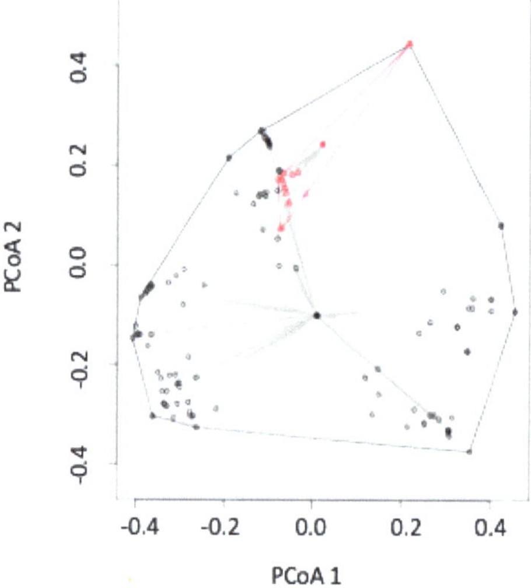

Figure 4. PCoA distances for fish species sampled at NEON sites from 2017 to 2022 and apportioned by Atlantic (black circle) vs. Pacific (red triangle) drainage.

3.3. Size Composition

Fish wet weights ranged from four orders of magnitude across all collections (Figure 5). Individual wet weights ranged from 0.3 g (n = >13,000 individuals from multiple species) to 1000 g (n = 7 *Oncorhynchus mykiss*). When averaged among sites, the median fish size varied from 0.3 to 22 g for wet weight and 31 to 144 mm for the total length (Table 3). Despite the variation in fish size among sites, there was strong consistency in fish sizes across years (Figure 6). For example, the grand median fish size at KING was 0.6 g, and yearly medians ranged only from 0.3 to 1 mg in wet weight. By comparison, the grand median at WLOU was 10 g, with yearly medians ranging from 6 to 13. In other words, fish size appeared to vary more among sites than across years within a site (Figure 6).

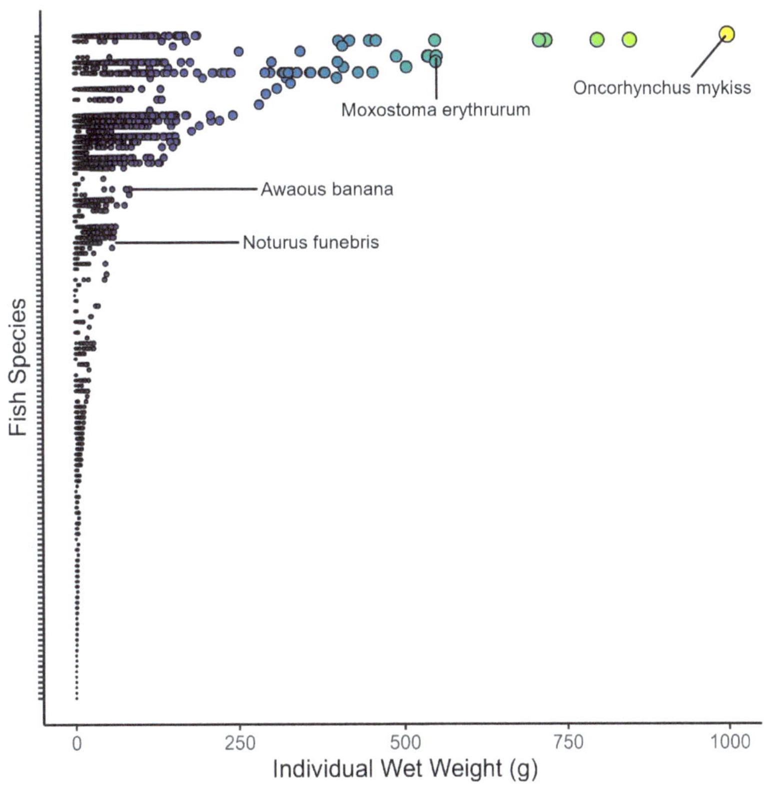

Figure 5. Distribution of 52,882 individual wet weights of fish measured in 23 NEON wadable stream sites. The data include all fish measured from 2015 to 2022. The y-axis represents 154 taxa ranked by the maximum fish size per taxon. Most taxon names are removed for clarity. Colors and sizes reflect the relative wet weights of fishes (yellow = largest, black = smallest).

Table 3. Median (and upper and lower 95%iles) of individual total lengths and wet weights summarized across all individuals collected between 2016 and 2022. N is the number of individual fish sizes recorded at a stream site.

Site	N	Total Length (mm)	Wet Weight (g)
CARI	186	144 (66 to 353)	22.1 (3 to 361)
REDB	242	142.5 (44 to 244)	27.25 (1 to 140)
TECR	876	123 (51 to 221)	17 (0 to 95)
BLDE	722	121 (65 to 215)	17 (3 to 89)
BIGC	2036	105 (31 to 215)	11 (0 to 96)
WLOU	840	104 (40 to 173)	10.5 (0 to 52)
MCRA	852	101 (41 to 158)	8.75 (1 to 37)
MART	1006	98 (57 to 154)	7.9 (2 to 32)
LECO	4456	75 (35 to 177)	3.8 (0 to 53)
HOPB	4144	62 (27 to 154)	2.2 (0 to 32)
LEWI	5641	59 (34 to 125)	2.4 (0 to 20)
MCDI	3488	55 (32 to 122)	1.6 (0 to 21)

Table 3. *Cont.*

Site	*N*	Total Length (mm)	Wet Weight (g)
WALK	4680	55 (23 to 91)	1.4 (0 to 8)
MAYF	417	52 (6 to 144)	1.2 (0 to 35)
POSE	6019	52 (24 to 89)	1.3 (0 to 8)
OKSR	180	50 (40 to 175)	0.9 (0 to 39)
BLUE	1519	45 (18 to 126)	1.1 (0 to 25)
KING	3652	42 (22 to 107)	0.7 (0 to 12)
PRIN	3861	42 (17 to 135)	0.8 (0 to 36)
ARIK	3201	41 (21 to 95)	0.3 (0 to 10)
CUPE	890	37 (14 to 220)	0.6 (0 to 118)
GUIL	1880	33 (13 to 82)	0.3 (0 to 4)
SYCA	2094	31 (17 to 67)	0.3 (0 to 4)

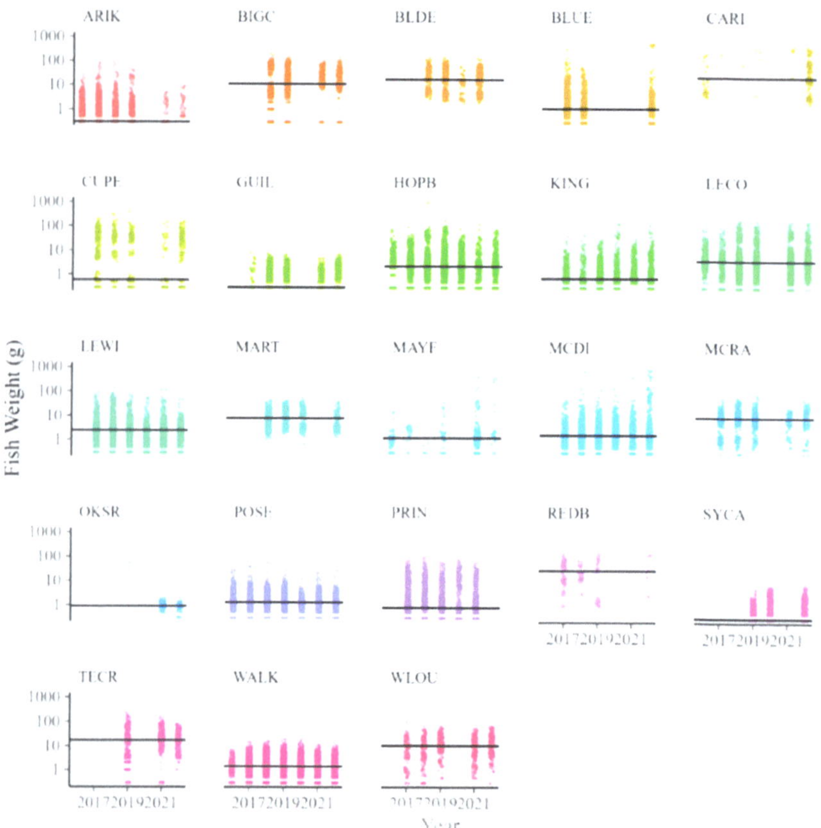

Figure 6. Individual fish wet weights (*n* = 52,882) collected across 23 NEON stream sites from 2017 to 2022. The horizontal line shows the grand median for each site.

4. Discussion

The site-level fish assemblage richness in wadable streams was determined by several site-specific features, but at North American temperate sites, one of the most important determinants was where in the continent the site was located [10,11]. In the U.S., primarily because of glacial history and historic post-glacial fish colonization patterns, we expected to see the highest alpha and beta diversities amongst Atlantic drainage sites, particularly in those sites found in warmer lowland river drainages [12,13]. Sites in Pacific drainages where post-glacial fish colonization was much more limited were expected to have lower freshwater diversity [20]. Our results confirmed this pattern.

The NEON program is designed to monitor anthropogenic change at a continental scale. Fish assemblages are determined by spatially extensive macroecological drivers, as well as by local natural (barriers, natural disturbance) and anthropogenic (dams and species introduction) factors (4). Many studies focus on more localized drivers of fish assemblage composition, but NEON's mission is to provide data that use measures of site-specific conditions with the intention of scaling these conditions to the continental scale.

Because of the ability of local conditions to create anomalous fish assemblages in comparison to regional fish assemblages, it is important to determine whether locally collected and analyzed NEON fish assemblages represent the expected fish assemblage distribution for wadable streams at a continental scale. If NEON selection selects regionally anomalous sites where Pacific draining sites have high biodiversity and Atlantic draining sites low biodiversity, future changes in regional biodiversity caused by climate and land and water use changes may not be detected.

4.1. Spatial Patterns of NEON Stream Fish Data

NEON's site selection is driven by several factors, including the need to have sites represented in 19 nationwide domains. NEON sites are installed to capture local conditions but then scale up to the continental scale. Fish assemblage composition was not a factor used to select sites; instead, site selection was driven by the need to distribute sites along major continental-scale ecological gradients [15]. Because of the need to place sites on such a broad continental scale, we assumed that spatial patterns of fish assemblages would follow continental-scale diversity patterns, with the largest number of species and the highest alpha diversity metrics found at sites in the Plains, Prairies, Gulf Coast, Atlantic, and Caribbean domains, opposite of the West Coast and Mountain West domains. As expected, we found that NEON fish diversity was distributed along the expected continental-scale patterns, with the highest diversity sites found in Atlantic drainages and the lowest diversity found in Pacific drainages farther west. Also, sites west of the divide are dominated by salmonids, except for SYCA in the Desert Southwest domain, where *Agosia chrysogaster*, longfin dace, dominates. Eastern sites are dominated primarily by *Cyprinidae*, *Poecilidae*, *Cottidae*, and *Cyprinidae*.

4.2. Occurence

The occurrence of fish species both seasonally and by year showed that at more diverse sites, there was a higher likelihood of collecting fish in one bout but not another. This was particularly true at BLUE and MAYF. Most species caught only in one bout are relatively rare seasonally because of migratory behavior (see below). Of the 49 collection times, one species was caught at a site during only one of the seasonal bouts, but 39 of those times, it represented five fish or fewer. When comparing fish caught in either the 2017–2019 group or the 2020–2022 group, 44 out of the 68 collection times represented occurrences of five fish or less. This indicates the difficulty of collecting rare species, especially singletons and doubletons (one or two individuals) (Table A3) [21,22]. Furthermore, spring and fall sampling periods are more likely to coincide with fish migration periods, meaning natural periods of presence and absence [22–25].

4.3. Spatial Patterns in Size Composition

In addition to taxonomic composition and abundance, body size provides critical ecological information in relation to the age–structure, size–abundance [26] metabolism [27], food web structure [28], and trophic transfer efficiencies [29]. Arranz et al. [30] used stream fish size spectra to detect responses to species invasion and eutrophication. Similar analyses are possible with NEON data. For example, Pomeranz et al. [31] used NEON macroinvertebrate body sizes to examine how size spectra scaled with temperature from Puerto Rico to Alaska. A major benefit of NEON size data is its collection of repeated measures over time. As shown in Figure 5, size data are consistent across the years, meaning that future disturbances to NEON sites may reveal shifts in the size structure of fish species relative to the baseline. Future data collected at NEON sites can reveal whether and how disturbances affect the size structure of fish species, yielding information not only on taxonomic persistence and abundance (from other NEON metrics) but also allowing the ecological functions of these sites to be correlated with body size [25].

5. Summary and Conclusions

This analysis confirms that fish assemblage patterns at NEON sites follow predicted continental-scale patterns, even though they are not selected using fish assemblage criteria. Speciose Atlantic sites were dominated by smaller-sized fish, with the most common fish representing at least five families. Low-diversity Pacific sites often contain only a single species and are dominated by salmonids. Sites representing expected spatial assemblage patterns are a good sign; potential changes to fish assemblages could be easier to detect and attribute to environmental drivers.

It is also important to learn why some fish are present in the first three years but not the last three years. Was this an accident of sampling, or are there other reasons (i.e., El Niño years vs. La Niña years)? Is it cyclical, or are those fish that were collected in the first three years gone forever? Are some the results of narrow window spawning migrations? Research in Brazil is currently examining this hypothesis at a large spatial scale (via ichthyoplankton sampling) and collaborating on an important question about how fish assemblages at a multi-continental level can maximize the benefits of NEON data [32]. The application of more sophisticated beta diversity metrics to spatial diversity questions could also be a valuable next step.

Author Contributions: D.M.: Introduction, Methods (Biodiversity), Results, Discussion, Conclusion. J.S.W.: Methods (Size Composition), Results (Size Composition), Discussion (Size Composition). S.M.P.: Methods (NEON Site Selection, Biological Sampling Windows). H.S.: Review documents and figures. All authors have read and agreed to the published version of the manuscript.

Funding: The National Ecological Observatory Network is a program sponsored by the National Science Foundation and operated under cooperative agreement by Battelle. This material is based in part upon work supported by the National Science Foundation through the NEON Program, as well as Grant No. 2106067 to JSW.

Data Availability Statement: Available online: https://data.neonscience.org/data-products/DP1.2 0107.001/RELEASE-2023 (accessed on 1 May 2023).

Conflicts of Interest: The authors declare no conflict of interest.

Appendix A

Table A1. Species richness per site, as described by the mean number of species, spring and fall bout means, the bout mean, highest and lowest scores, most common species (the total number caught), and domain. Fish species are written in 6-letter code, with a star next to it indicating that the fish in question in not native to that site. Codes are described at the bottom of the table.

Site	Mean	Spring Bout	Fall Bout	Highest	Lowest	Spring Bout Most Com. Spec.	Fall Bout Most Com Spec.	Domain
BLUE	20.75	17.67	30	30	5	ETHRAD (786)	ETHRAD (644)	Southern Plains

Table A1. *Cont.*

Site	Mean	Spring Bout	Fall Bout	Highest	Lowest	Spring Bout Most Com. Spec.	Fall Bout Most Com Spec.	Domain
MAYF	17	17	17	23	13	NOTBAI (561)	NOTBAI (1053)	Ozarks Complex
PRIN	9.5	10	9	11	7	GAMAFF (654)	GAMAFF (1954)	Southern Plains
MCDI	9.17	7.5	10.83	13	6	CAMANO (1652)	CAMANO (566)	Prairie Peninsula
ARIK	7.13	6.4	8.33	10	5	ETHSPE (552)	* GAMAFF (1914)	Central Plains
KING	6.42	5.33	7.5	9	3	ETHSPE (790)	CHRERY (4911)	Prairie Peninsula
LEWI	5.81	5.2	5.17	7	4	COTGIR (1955)	COTGIR (2930)	Mid-Atlantic
CUPE	4.4	4.8	4	7	2	POERET (265)	POERET (111)	Atlantic NeoTropical
HOPB	4.19	4.6	3	7	2	RHIATR (674)	RHIATR (1382)	Northeast
POSE	4.19	4.2	4.17	5	4	RHIATR (1351)	RHIATR (2181)	Mid-Atlantic
LECO	3.44	3.5	3.4	4	3	RHIATR (755)	RHIATR (1965)	Appalachian and Cumberland Plateau
WALK	2.46	2.4	2.5	4	2	RHIATR (1221)	RHIATR (2811)	Appalachian and Cumberland Plateau
GUIL	2.3	2.4	2.2	3	2	POERET (3698)	POERET (4199)	Atlantic NeoTropical
BIGC	2.29	2.75	1.667	3	1	* SALTRU (645)	* SALTRU (695)	Pacific Southwest
SYCA	2.25	2.33	2	3	1	AGOCHR (2410)	AGOCHR (699)	Desert Southwest
OKSR	1.33	2	1	2	1	THYARC (1)	THYARC (64)	Tundra
CARI	1.14	0.75	1.67	3	1	THYARC (8)	THYARC (87)	Taiga
BLDE	1	NA	1	1	1	NA	* SALFON (587)	Northern Rockies
MART	1	NA	1	1	1	NA	SALSP (785)	Pacific Northwest
MCRA	1	NA	1	1	1	NA	ONCCLA (720)	Pacific Northwest
REDB	1	1	1	1	1	ONCCLA (24)	ONCCLA (154)	Great Basin
TECR	1	1	1	1	1	* SALFON (529)	* SALFON (387)	Pacific Southwest
WLOU	1	1	1	1	1	* SALFON (74)	* SALFON (494)	Southern Rockies and Colorado Plateau

Etheostoma radiosum = ETHRAD, Notropis baileyi = NOTBAI, Gambusia affinis = GAMAFF, Campostoma anomalum = CAMANO, Campostoma anomalum = ETHSPE, Chrosomus erythrogaster = CHRERY, Cottus girardi = COTGIR, Poecilia reticulata = POERET, Rhinichthys atratulus = RHIATR, Salmo trutta = SALTRU, Agosia chrysogaster = AGOCHR, Thymallus arcticus = THYARC, Salvelinus fontinalis = SALFON, Oncorhynchus clarki = ONCCLA.

Table A2. Species caught during either the spring sampling or the fall sampling bout but not the other per-site since 2017.

Site	Spring/Fall	Species	Count
ARIK	Spring	*Etheostoma exile*	206
BLUE	Fall	*Ameiurus melas*	1
BLUE	Spring	*Cyprinella camura*	1
BLUE	Fall	*Lythrurus umbratilis*	6
BLUE	Spring	*Micropterus punctulatus*	1
BLUE	Spring	*Micropterus salmoides*	1
BLUE	Spring	*Notropis boops*	13
BLUE	Spring	*Notropis buchanani*	6
BLUE	Fall	*Pylodictis olivaris*	1
CUPE	Spring	*Anguilla rostrata*	3
GUIL	Fall	*Tilapia rendalli*	1
HOPB	Spring	*Ameiurus nebulosus*	1

Table A2. *Cont.*

Site	Spring/Fall	Species	Count
HOPB	Spring	*Notemigonus crysoleucas*	1
HOPB	Spring	*Noturus gyrinus*	1
HOPB	Fall	*Notemigonus crysoleucas*	6
KING	Spring	*Cyprinella lutrensis*	2
KING	Spring	*Etheostoma pseudovulatum*	4
KING	Spring	*Etheostoma tennesseense*	1
KING	Fall	*Lepomis macrochirus*	1
KING	Fall	*Luxilus cornutus*	1
KING	Fall	*Moxostoma pisolabrum*	1
KING	Fall	*Phoxinus erythrogaster*	78
LECO	Spring	*Campostoma anomalum*	1
LEWI	Fall	*Cyprinella spiloptera*	1
LEWI	Spring	*Etheostoma flabellare*	1
LEWI	Spring	*Lepomis cyanellus*	1
LEWI	Spring	*Lepomis macrochirus*	4
MAYF	Fall	*Elassoma zonatum*	1
MAYF	Fall	*Erimyzon oblongus*	1
MAYF	Spring	*Lepomis auritus*	2
MAYF	Spring	*Lepomis cyanellus*	1
MAYF	Fall	*Etheostoma histrio*	1
MAYF	Spring	*Lepomis macrochirus*	3
MAYF	Fall	*Lythrurus bellus*	2
MAYF	Spring	*Minytrema melanops*	2
MAYF	Fall	*Micropterus henshalli*	4
MAYF	Fall	*Micropterus warriorensis*	1
MAYF	Spring	*Moxostoma poecilurum*	10
MAYF	Fall	*Notropis stilbius*	65
MAYF	Spring	*Pteronotropis hypselopterus*	1
MCDI	Fall	*Catostomus commersonii*	1
MCDI	Fall	*Etheostoma nigrum*	133
MCDI	Fall	*Lepomis megalotis*	3
MCDI	Fall	*Pimephales vigilax*	5
PRIN	Fall	*Cyprinus carpio*	1
PRIN	Spring	*Micropterus salmoides*	2
PRIN	Spring	*Notropis volucellus*	71
PRIN	Spring	*Pimephales vigilax*	1
WALK	Fall	*Notropis atherinoides*	1

Table A3. Species caught during either 2017–2019 bouts or the 2020–2022 bouts but not over three-year periods.

Site	Species	Years Caught	Count
ARIK	*Ameiurus melas*	2017–2019	16
ARIK	*Etheostoma exile*	2017–2019	206
ARIK	*Fundulus zebrinus*	2017–2019	20
ARIK	*Lepomis cyanellus*	2017–2019	203
BLUE	*Ameiurus melas*	2017–2019	1
BLUE	*Lythrurus umbratilis*	2017–2019	6
BLUE	*Cyprinella camura*	2020–2022	1
BLUE	*Micropterus salmoides*	2017–2019	1
BLUE	*Micropterus punctulatus*	2017–2019	1
BLUE	*Nocomis asper*	2017–2019	10
BLUE	*Notropis buchanani*	2020–2022	6
BLUE	*Notropis nubilus*	2020–2022	1
BLUE	*Notropis suttkusi*	2017–2019	61
BLUE	*Notropis volucellus*	2017–2019	99

Table A3. *Cont.*

Site	Species	Years Caught	Count
BLUE	*Pimephales notatus*	2017–2019	79
BLUE	*Pylodictis olivaris*	2017–2019	1
CUPE	*Anguilla rostrata*	2017–2019	3
CUPE	*Gobiomorus dormitor*	2020–2022	4
CUPE	*Sicydium punctatum*	2017–2019	30
CUPE	*Sicydium plumieri*	2017–2019	45
GUIL	*Gambusia affinis*	2017–2019	43
GUIL	*Tilapia rendalli*	2020–2022	1
HOPB	*Ameiurus nebulosus*	2017–2019	1
HOPB	*Noturus gyrinus*	2017–2019	1
HOPB	*Salmo trutta*	2017–2019	57
KING	*Cyprinella lutrensis*	2020–2022	2
KING	*Etheostoma pseudovulatum*	2017–2019	4
KING	*Etheostoma tennesseense*	2017–2019	1
KING	*Luxilus cornutus*	2020–2022	1
KING	*Lepomis macrochirus*	2017–2019	1
KING	*Moxostoma pisolabrum*	2020–2022	1
KING	*Notropis percobromus*	2020–2022	1
KING	*Phoxinus erythrogaster*	2017–2019	78
KING	*Noturus exilis*	2020–2022	2
LEC0	*Campostoma anomalum*	2017–2019	1
LEWI	*Gambusia holbrooki*	2020–2022	46
LEWI	*Lepomis cyanellus*	2020–2022	1
LEWI	*Nocomis leptocephalus*	2020–2022	3
LEWI	*Etheostoma flabellare*	2017–2019	1
LEWI	*Lepomis macrochirus*	2017–2019	4
MAYF	*Elassoma zonatum*	2017–2019	1
MAYF	*Lythrurus bellus*	2017–2019	2
MAYF	*Minytrema melanops*	2017–2019	2
MAYF	*Notropis ammophilus*	2017–2019	4
MAYF	*Erimyzon oblongus*	2020–2022	1
MAYF	*Etheostoma chlorosomum*	2020–2022	2
MAYF	*Etheostoma histrio*	2020–2022	1
MAYF	*Etheostoma nigrum*	2020–2022	7
MAYF	*Lepomis cyanellus*	2020–2022	1
MAYF	*Lepomis macrochirus*	2020–2022	3
MAYF	*Micropterus warriorensis*	2020–2022	1
MAYF	*Notropis volucellus*	2020–2022	30
MAYF	*Pteronotropis hypselopterus*	2020–2022	1
MCDI	*Ameiurus natalis*	2020–2022	4
MCDI	*Catostomus commersonii*	2020–2022	1
MCDI	*Notropis atherinoides*	2017–2019	15
MCDI	*Notropis shumardi*	2017–2019	4
MCDI	*Micropterus punctulatus*	2017–2019	4
MCDI	*Etheostoma nigrum*	2017–2019	133
POSE	*Cottus bairdii*	2017–2019	1010
PRIN	*Cyprinus carpio*	2017–2019	1
PRIN	*Micropterus salmoides*	2017–2019	2
PRIN	*Notropis stramineus*	2017–2019	1
PRIN	*Notropis volucellus*	2017–2019	71
PRIN	*Pimephales vigilax*	2017–2019	1
WALK	*Cottus caeruleomentum*	2017–2019	55
WALK	*Notropis atherinoides*	2017–2019	1

References

1. Zogaris, S.; Koutsikos, N.; Chatzinikolaou, Y.; Őzeren, S.C.; Yence, K.; Vlami, V.; Kohlmeier, P.G.; Akyildiz, G.K. Fish assemblages as ecological indicators in the Büyük Menderes (Great Meander) River, Turkey. *Water* **2023**, *15*, 2292. [CrossRef]
2. Pont, D.; Hughes, R.M.; Whittier, T.R.; Schmutz, S. A predictive index of biotic integrity model for aquatic-vertebrate assemblages of western U.S. streams. *Trans. Am. Fish. Soc.* **2009**, *138*, 292–305. [CrossRef]
3. Pont, D.; Hugueny, B.; Beier, U.; Goffaux, D.; Melcher, A.; Noble, R.; Rogers, C.; Roset, N.; Schmutz, S. Assessing river biotic condition at a continental scale: European approach using functional metrics and fish assemblages. *J. Appl. Ecol.* **2006**, *43*, 70–80. [CrossRef]
4. Vadas, R.L., Jr.; Hughes, R.M.; Bello-Gonzales, O.; Callisto, M.; Carvalho, D.; Chen, K.; Davies, P.E.; Ferreira, M.T.; Fierro, P.; Harding, J.S.; et al. Assemblage-based biomonitoring of freshwater ecosystem health via multimetric indices: A critical review and suggestions for improving their applicability. *Water Biol. Secur.* **2022**, *1*, 100054. [CrossRef]
5. Cawley, K.M.; Parker, S.; Utz, R.; Goodman, K.; Scott, C.; Fitzgerald, M.; Vance, J.; Jensen, B.; Bohall, C.; Baldwin, T. NEON aquatic sampling strategy 2022, *NEON.DOC.001152vB*. NEON (National Ecological Observatory Network). Available online: https://data.neonscience.org/api/v0/documents/NEON.DOC.001152vA (accessed on 1 May 2023).
6. McDowell, W.H. NEON and STREON: Opportunities and challenges for the aquatic sciences. *Freshw. Sci.* **2015**, *34*, 386–391. [CrossRef]
7. Tesfaye, C.T.; Souza, A.T.; Barton, D.; Blabolil, P.; Cech, M.; Drastik, V.; Jaroslava, F.; Holubova, M.; Kocvara, L.; Kolarik, T.; et al. Long-term monitoring of fish in a freshwater reservoir: Different ways of weighting complex spatial samples. *Front. Environ. Sci.* **2022**, *6*, 303. [CrossRef]
8. da Silva, V.E.L.; Silva-Firmiano, L.P.S.; Teresa, F.B.; Batista, V.S.; Ladle, R.J.; Fabré, N.N. Functional traits of fish species: Adjusting resolution to accurately express resource partitioning. *Front. Mar. Sci.* **2019**, *6*, 303. [CrossRef]
9. Dias, M.S.; Zuanon, J.; Couto, T.B.; Carvalho, M.; Carvalho, L.N.; Espírito-Santo, H.M.; Frederico, R.; Leitão, R.P.; Mortati, A.F.; Pires, T.H.; et al. Trends in studies of Brazilian stream fish assemblages. *Nat. Conserv.* **2016**, *14*, 106–111. [CrossRef]
10. Tonn, W.M. Climate change and fish communities: A conceptual framework. *Trans. Am. Fish. Soc.* **1990**, *119*, 337–352. [CrossRef]
11. Azaele, S.; Muneepeerakul, R.; Maritan, A.; Rinaldo, A.; Rodriguez-Iturbe, I. Predicting spatial similarity of freshwater fish biodiversity. *Proc. Natl. Acad. Sci. USA* **2009**, *106*, 7058–7062. [CrossRef]
12. Smith, G.R.; Badgley, C.; Eiting, T.P.; Larson, P.S. Species diversity gradients in relation to geological history in North American freshwater fishes. *Evolut. Ecol. Res.* **2010**, *2*, 693–726.
13. Hocutt, C.H.; Wiley, E.O. *The Zoogeography of North American Freshwater Fishes*; Wiley: New York, NY, USA, 1986.
14. Parker, S.M.; Utz, R.M. Temporal design for aquatic organismal sampling across the National Ecological Observatory Network. *Methods Ecol. Evol.* **2022**, *13*, 1834–1848. [CrossRef]
15. A NEON (National Ecological Observatory Network). Aquatic Field Sites—Excluding Lake Sites—Across the NEON Domains. Courtesy of National Ecological Observatory Network/Battelle. Available online: https://www.neonscience.org/field-sites/explore-field-sites (accessed on 1 November 2023).
16. Monahan, D.; Jensen, B.; Parker, S.; Fischer, J.R. AOS Protocol and Procedure: FSS-Fish Sampling in Wadeable Streams. *NEON.DOC.001295vG.NEON*2021 (National Ecological Observatory Network). Available online: https://data.neonscience.org/documents/10179/1883159/NEON.DOC.001295vG (accessed on 1 November 2023).
17. Froese, R.; Pauly, D. FishBase. World Wide Web Electronic Publication. Available online: https://www.fishbase.de/home.htm (accessed on 1 October 2016).
18. NEON (National Ecological Observatory Network). Fish Electrofishing, Gill Netting, and Fyke Netting Counts (DP1.20107.001), RELEASE-2023. Available online: https://data.neonscience.org/data-products/DP1.20107.001/RELEASE-2023 (accessed on 16 April 2023). [CrossRef]
19. Oksanen, J.; Blanchet, F.G.; Kindt, R.; Legendre, P.; O'hara, R.B.; Simpson, G.L.; Solymos, P.; Stevens, M.H.; Wagner, H. Vegan: Community Ecology Package. R Package Version 2.4-6. 2022. Available online: https://CRAN.R-project.org/package=vegan (accessed on 1 May 2023).
20. Hughes, R.M.; Herlihy, A.T.; Comeleo, R.; Peck, D.V.; Mitchell, R.M.; Paulsen, S.G. Patterns in and predictors of stream and river macroinvertebrate genera and fish species richness across the conterminous USA. *Knowl. Manag. Aquat. Ecosyst.* **2023**, *424*, 1–16. [CrossRef]
21. Kanno, Y.; Vokoun, J.C.; Dauwalter, D.C.; Hughes, R.M.; Herlihy, A.T.; Maret, T.R.; Patton, T.M. Influence of rare species on electrofishing distance–species richness relationships at stream sites. *Trans. Am. Fish. Soc.* **2009**, *138*, 1240–1251. [CrossRef]
22. Hughes, R.M.; Herlihy, A.T.; Peck, D.V. Sampling effort for estimating fish species richness in western USA river sites. *Limnologica* **2021**, *87*, 125859. [CrossRef] [PubMed]
23. Schlosser, I.J.; Angermeier, P.L. Spatial variation in demographic processes in lotic fishes: Conceptual models, empirical evidence, and implications for conservation. *Am. Fish. Soc. Symp.* **1995**, *17*, 360–370.
24. Fausch, K.D.; Torgersen, C.E.; Baxter, C.V.; Li, H.W. Landscapes to riverscapes: Bridging the gap between research and conservation of stream fishes. *BioScience* **2002**, *52*, 483–498. [CrossRef]
25. Vadas, R.L., Jr. Seasonal habitat use, species associations, and assemblage structure of forage fishes in Goose Creek, northern Virginia. I. Macrohabitat patterns. *J. Freshw. Ecol.* **1991**, *6*, 403–417. [CrossRef]

26. White, E.P.; Ernest, S.M.; Kerkhoff, A.J.; Enquist, B.J. Relationships between body size and abundance in ecology. *Trends Ecol. Evol.* **2007**, *22*, 323–330. [CrossRef]
27. Brown, J.H.; Gillooly, J.F.; Allen, A.P.; Savage, V.M.; West, G.B. Toward a Metabolic Theory of Ecology. *Ecology* **2004**, *85*, 1771–1789. [CrossRef]
28. Jennings, S. Size-based analyses of aquatic food webs. In *Aquatic Food Webs: An Ecosystem Approach*; Oxford University Press: Oxford, UK, 2005; pp. 86–97.
29. Sprules, W.G.; Barth, L.E. Surfing the biomass size spectrum: Some remarks on history, theory, and application. *Can. J. Fish. Aquat. Sci.* **2016**, *73*, 477–495. [CrossRef]
30. Arranz, I.; Grenouillet, G.; Cucherousset, J. Biological invasions and eutrophication reshape the spatial patterns of stream fish size spectra in France. *Divers. Distrib.* **2023**, *29*, 590–597. [CrossRef]
31. Pomeranz, J.P.F.; Junker, J.R.; Wesner, J.S. Individual size distributions across North American streams vary with local temperature. *Glob. Chang. Biol.* **2021**, *28*, 848–858. [CrossRef] [PubMed]
32. Pompeu, P.S.; Wouters, L.; Hilário, H.O.; Loures, R.C.; Peressin, A.; Prado, I.G.; Suzuki, F.M.; Carvalho, D.C. Inadequate Sampling Frequency and Imprecise Taxonomic Identification Mask Results in Studies of Migratory Freshwater Fish Ichthyoplankton. *Fishes* **2023**, *8*, 518. [CrossRef]

Article

The Status of South Africa's Freshwater Fish Fauna: A Spatial Analysis of Diversity, Threat, Invasion, and Protection

Mohammed Kajee [1,2,3,*], Helen F. Dallas [2,4], Charles L. Griffiths [1], Cornelius J. Kleynhans [5] and Jeremy M. Shelton [2]

[1] Department of Biological Sciences, University of Cape Town, Cape Town 7700, South Africa; charles.griffiths@uct.ac.za

[2] Freshwater Research Centre, Cape Town 7975, South Africa; helen@frcsa.org.za (H.F.D.); jeremy@frcsa.org.za (J.M.S.)

[3] DSI/NRF Research Chair in Inland Fisheries and Freshwater Ecology, South African Institute for Aquatic Biodiversity (SAIAB), Makhanda 6140, South Africa

[4] Faculty of Science, Nelson Mandela University, Gqeberha 6031, South Africa

[5] Department of Water Affairs and Forestry (Retired), 8th Avenue 771, Wonderboom South, Pretoria 0084, South Africa; kneria@gmail.com

* Correspondence: kjxmoh007@myuct.ac.za

Abstract: In South Africa, freshwater habitats are among the most threatened ecosystems, and freshwater fishes are the most threatened species group. Understanding patterns in freshwater fish diversity, threat, invasion, and protection status are vital for their management. However, few studies have undertaken such analyses at ecologically and politically appropriate spatial scales, largely because of limited access to comprehensive biodiversity data sets. Access to freshwater fish data for South Africa has recently improved through the advent of the Freshwater Biodiversity Information System (FBIS). We used occurrence records downloaded from the FBIS to evaluate spatial patterns in distribution, diversity, threat, invasion, and protection status of freshwater fishes in South Africa. Results show that record density varies spatially, at both primary catchment and provincial scales. The diversity of freshwater fishes also varied spatially: native species hotspots were identified at a provincial level in the Limpopo, Mpumalanga, and KwaZulu-Natal provinces; endemic species hotspots were identified in the Western Cape; and threatened species hotspots in the Western Cape, Mpumalanga, Eastern Cape, and KwaZulu-Natal. Non-native species distributions mirrored threatened species hotspots in the Western Cape, Mpumalanga, Eastern Cape, and KwaZulu-Natal. Some 47% of threatened species records fell outside of protected areas, and 38% of non-native species records fell within protected areas. Concerningly, 58% of the distribution ranges of threatened species were invaded by non-native species.

Keywords: freshwater fishes; South Africa; biodiversity data; FBIS; species richness; threatened species; occurrence data

Key Contribution: This study uses occurrence records downloaded from the FBIS to evaluate spatial patterns in distribution, diversity, threat, invasion, and protection status of freshwater fishes in South Africa. There is an urgent need for better monitoring of freshwater fishes in South Africa, so that large-scale assessments of the status of the country's freshwater fish fauna can be more accurately assessed.

Citation: Kajee, M.; Dallas, H.F.; Griffiths, C.L.; Kleynhans, C.J.; Shelton, J.M. The Status of South Africa's Freshwater Fish Fauna: A Spatial Analysis of Diversity, Threat, Invasion, and Protection. *Fishes* **2023**, *8*, 571. https://doi.org/10.3390/fishes8120571

Academic Editors: Robert L. Vadas, Jr. and Robert M. Hughes

Received: 19 October 2023
Revised: 15 November 2023
Accepted: 17 November 2023
Published: 22 November 2023

1. Introduction

Freshwater ecosystems across the world are under threat because they face major impacts from human activities, including land-use change, non-native invasives, water over-abstraction, and the climate crisis [1–4]. Consequently, many organisms that rely on these habitats are threatened with extinction [5,6]. Recent reports indicate that almost one-in-three freshwater taxa are threatened with extinction globally [7,8]. This is of major concern, given that freshwater ecosystems account for ~10% of global biodiversity and 51% of all fish species, despite only covering 1% of the earth's surface [8–10].

In South Africa, freshwater habitats (wetlands and rivers) are among the most threatened ecosystem types, with freshwater fishes the country's most threatened species group [11]. Of 105 formally described native species, 25 are classified as threatened (Vulnerable, Endangered, or Critically Endangered) by the International Union for the Conservation of Nature (IUCN) Red List of Threatened Species [12]. For the 40 South African-endemic freshwater fish species, the threat level is even higher, with two-thirds of endemics being currently classified as threatened [13]. It is thus critical that trends in diversity, distribution, threat status, and protection status of freshwater fishes in South Africa are comprehensively assessed to support management approaches and conservation action.

South Africa has a rich history of freshwater fish research dating back more than 200 years [14]. The first comprehensive catalogue of freshwater fishes for southern Africa was written by G.D.F. Gilchrist and W.W. Thompson between 1913 and 1917 [14]. Skelton's [14] field guide—*A Complete Guide to the Freshwater Fishes of Southern Africa*—followed with updated information on species diversity and distribution [14]. Skelton et al. [15] undertook the first broad-scale analysis of distribution, richness, endemism, and conservation status of freshwater fishes in South Africa, Lesotho, and the Kingdom of eSwatini, including 94 native fish taxa and 18 non-native species [15]. In 2011, an atlas of National Freshwater Ecosystem Priority Areas for South Africa (NFEPA) was published, which summarised data and expert knowledge of regional freshwater ecosystems and fish distributions [16].

The most recent national-scale assessment of freshwater fishes in South Africa, by Chakona et al. [13], summarised the diversity, distribution, and extinction risk of native freshwater fishes. They assessed the extinction risk for 101 valid species and 18 unique genetic lineages of native fishes, finding that 36% of South Africa's freshwater fishes are threatened with extinction. Important diversity and threat hotspots were also assessed, with the Cape Fold Ecoregion [17] being identified as both a high-diversity and high-threat region [13].

In addition to peer-reviewed research articles, the South African National Biodiversity Institute (SANBI), which is mandated to assess and monitor the state of South Africa's biodiversity through the National Environmental Management: Biodiversity Act: Act 10 of 2004 [18], also assessed the status of South Africa's freshwater fishes in their National Biodiversity Assessments (NBA) conducted in 2004, 2011, and 2018 [11]. The most recent NBA [11] assessed the threat status of 118 native freshwater fishes and found that freshwater fishes contained the highest percentage of threatened taxa of any species group in the country [11]. Whilst this assessment has been an important update to the status of freshwater fishes in the country, no specific spatio-temporal analyses of the available fish occurrence data were conducted.

Regardless of this well-established research infrastructure, long-term monitoring data sets for South Africa's freshwater fishes are limited [13,19], with no formalised national or even provincial monitoring programmes currently being undertaken. Scott [20], along with Skelton et al. [21], presented the only known Atlas of southern African freshwater fishes, which contained 35,180 georeferenced specimen records covering 735 species from across the region [20,21]. Data from the atlas have since been uploaded to the GBIF and used to evaluate fish distributions and links with environmental gradients at regional scales [21,22]. More recently, however, platforms such as FishBase [23] and GBIF [24] have facilitated the storage of, and access to, large databases via online, open access platforms, which has allowed government and research organisations, as well as private individuals, to share and access data freely online. For example, the South African Institute for Aquatic Biodiversity (SAIAB) uploaded its entire freshwater fish database to the GBIF platform [25], thereby greatly improving access to this data set [26].

Despite the wealth of freshwater research and biodiversity information available in South Africa, there was no central database for housing freshwater biodiversity data until the recent development of the Freshwater Biodiversity Information System (FBIS;

freshwaterbiodiversity.org) [19]. The FBIS is a data-rich, open access online platform that hosts, analyses, and serves freshwater biodiversity data [19], and aims to serve as a platform for the inventory and maintenance of freshwater data, improving access to comprehensive and reliable freshwater biodiversity data [19]. Consequently, the FBIS functions as a repository for freshwater biodiversity data in South Africa and has been populated with data from a variety of key sources, including published scientific literature, government organisations, and online databases [19]—making it the first comprehensive, accessible national-level resource for freshwater biodiversity data in the country [19]. The database currently hosts more than 57,000 occurrence records for freshwater fishes in South Africa.

Given growing anthropogenic pressures on freshwater ecosystems and fish, and recent improvements in freshwater fish data access in South Africa, there is now both an urgent need and new opportunity for a data-driven assessment of trends in diversity, distribution, and threat status of South Africa's freshwater fish fauna to support improved management and conservation decisions. Freshwater fishes in South Africa face pressure from water abstraction [15], climate change [27–30], and the introduction of non-native species [31–33]. Given the threats to, and observed declining trends in, South Africa's freshwater fish fauna, assessing the conservation status and effectiveness of conservation areas at protecting threatened fishes is both imperative and urgent [13,34].

South Africa has a relatively extensive terrestrial protected area network covering an area of approximately 270,000 km^2 [35,36]. It includes both statutory conservation areas (e.g., National Parks) and non-statutory conservation areas (e.g., Private Nature Reserves) [35–37]. Whilst these protected areas offer some protection to freshwater fishes, few conservation areas and reserves protect entire catchments [15,38], with freshwater systems, in general, being especially neglected [39]. Consistent with global trends [40–42], South Africa's network of protected areas is severely lacking in terms of adequately conserving freshwater ecosystems, with over 90% of the country's main rivers falling outside protected areas [43,44]. The current network only includes small components of protected river areas that form part of much larger, degraded aquatic systems further upstream and downstream of the parks [15,45]. Recent studies have shown that the current protected area network does not adequately protect native freshwater fish [34,45,46], with 84% of taxa regarded as under-protected [34]. Given that approximately 90% of freshwater species listed as Critically Endangered, Endangered, or Vulnerable on the IUCN Red List [12] are threatened by human-induced habitat loss [44], providing adequate protection and interventions to prevent further habitat loss and degradation is of utmost importance. Incorporating spatial freshwater biodiversity data into both protected area planning and management in South Africa will improve the role of protected areas in conserving freshwater ecosystems [47–50].

We used historic-to-present-day freshwater fish data currently available on the FBIS database to assess spatial patterns of distribution, diversity, invasion, and threat status in South Africa at provincial and primary catchment scales. Additionally, we assessed how well South Africa's protected area network protects threatened species and fish diversity hotspots.

2. Materials and Methods

2.1. Study Site

The geographic scope of this study was restricted to the Republic of South Africa (Figure 1). Occurrence records were also limited to rivers, dams, and freshwater lakes, with marine systems in South Africa excluded. South Africa's freshwater fish fauna are managed at provincial level (Figure 1B) via individual provincial conservation authorities. However, primary hydrological catchments (Figure 1C) and freshwater ecoregions (Figure 1D) represent more ecologically relevant assessment and management extents, given that the distribution of fishes are strongly impacted by the climate, geomorphological history, and topography of each region [14]. Analyses for this study were therefore conducted at both provincial and primary catchment scales.

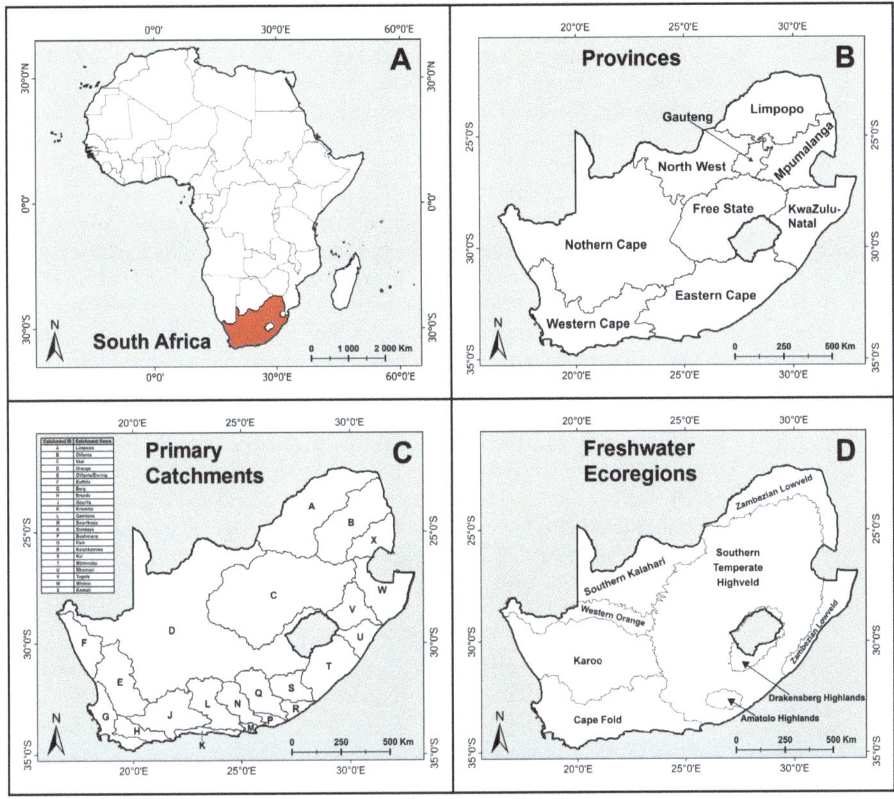

Figure 1. Republic of South Africa (**A**) showing provincial boundaries (**B**), primary catchments (**C**), and freshwater ecoregions (**D**). Catchment IDs are represented, where: A = Limpopo; B = Olifants; C = Vaal; D = Orange; E = Olifants/Doring; F = Buffels; G = Berg; H = Breede; J = Gourits; K = Kromme; L = Gamtoos; M = Swartkops; N = Sundays; P = Bushmans; Q = Fish; R = Keishkamma; S = Kei; T = Mzimvubu; U = Mkomazi; V = Tugela; W = Mfolozi; and X = Komati.

2.2. Data Collection and Cleaning

Occurrence records for all freshwater fish (all known primary and secondary freshwater fish, as well as catadromous species) occurring in South Africa were downloaded from the Freshwater Biodiversity Information System (FBIS) on 30 July 2023 [51] for further analysis.

Preliminary data cleaning was conducted in R [52]; Version 4.2.3 and ArcGIS Pro [53] as follows. First, the data set was clipped to the political boundary of the Republic of South Africa (Figure 1), ensuring that all records in the ocean were removed from the data set. Next, a list of native species that have been translocated outside of their native range (see Supplementary Material S1) was extracted from Ellender and Weyl [32]. For species known to be native but extralimital [32], records occurring outside of the native range of the species—based on IUCN range maps [54] and expert knowledge—were flagged. These records were not included in the species richness counts for native, threatened, and endemic analyses. The R package, *biogeo* [55] was then used to 'autoclean' the data set prior to final analyses. This was done using the '*quickclean*' function in *biogeo* [56], which performs a check to determine if records are at appropriate spatial resolution, removes records that are deemed to be erroneous, and flags duplicate records per species per grid cell. Supplementary Material S4 provides a list of data sources included in the final data set.

2.3. Data Analysis

All data analyses were conducted using R ([52]; Version 4.2.3) and ArcGIS Pro [53]. Using the *richnessmap* function from the R package, *biogeo* [55], we produced species richness maps at a quarter-degree square (QDS) spatial scale (15′ resolution). We used a QDS resolution since accurate distribution maps for all freshwater fishes occurring in South Africa do not currently exist [13]. This spatial scale was also used by Skelton et al. [21], Scott et al. [20], and Skelton et al. [15]. We produced richness maps for all native, non-native, and threatened (listed as Vulnerable, Endangered, or Critically Endangered according to the IUCN Red List of Threatened Species) native species. Species richness was also assessed for species endemic to a single freshwater ecoregion [19,26]. We calculated species counts and the number of records for each species for each province and within each primary catchment, these being considered politically and ecologically useful scales, respectively.

We conducted protected area analyses by overlaying occurrence records and species richness maps at QDS resolution with the protected areas spatial layer downloaded from the South African Protected and Conservation Areas Database on 10 September 2023 [35,36]. The intersections were then used to calculate number of records and distribution area found within protected and conservation areas for native, non-native, threatened, and endemic species. Additionally, we used species richness maps at QDS resolution for threatened and non-native species to assess spatial overlap between these two species groups. Final maps were produced in ArcGIS Pro [53].

3. Results

A total of 57,485 records for freshwater fish were downloaded from the FBIS. After data cleaning, these were reduced to 55,215 records, comprising both native (n = 50,927) and non-native (n = 4288) fishes occurring in South Africa (Figure 2). The final, cleaned data set spans 184 years (1839–2023) and represents 129 species, of which 105 (81%) were native and 24 (19%) were non-native. A list of species with occurrence records in South Africa is available in Kajee et al. [26]. The data vary in time, with 89% of records being collected between 1975 and present.

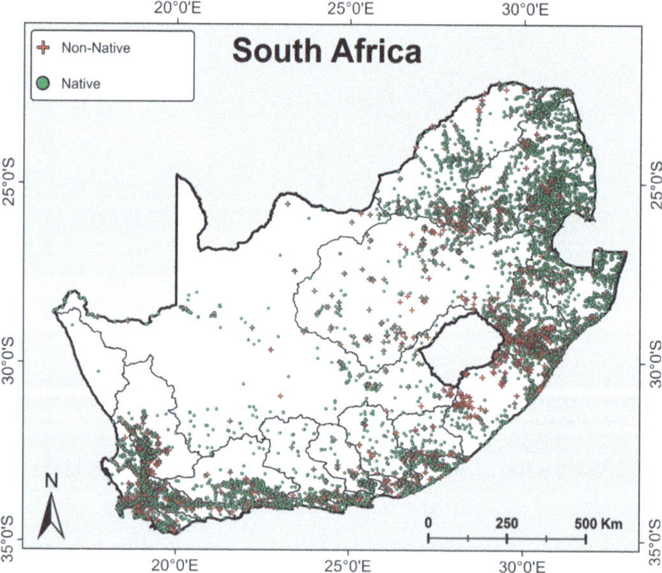

Figure 2. Republic of South Africa, showing primary catchment boundaries and the distribution of available, uncleaned freshwater fish occurrence records for native (green circles) and non-native (red crosses) species (n = 57,485 records).

3.1. Spatial Distribution of Records

The density of records per QDS grid cell shows that data were very unevenly distributed across the country (Figure 3). At the provincial level, fish occurrence records varied substantially between provinces (Table 1). Limpopo (n = 14,353), KwaZulu-Natal (n = 11,411), and Mpumalanga (n = 11,411) each had greater than 10,000 records, whilst Gauteng (n = 907) and Free State (n = 890) both had fewer than 1000 records, respectively. Similarly, the distribution of records among primary catchments varied widely (Table 2). Areas of relatively high density of occurrences (>100 records per QDC grid cell) were in the northeast (Limpopo, Olifants, Komati, Mfolozi, Tugela, and Mkomazi Primary Catchments) and southwest (Olifants/Doring, Berg, Breede, and Gourits Primary Catchments) of South Africa. Conversely, large areas with no records were observed within the central (Orange and Vaal Primary Catchments) and western (Buffels Primary Catchment) parts of the country. On a finer scale, there were noticeable gaps in data in the northern part of the Olifants Doring, Gourits, Mzimvubu, and Limpopo Primary Catchments, respectively. The vast majority of grid cells contained fewer (between 1–10) records (represented in violet; Figure 3).

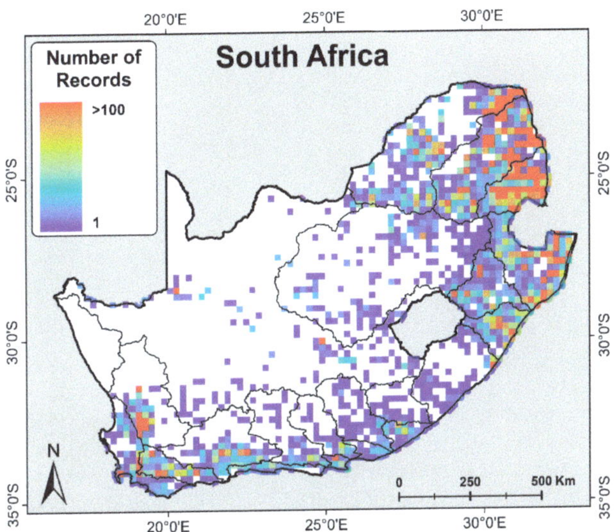

Figure 3. Density of native freshwater fish occurrence records across South Africa per quarter-degree square (QDS—15′) grid cell. Primary catchment boundaries are shown.

Table 1. Number of records and native, extralimital, non-native, endemic, and threatened species occurring in each province of South Africa.

Province	Records	Native Species	Extralimital Species *	Non-Native Species	Endemic Species +	Threatened Species
Eastern Cape	2725	35	10	14	13	8
Free State	890	19	4	7	8	0
Gauteng	907	38	3	9	5	1
KwaZulu-Natal	11,411	74	8	20	13	8
Limpopo	14,353	61	6	11	7	2
Mpumalanga	11,411	65	7	12	7	6
North West	1058	37	5	7	6	1
Northern Cape	1285	39	9	9	13	1
Western Cape	8159	30	10	16	19	11

* As defined by Ellender and Weyl [32] and occurring outside of their home range. + Endemic refers only to Regional endemic level 2, Regional endemic level 1, Micro-endemic level 2, and Micro-endemic level 1 species, as defined by Dallas et al. [19]. Nationally endemic species were omitted.

Table 2. Number of records and native, extralimital, non-native, endemic, and threatened species occurring in each Primary Catchment of South Africa.

Catchment ID	Catchment Name	Records	Native Species	Extralimital Species *	Non-Native Species	Endemic Species +	Threatened Species
Region A	Limpopo	8355	60	7	9	7	1
Region B	Olifants	9823	59	7	10	6	4
Region C	Vaal	1240	34	6	10	7	1
Region D	Orange	1100	37	6	8	11	0
Region E	Olifants/Doring	3443	16	7	7	13	6
Region F	Buffels	6	0	3	0	0	0
Region G	Berg	1690	15	8	14	10	4
Region H	Breede	1902	13	5	10	10	3
Region J	Gourits	1269	12	7	7	7	2
Region K	Kromme	488	8	4	9	6	4
Region L	Gamtoos	433	10	4	6	6	2
Region M	Swartkops	309	10	3	6	4	1
Region N	Sundays	342	13	6	4	2	2
Region P	Bushmans	243	14	5	6	3	2
Region Q	Fish	293	11	5	7	2	2
Region R	Keishkamma	444	12	4	11	4	3
Region S	Kei	176	6	5	9	0	1
Region T	Mzimvubu	713	18	5	13	3	3
Region U	Mkomazi	2359	27	4	17	4	3
Region V	Tugela	2089	29	3	11	5	3
Region W	Mfolozi	7017	68	6	12	8	6
Region X	Komati	8471	58	3	12	4	5

* As defined by Ellender and Weyl [32] and occurring outside of their home range. + Endemic refers only to Regional endemic level 2, Regional endemic level 1, Micro-endemic level 2, and Micro-endemic level 1 species. Nationally endemic species were omitted.

3.2. Species Richness

Native species richness per QDS grid cell (Figure 4A) followed a similar pattern to the record density per grid cell (Figure 3) across South Africa. There were noticeable areas of high species richness in Mpumalanga, Limpopo, and northern KwaZulu-Natal (corresponding with the Limpopo, Olifants, Komati, and Mfolozi Primary Catchments). Additionally, species richness was relatively high (between 10–15 species per QDS grid cell) along the east coast of KwaZulu-Natal (Tugela and Mkomazi Primary Catchments) and in a small cluster in the Western Cape (along the Berg-Olfiants/Doring Primary Catchment boundary). As expected, based on the recorded occurrence density (Figure 3), species richness was lowest in the Free State, Northern Cape, and North West Province (Orange, Vaal and Buffles Primary Catchments).

Endemic species richness per QDS grid cell (Figure 4B) was highest in the Western Cape Province (Olfiants/Doring, Berg, and Bree Primary Catchments) and along the south coast of the Eastern Cape (Swartkops, Kromme, and Gamtoos Primary Catchments). There were small clusters of higher endemic species richness in the northeastern part of the country (Komati, Mfolozi, and Vaal Primary Catchments).

Threatened species were distributed fairly consistently in Limpopo and Mpumalanga (Limpopo, Olifants, and Komati Primary Catchments), along the east coast of KwaZulu-Natal (Mfolozi, Tugela, and Mkomazi Primary Catchments), and south coast of the Eastern Cape (Keishkamma, Bushmans, Swartkops, Fish, Sundays, Gamtoos, Kromme, and Gourits Primary Catchments) (Figure 4C). The Western Cape (Olifants/Doring, Berg, and Bree Primary Catchments) contained the highest concentration of threatened species, followed by the Komati Primary Catchment in the KwaZulu-Natal. There were also notable hotspots within the Kromme and Keishkamma Primary Catchments, as well as within the Mkomazi and Mfolozi Primary Catchments (Figure 4C).

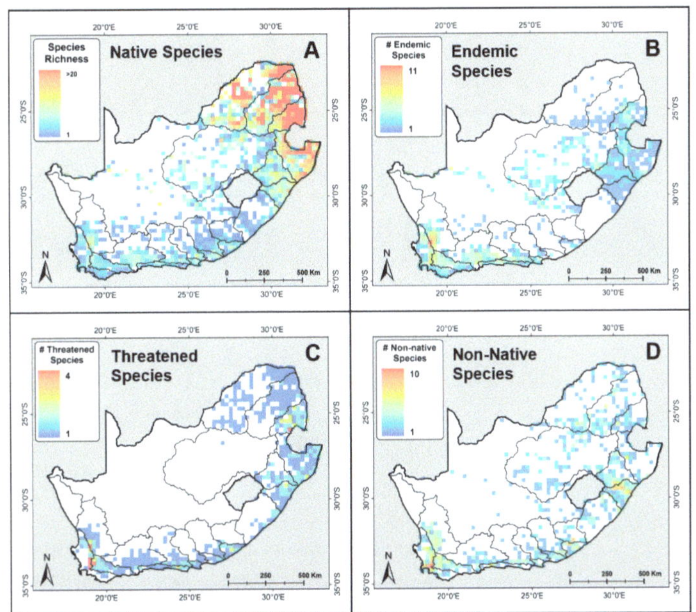

Figure 4. Species richness per quarter-degree square (QDS—15′) grid cell for all native (**A**), endemic* (**B**), threatened⁺ (**C**), and non-native (**D**) freshwater water fish species occurring in South Africa. Primary catchment boundaries are shown. Endemic refers only to Regional endemic level 2, Regional endemic level 1, Micro-endemic level 2, and Micro-endemic level 1 species, as defined by Dallas et al. [19] Nationally endemic species were omitted. Threatened species refers to freshwater water fish listed as threatened according to the IUCN Red List of Threatened Species (Vulnerable, Endangered, or Critically Endangered).

Non-native species richness per QDS grid cell (Figure 4D) was relatively high (4–7 species) along the east coast of KwaZulu-Natal (Mkomazi and Keishkamma Primary Catchments) and highest in the Western Cape (Olifants/Doring, Berg, and Bree Primary Catchments). In general, non-native species were present in all provinces and catchments where records were available (Figure 4D). Importantly, 'species richness' hotspots for non-native species indicate grid cells where several non-native species have established successful, self-sustaining (often invasive) populations.

3.3. Protected Areas

Of the 47,946 records for native fish species analysed in this study, 47% (*n* = 22,756) occurred outside of protected areas, whilst 53% (*n* = 25,190) were located inside protected areas (Figure 5). Of these, 28% (*n* = 6984) were located within a formally protected area, whereas the remaining 72% (*n* = 18,206) were located within a conservation area (Figure 5), as defined by the South Africa Protected and Conservation Areas Database [35,36]. When assessing records for threatened species, a total of 5740 records were used in the final analyses. Of these, 43% (*n* = 2464) were located outside a protected area, and 57% (*n* = 3276) within either a formally protected area (32%; *n* = 1060) or a conservation area (68%; *n* = 2216) (Figure 6A). Based on the QDS grid cell combined distribution, threatened species covered an area of 328,319 km², with only 36% of this range overlapping with South Africa's protected area network (Figure 6B). However, all QDS grid cells that contained more than three threatened species (i.e., the highest density of threatened species) overlapped with a protected area (Figure 6C). Concerningly, 58% of threatened species co-occurred in the same grid cells as non-native species (Figure 6D).

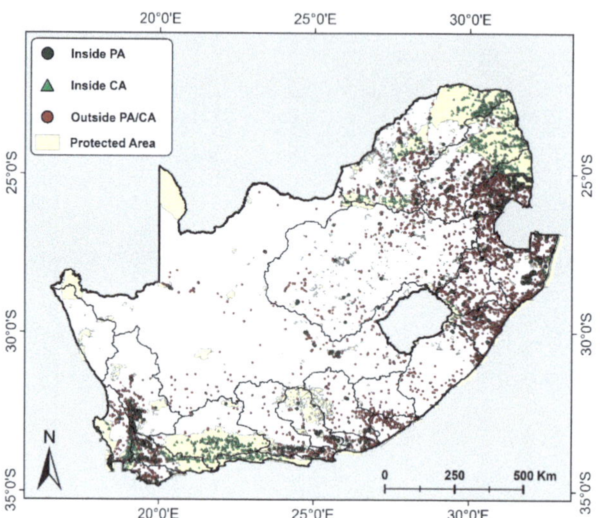

Figure 5. Freshwater fish occurrence records for native freshwater fish occurring within protected areas (green circles), conservation areas (green triangles), and outside of either protected or conservation areas (red circles) in South Africa. Primary catchment boundaries are shown.

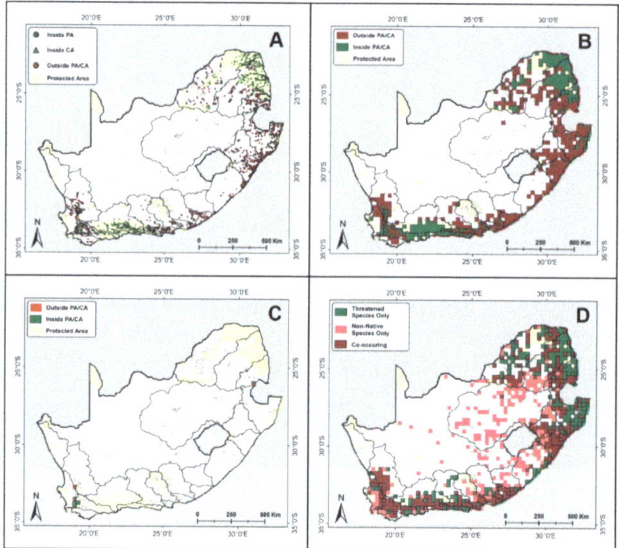

Figure 6. Primary Catchment map of the Republic of South Africa, showing the distribution of threatened freshwater fish (listed as either Vulnerable, Endangered, or Critically Endangered according to the IUCN Red List of Threatened Species) within South Africa's protected areas network. Panel (**A**) shows threatened occurrence records within protected areas (green circles), conservation areas (green triangles), and outside of either protected or conservation areas (red circles). Panel (**B**) shows QDS grid cell distribution of threatened freshwater fish occurring within protected or conservation areas (green) and outside of protected or conservation areas (red). Panel (**C**) shows QDS grid cells that contain the most (*n* = 4) threatened species. Panel (**D**) shows QDS grid cell co-distributions for threatened and non-native species, indicating where threatened species do not overlap with non-native species (green), where threatened species overlap with non-native species (red), and where non-native species do not overlap with threatened species (pink).

A total of 4244 records for non-native species were used in the final analyses for non-native species occurring in South Africa. Of these 62% (*n*= 2625) were recorded outside a formally protected area, whilst 38% (*n*= 1619) were recorded inside either a protected area (53%; *n* = 870) or a conservation area (47%; *n* = 749) (Figure 7A). South Africa's total terrestrial protected area network covers 272,485 km^2, of which 100,815 km^2 (37%) overlaps with non-native species distributions based on species richness grids generated at quarter-degree square (QDS—15′) resolution (Figure 7B).

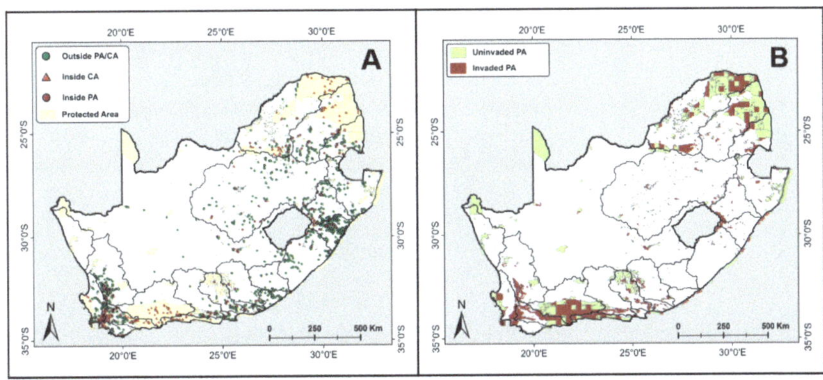

Figure 7. Primary Catchment map of the Republic of South Africa showing (**A**) records for non-native freshwater fish occurring within protected areas (red circles), conservation areas (red triangles), and outside of either protected or conservation areas (green circles); and (**B**) QDS grid cell distribution of non-native freshwater fish within protected or conservation areas (red).

4. Discussion

We found that record density varied spatially, at both primary catchment and provincial scales. The diversity of freshwater fishes also varied spatially. Native fish hotspots were identified in Limpopo, Mpumalanga, and KwaZulu-Natal; an endemic species hotspot was identified in the Western Cape, and threatened species hotspots were identified in the Western Cape, Mpumalanga, Eastern Cape, and KwaZulu-Natal. Non-native species hotspots mirrored threatened species hotspots in the Western Cape, Mpumalanga, Eastern Cape, and KwaZulu-Natal. A total of 43% of threatened species records fell outside of protected areas, and 38% of non-native species records fell within protected areas.

4.1. Spatial Distribution of Records

The spatial distribution of freshwater fish records was uneven across South Africa, indicating that sampling across the country has been biased and concentrated in a few regions. There are several factors that can influence the distribution of these species occurrence data. These include the spatial extent of the original study [56,57], how data from individual studies were stored [56], accessibility of sites [58], and proximity to man-made infrastructure such as roads [58,59], cities [60], and research institutions [59]. There is also evidence that a higher proportion of occurrence records are collected within protected areas, or hotspots of species richness [61,62]. Whilst there have been numerous studies [59,60,63–67] that have sought to identify and quantify inherent bias when using large, publicly available species occurrence data sets, adequately dealing with these biases has proven difficult, and in some cases, even impossible [63]. One concern with working with large, historic data sets is how these underlying biases influence analyses and ultimately the conclusions that can be drawn. Large species occurrence data sets that are prone to bias can have far-reaching implications for the perception of, and inferences about, macroecological patterns [66], which limits the usefulness of these outputs. Another major issue with large historic data sets is that species absences are very rarely reported [68]. This obviously adds additional uncertainty when interpreting analyses based on these data. For example, a data gap could mean that: (i) a site was not sampled; (ii) a site was sampled and did not contain any fish species; or (iii) a site was sampled but there was a failure to detect all fish species present. This is especially the case for rare species, i.e., those likely to be represented by only one or two individuals per site [69,70]. In this regard, interpretation of results from large, historic data sets should be approached with caution.

Whilst the scope of this study did not include assessing the level of bias and spatial autocorrelation between the distribution of man-made landmarks and the FBIS fish data set, this is an important next step that should be carried out when conducting future analyses at national, provincial, or catchment scales. Quantifying this will allow for better contextualisation of these results. Additionally, the unevenness in available freshwater fish data in South Africa speaks to the urgent need for a coordinated national and provincial fish monitoring programme. Unfortunately, however, such biodiversity data depend on a wealth of scientific, human, and financial resources [65], which can be a limiting factor in South Africa's current socio-political and economic landscape [71]. However, given the importance of comprehensive and up-to-date biodiversity data for making informed conservation and management decisions, adequately and strategically monitoring river systems across the country should be prioritised.

Additionally, it is important to acknowledge that using a QDS resolution for this assessment provided a useful, but generalised, insight into freshwater fish spatial patterns in South Africa at a national scale. River habitats are linear systems that are not equally distributed across the geographic landscape. As such, many QDS grid cells are simply not sampled because there are no surface water or drainage regions with regular enough surface water to sustain permanent fish populations.

4.2. Species Richness

The patterns of species richness, endemism, and threatened species are congruent with previously published descriptions of South Africa's freshwater fish diversity. Primary catchments containing the highest freshwater fish species richness in South Africa included the Limpopo, Olifants, Komati, Mfolozi, Tugela, and Mkomazi Primary Catchments, as well as along the Berg-Olfiants/Doring Primary Catchment boundary. Skelton [15] also identified richness hotspots in the northeastern region of the country, with endemism and threat hotspots in the Western and Eastern Cape [15]. Interestingly, Skelton [15] attributed these hotspots to the relatively high topographical relief characteristic of the Cape Fold Mountains. This hypothesis was supported by more recent work [72,73], which found that South Africa's complex geological and climate history, characterised by tectonism and sea-level fluctuations, created unique biogeographic conditions that allowed for the diversification of stream-dwelling taxa, particularly obligate freshwater fishes [72,74]. More recently, Chakona et al. [13] assessed the distribution of freshwater fishes in South Africa, using distribution records from SAIAB's National Fish Collection only (a subset representing ~40% of the raw data included in our study). Whilst they assessed richness at the ecoregion scale, broad patterns in species richness are similar to those presented here. Consequently, it is safe to conclude that the broad patterns of richness identified are an accurate representation of the current state of freshwater fishes in South Africa—despite the concerns regarding the bias contained within the currently available data set. Similarly, endemic species hotspots were found in the Olfiants/Doring, Berg, and Bree Primary Catchments, and, to a lesser degree, the Swartkops, Kromme, Gamtoos, Komati, Mfolozi, and Vaal Primary Catchments. Threatened species hotspots were identified in the Olifants/Doring, Berg, and Bree Primary Catchments, as well as the Mkomazi and Mfolozi catchments, respectively. It is thus recommended that all QDS grid cells identified as having high levels of species richness, endemism, and threatened species be prioritised for resampling and monitoring, to better inform the conservation interventions required in these catchments. Focussing effort and resources in this targeted manner could provide the most efficient use of the limited national, provincial, and scientific resources available in the country.

Of particular concern was the large overlap between the distributions of threatened species and non-native species in South Africa, with the majority (58%) of threatened species co-occurring with non-native species (Figure 6D). Non-native species can have profound and devastating impacts on both native threatened species and, more broadly, freshwater habitats as a whole, and are considered a top threat to native freshwater fish in South Africa [13,32–34,75,76]. The presence of non-native species has resulted in the widescale extirpation of many native species from their historic distributions (especially in mainstem rivers), with many native species now relegated to the upper reaches of tributaries, surviving in small, fragmented populations above waterfalls or other physical barriers that have prevented invasion by non-native species [32,34,77].

It is also important to note that this study was limited to formally described primary and secondary freshwater, as well as catadromous, fishes based on the GBIF taxonomic backbone and available IUCN Red List assessments. As such, the analyses presented will likely need to be updated in the near future once: (i) experts settle on an updated species list for freshwater fishes in the country; (ii) all species in South Africa have had their threat status assessed; and (iii) ongoing taxonomic revisions for several species suites are formalised [13,78–82].

4.3. Protected Areas

We found that South Africa's protected area network [35,36] does not adequately cover the distributions of threatened species in the country. That 57% of South Africa's freshwater fish records were located within a protected or conservation area suggests a strong bias towards sampling in those areas, because they account for only ~20% of South Africa's total land surface area (Figure 6A). Moreover, only 36% of the total area occupied by threatened species occurs within a protected or conservation area (Figure 6B). On the one hand, this indicates that the majority of threatened species distributions are not under formal protection and at heightened risk of extinction. Conversely, however, this level of protection is relatively high, when compared to the percentage of terrestrial land area under protection globally [38,49]. Furthermore, 37% of the country's formal protected area network is invaded by non-native species, with a high percentage over-lap between threatened and non-native species at the QDS scale (Figure 6D). As such, threatened freshwater fish in South Africa still face direct threats from non-native species, even though it may appear that these species are well-considered in the protected area network. Thus, our findings add further evidence to the growing body of research that considers South Africa's protected area network to provide inadequate protection for sensitive freshwater species. Kleynhans [83], Nel et al. [43], Abell et al. [44], and Nel at al. [49] all concluded that South Africa's protected areas did not adequately conserve freshwater ecosystems, with the majority of rivers falling outside the protected area network. Of the rivers located in South Africa, 70% are classified as either Not Protected or Poorly Protected [39]. Furthermore, of the river systems that are formally protected, almost half of these have already been degraded by human activities upstream of the protected area [49,83].

For example, a comprehensive assessment of the 19 National Parks managed by SAN-Parks found that the National Parks protected network only includes small components of protected river areas that form part of much larger, degraded aquatic systems further up- and downstream of the parks [45]. Consequently, very few sites within National Parks contain freshwater fishes that are not under direct threat from land-use change, habitat loss, and non-native species impacts [45]. Similarly, an analysis of the National Parks and nature reserves within the Cape Floral Kingdom (roughly the same geographic range as the Western Cape Province) found that, whilst protected areas do contain populations of most native fish, actual protection was impaired because species ranges extended beyond the boundaries of protection or were protected in rivers with substantial invasion by non-native species [44,46]. More recently, Jordaan et al. [34] also found that protected areas in the Cape Fold Ecoregion did not adequately protect native freshwater fish, with 84% of taxa regarded as under-protected [34].

However, there has been some improvement to the level of protection afforded to South Africa's threatened fish populations. Firstly, the NFEPA project developed a series of strategic spatial maps, prioritising the conservation of the country's freshwater ecosystems [16]. Importantly, the NFEPA provides for Fish Sanctuaries and associated Fish Support Areas, which includes rivers that are essential for protecting threatened and near-threatened freshwater fish native to South Africa [16]. Furthermore, the NFEPA also highlights important Upstream Management Areas, which flag sub-quaternary catchments where human activities need to be carefully managed to prevent degradation of downstream river Fish Sanctuaries and Fish Support Areas [16]. More recently, Kajee et al. [82] reported on the first inclusion of freshwater fishes into the DFFE National Environmental Screening Tool [84]. This process provided an additional layer of protection for South Africa's threatened freshwater fish species, along roughly 50,000 km of river habitat [82]. However, it is also important to acknowledge that South Africa's freshwater fish fauna have the potential to serve as a vital food and income source for rural communities that face extreme levels of poverty and food insecurity [85]. Subsistence fishing activity, in response to modern socio-economic circumstances, was recorded at 77% of dams in South Africa, with recent studies indicating that more than 1.5 million people are involved in freshwater angling activities, in an industry worth approximately ZAR 9 billion annually [85,86]. Consequently, there is a need to reimagine the country's protected area network to better safeguard freshwater fish, and freshwater habitats in general, whilst also accounting for the socio-economic needs of rural communities in South Africa.

4.4. Limitations

Whilst these results will no doubt be useful for researchers, catchment and provincial conservation managers, and policymakers, there are limitations to our study. Given that the basis of this assessment is occurrence records sourced from an open access biodiversity database [19,51], analyses at national (and even sub-national) scales were complicated by a lack of consistency in the types of metadata available for each record. For example, inconsistent abundance and effort data limited our ability to undertake additional diversity analyses. Additionally, working with a large data set (in our case, more than 50000 records) can make manual cleaning of raw data unsustainably time-consuming. In this study, we chose to follow the methods of Robertson et al. [55] and used the biogeo R package to 'autoclean' our data prior to analysis. Whilst this approach seems to have worked well when assessing species richness patterns across the country, further scrutiny of species lists at provincial and catchment levels revealed some inaccuracies.

For example, when reviewing the final species lists for the Olifants/Doring Primary Catchment (a catchment known to the authors), we found that there were inaccuracies in the final number of native, endemic, and threatened species reported in the catchment. Most notably, our data indicated that there were 16 native species occurring in the catchment. However, a literature search, along with expert consultations, revealed 11 such species in this catchment. Further scrutiny revealed that five species (*Labeo rosae*, *Pseudobarbus burchelli*, *Pseudobarbus burgi*, *Pseudobarbus capensis*, and *Sandelia capensis*) had occurrence records in the catchment, when in fact none of these species are known to occur in the area. These records ($n < 10$) are likely a result of misidentifications or inaccurate GPS coordinates. Regardless, their existence undermines the potential benefits of using large, historic data sets to assess species diversity at large spatial extents. Accurate cleaning of this dataset prior to analysis was beyond the scope of this study but should be a priority for such analyses in the future. This would need to be expert-driven and would require substantial time and resources. We suggest that such a project be planned and coordinated by a national body, such as the SANBI or the SAIAB. As such, we chose to include these occurrences in our assessment, but to flag them in our final species list reports (see Supplementary Materials S2 and S3).

Additionally, our study focussed only on the spatial patterns of freshwater fishes, without conducting a detailed temporal analysis of the data, which fell outside the scope of our study. It is important to acknowledge that these data could, and likely are, influenced by inter-specific temporal variation in the data. Consequently, this study, and the published species list, should act only as a first step towards understanding species richness at national, provincial, and catchment scales. A logical next step would be to repeat this study, after having spent the time to thoroughly clean the data set (as was done for threatened fish taxa only by Kajee et al. [82]) and conduct a comprehensive temporal analysis of the data.

5. Conclusions

We present the first assessment of the status of freshwater fish distributions in South Africa using all available data sources from the FBIS. While acknowledging the shortcomings of working with a large, historic data set is important, the patterns emerging from this assessment do adequately identify key fish richness, endemism, and threat hotspots. These patterns are broadly aligned with historic and current expert-driven assessments. This provides a much-needed snapshot of the most important geographic areas for freshwater fishes, and a valuable resource for identifying scientific, conservation, and management priorities in South Africa. We also present the first national-scale assessment of the effectiveness (in terms of geographic coverage and invasion status) of South Africa's protected area network in protecting threatened freshwater fishes. We concluded that the current protected area network is not sufficient to functionally conserve threatened species and prevent future population or species extinctions, given that the majority of the distributions of these species are either outside of the protected area network, invaded by non-native species, or do not have adequate upstream protection. Future interventions should prioritise systematically sampling river systems to fill in identified data gaps, developing strategic long-term monitoring programmes in key hotspot catchments, the comprehensive cleaning of available data to produce accurate distribution maps for all species, and reimagining protected areas to better conserve freshwater fishes.

Supplementary Materials: The following supporting information can be downloaded at: https://www.mdpi.com/article/10.3390/fishes8120571/s1, Supplementary Material S1: List of native species that have been translocated outside of their native range, extracted from Ellender and Weyl [32].; Supplementary Material S2: List of all native freshwater fishes occuring in each province in South Africa; Supplementary Material S3: List of all native freshwater fishes occuring in each primary catchment in South Africa. Supplementary Material S4: List of data sources on which the analyses presented in this paper are based.

Author Contributions: Conceptualization, M.K., H.F.D., C.L.G., C.J.K. and J.M.S.; methodology, M.K. and J.M.S.; software, M.K.; validation, M.K., H.F.D., J.M.S. and C.J.K.; formal analysis, M.K.; investigation, M.K.; resources, H.F.D., J.M.S. and C.L.G.; data curation, M.K., H.F.D. and J.M.S.; writing—original draft preparation, M.K. and J.M.S.; writing—review and editing, M.K., H.F.D., C.L.G., C.J.K. and J.M.S.; visualization, M.K.; supervision, H.F.D., C.L.G. and J.M.S.; project administration, C.L.G.; funding acquisition, H.F.D. and J.M.S. All authors have read and agreed to the published version of the manuscript.

Funding: This research was funded by the JRS Biodiversity Foundation, grants 60606 and 60919, and the South African National Biodiversity Institute (SANBI). This work is based on research supported in part by the National Research Foundation (NRF) of South Africa, grant number MND2000621534710–UID: 133692, and the NRF-SAIAB DSI/NRF Research Chair in Inland Fisheries and Freshwater Ecology (UID 110507). Student funding, in the form of an MSc bursary, was also provided by the Freshwater Biodiversity Unit of the South African National Biodiversity Institute (SANBI).

Institutional Review Board Statement: All the data come from previous research or other databases. In this case, ethical approval is not needed for this article.

Data Availability Statement: The raw data presented in this study are openly available on the Freshwater Biodiversity Information System (FBIS) at https://freshwaterbiodiversity.org/ The data set can also be accessed via GBIF at Freshwater Biodiversity Information System (FBIS) Fish Data. Version 1.6. Freshwater Research Centre [https://doi.org/10.15468/gmk6hg]. All additional data and analyses are available on request from the corresponding author.

Acknowledgments: We would like to thank the FRC staff and interns, specifically Aneri Swanepoel, who provided invaluable assistance with data collection and cleaning. We express our gratitude to the Kartoza team for assistance with the technical development and implementation of the FBIS platform. Lastly, we would like to acknowledge the contributions of Dominic Henry, Res Altwegg, and Vernon Visser, who all provided invaluable advice with aspects of the data analyses. The authors acknowledge the individuals and organisations whose data have been used in the analyses presented in this paper.

Conflicts of Interest: The authors declare no conflict of interest.

References

1. Darwall, W.; Smith, K.; Allen, D.; Seddon, M.; Reid, G.M.; Clausnitzer, V.; Kalkman, V.J. Freshwater Biodiversity: A Hidden Resource under Threat. In *Wildlife in a Changing World–An Analysis of the 2008 IUCN Red List of Threatened Species*; IUCN: Gland, Switzerland, 2009; Volume 43.
2. Best, J. Anthropogenic Stresses on the World's Big Rivers. *Nat. Geosci.* **2019**, *12*, 7–21. [CrossRef]
3. Dudgeon, D.; Arthington, A.H.; Gessner, M.O.; Kawabata, Z.-I.; Knowler, D.J.; Lévêque, C.; Naiman, R.J.; Prieur-Richard, A.-H.; Soto, D.; Stiassny, M.L.J. Freshwater Biodiversity: Importance, Threats, Status and Conservation Challenges. *Biol. Rev.* **2006**, *81*, 163–182. [CrossRef] [PubMed]
4. Capon, S.J.; Stewart-Koster, B.; Bunn, S.E. Future of Freshwater Ecosystems in a 1.5 °C Warmer World. *Front. Environ. Sci.* **2021**, *9*, 596–602. [CrossRef]
5. He, F.; Zarfl, C.; Bremerich, V.; David, J.N.W.; Hogan, Z.; Kalinkat, G.; Tockner, K.; Jähnig, S.C. The Global Decline of Freshwater Megafauna. *Glob. Chang. Biol.* **2019**, *25*, 3883–3892. [CrossRef] [PubMed]
6. Reid, A.J.; Carlson, A.K.; Creed, I.F.; Eliason, E.J.; Gell, P.A.; Johnson, P.T.J.; Kidd, K.A.; MacCormack, T.J.; Olden, J.D.; Ormerod, S.J.; et al. Emerging Threats and Persistent Conservation Challenges for Freshwater Biodiversity. *Biol. Rev.* **2019**, *94*, 849–873. [CrossRef]
7. Collen, B.; Whitton, F.; Dyer, E.E.; Baillie, J.E.M.; Cumberlidge, N.; Darwall, W.R.T.; Pollock, C.; Richman, N.I.; Soulsby, A.M.; Böhm, M. Global Patterns of Freshwater Species Diversity, Threat and Endemism. *Glob. Ecol. Biogeogr.* **2014**, *23*, 40–51. [CrossRef]
8. Tickner, D.; Opperman, J.J.; Abell, R.; Acreman, M.; Arthington, A.H.; Bunn, S.E.; Cooke, S.J.; Dalton, J.; Darwall, W.; Edwards, G.; et al. Bending the Curve of Global Freshwater Biodiversity Loss: An Emergency Recovery Plan. *Bioscience* **2020**, *70*, 330–342. [CrossRef]
9. Strayer, D.L.; Dudgeon, D. Freshwater Biodiversity Conservation: Recent Progress and Future Challenges. *J. N. Am. Benthol. Soc.* **2010**, *29*, 344–358. [CrossRef]
10. Hughes, K.; Harrison, I.; Darwall, W.; Lee, R.; Muruven, D.; Revenga, C.; Claussen, J.; Lynch, A.; Pinder, A.; Abell, R.; et al. *The World's Forgotten Fishes*; World Wide Fund for Nature: Gland, Switzerland, 2021.
11. Skowno, A.L.; Poole, C.J.; Raimondo, D.C.; Sink, K.J.; Van Deventer, H.; van Niekerk, L.; Harris, L.; Smith-Adao, L.B.; Tolley, K.A.; Zangeya, T.A.; et al. *National Biodiversity Assessment 2018: The Status of South Africa's Ecosystems and Biodiversity*; National Biodiversity Assessment: Pretoria, South Africa, 2019; Volume 53.
12. International Union for Conservation of Nature The IUCN Red List. Available online: https://www.iucnredlist.org/ (accessed on 14 September 2022).
13. Chakona, A.; Jordaan, M.S.; Raimondo, D.C.; Bills, R.I.; Skelton, P.H.; van Der Colff, D. Diversity, Distribution and Extinction Risk of Native Freshwater Fishes of South Africa. *J. Fish. Biol.* **2022**, *100*, 1044–1061. [CrossRef]
14. Skelton, P.H. *A Complete Guide to the Freshwater Fishes of Southern Africa*, 1st ed.; Southern Book Publishers (Pty) Ltd.: Johannesburg, South Africa, 2001; ISBN 1868726436.
15. Skelton, P.H.; Cambray, J.A.; Lombard, A.; Benn, G.A. Patterns of Distribution and Conservation Status of Freshwater Fishes in South Africa. *South Afr. J. Zool.* **1995**, *30*, 71–81. [CrossRef]
16. Nel, J.L. *Technical Report for the National Freshwater Ecosystem Priority Areas Project: Report to the Water Research Commission*; Water Research Commission: Pretoria, South Africa, 2011.
17. Abell, R.; Thieme, M.L.; Revenga, C.; Bryer, M.; Kottelat, M.; Bogutskaya, N.; Coad, B.; Mandrak, N.; Balderas, S.C.; Bussing, W. Freshwater Ecoregions of the World: A New Map of Biogeographic Units for Freshwater Biodiversity Conservation. *Bioscience* **2008**, *58*, 403–414. [CrossRef]
18. Republic of South Africa. *National Environmental Management: Biodiversity Act (10/2004): Threatened or Protected Species Regulations*; South African Government: Pretoria, South Africa, 2004.

19. Dallas, H.; Shelton, J.; Sutton, T.; Tri Cuptura, D.; Kajee, M.; Job, N. The Freshwater Biodiversity Information System (FBIS)—Mobilising Data for Evaluating Long-Term Change in South African Rivers. *Afr. J. Aquat. Sci.* **2022**, *47*, 291–306. [CrossRef]

20. Scott, L.E.P.; Skelton, P.H.; Booth, A.J.; Verheust, L. The Development of a GIS Atlas of Southern African Freshwater Fish. *S. Afr. J. Aquat. Sci.* **2000**, *25*, 43–48. [CrossRef]

21. Skelton, P.; Verheust, L.; Scott, L. A Collection-Based GIS Atlas of Fish Distribution in South Africa: Collections and Research. *S. Afr. Mus. Assoc. Bull.* **2000**, *25*, 11–15.

22. Scott, L.E.P. *The Development of a Geographic Information Systems Based Atlas of Southern African Freshwater Fish, and Its Application to Biogeographic Analysis*; Rhodes University: Makhanda, South Africa, 2000.

23. Froese, R.; Pauly, D. FishBase. Available online: www.fishbase.org (accessed on 29 July 2023).

24. GBIF.org GBIF Home Page. Available online: https://www.gbif.org (accessed on 1 December 2020).

25. *SAIAB National Fish Collection of South Africa 2023*; South African Institute for Aquatic Biodiversity: Makhanda, South Africa, 2023.

26. Kajee, M.; Dallas, H.F.; Swanepoel, A.; Griffiths, C.L.; Shelton, J.M. The Freshwater Biodiversity Information System (FBIS) Fish Data: A Georeferenced Dataset of Freshwater Fishes Occurring in South Africa. *J. Limnol.* **2023**, *82*, 102–110. [CrossRef]

27. Dallas, H.; Rivers-Moore, N. Ecological Consequences of Global Climate Change for Freshwater Ecosystems...: Discovery Service for Rhodes University Library. *S. Afr. J. Sci.* **2014**, *110*, 48–58. [CrossRef]

28. Ziervogel, G.; New, M.; Archer van Garderen, E.; Midgley, G.; Taylor, A.; Hamann, R.; Stuart-Hill, S.; Myers, J.; Warburton, M. Climate Change Impacts and Adaptation in South Africa. *Wiley Interdiscip. Rev. Clim. Chang.* **2014**, *5*, 605–620. [CrossRef]

29. Niang, I.; Ruppel, O.C.; Abdrabo, M.A.; Essel, A.; Lennard, C.; Padgham, J.; Urquhart, P. *Climate Change 2014: Impacts, Adaptation, and Vulnerability. Part B: Regional Aspects*; Cambridge University Press: Cambridge, UK, 2017.

30. Reizenberg, J.-L. *The Thermal Tolerances and Preferences of Native Fishes in the Cape Floristic Region: Towards Understanding the Effect of Climate Change on Native Fish Species*; University of Cape Town: Cape Town, South Africa, 2017.

31. Van Rensburg, B.J.; Weyl, O.L.F.; Davies, S.J.; van Wilgen, N.J.; Spear, D.; Chimimba, C.T.; Peacock, F. *Invasive Vertebrates of South Africa. Biological Invasions: Economic and Environmental Costs of Alien Plant, Animal, and Microbe Species*; CRC Press: Boca Raton, FL, USA, 2011; pp. 326–378.

32. Ellender, B.; Weyl, O. A Review of Current Knowledge, Risk and Ecological Impacts Associated with Non-Native Freshwater Fish Introductions in South Africa. *Aquat. Invasions* **2014**, *9*, 117–132. [CrossRef]

33. Weyl, O.L.F.; Ellender, B.R.; Wassermann, R.J.; Truter, M.; Dalu, T.; Zengeya, T.A.; Smit, N.J. Alien Freshwater Fauna in South Africa. In *Biological Invasions in South Africa*; van Wilgen, B.W., Measey, J., Richardson, D.M., Wilson, J.R., Zengeya, T.A., Eds.; Springer: Cham, Switzerland, 2020; pp. 153–183. ISBN 978-3-030-32394-3.

34. Jordaan, M.S.; Chakona, A.; Colff, D. Van Der Protected Areas and Endemic Freshwater Fishes of the Cape Fold Ecoregion: Missing the Boat for Fish Conservation? *Front. Environ. Sci.* **2020**, *8*, 1–13. [CrossRef]

35. Department of Forestry, Fisheries and the Environment (DFFE). *Department of Forestry Fisheries and the Environment South Africa Protected Areas Database (SAPAD_OR_2023_Q1)*; Department of Forestry, Fisheries and the Environment (DFFE): Pretoria, South Africa, 2023.

36. Department of Forestry, Fisheries and the Environment (DFFE). *Department of Forestry Fisheries and the Environment South Africa Conservation Areas Database (SACAD_OR_2023_Q1)*; Department of Forestry, Fisheries and the Environment (DFFE): Pretoria, South Africa, 2023.

37. Rouget, M.; Richardson, D.M.; Cowling, R.M. The Current Configuration of Protected Areas in the Cape Floristic Region, South Africa—Reservation Bias and Representation of Biodiversity Patterns and Processes. *Biol. Conserv.* **2003**, *112*, 129–145. [CrossRef]

38. Acreman, M.; Hughes, K.A.; Arthington, A.H.; Tickner, D.; Dueñas, M.A. Protected Areas and Freshwater Biodiversity: A Novel Systematic Review Distils Eight Lessons for Effective Conservation. *Conserv. Lett.* **2019**, *13*, e12684. [CrossRef]

39. Department of Environmental Affairs. *National Protected Area Expansion Strategy for South Africa 2016*; Department of Environmental Affairs: Pretoria, South Africa, 2016.

40. Nel, J.L.; Roux, D.J.; Abell, R.; Ashton, P.J.; Cowling, R.M.; Higgins, J.V.; Thieme, M.; Viers, J.H. Progress and Challenges in Freshwater Conservation Planning. *Aquat. Conserv.* **2009**, *19*, 474–485. [CrossRef]

41. Adams, V.M.; Setterfield, S.A.; Douglas, M.M.; Kennard, M.J.; Ferdinands, K. Measuring Benefits of Protected Area Management: Trends across Realms and Research Gaps for Freshwater Systems. *Philosophical Transactions of the Royal Society B: Biological Sciences* **2015**, *370*, 20140274. [CrossRef] [PubMed]

42. Saunders, D.L.; Meeuwig, J.J.; Vincent, A.C. Freshwater Protected Areas: Strategies for Conservation. *Cons. Biol.* **2002**, *16*, 30–41. [CrossRef] [PubMed]

43. Nel, J.; Maree, G.; Roux, D.; Moolman, J.; Kleynhans, N.; Silberbauer, M.; Driver, A. *South African National Spatial Biodiversity Assessment 2004: Technical Report, Volume 2: River Component*; South Africa's National Biodiversity Assessment: Pretoria, South Africa, 2004.

44. Abell, R.; Allan, J.D.; Lehner, B. Unlocking the Potential of Protected Areas for Freshwaters. *Biol. Conserv.* **2007**, *134*, 48–63. [CrossRef]

45. Russell, I.A. Conservation Status and Distribution of Freshwater Fishes in South African National Parks. *Afr. Zool.* **2011**, *46*, 117–132. [CrossRef]

46. Impson, N.D.; Bills, I.R.; Cambray, J.A. A Conservation Plan for the Unique and Highly Threatened Freshwater Fishes of the Cape Floral Kingdom. In *Conservation of Freshwater Fishes: Options for the Future*; Blackwell Science: Oxford, UK, 2002; pp. 432–440.
47. Roux, D.J.; Nel, J.L.; Ashton, P.J.; Deacon, A.R.; de Moor, F.C.; Hardwick, D.; Hill, L.; Kleynhans, C.J.; Maree, G.A.; Moolman, J.; et al. Designing Protected Areas to Conserve Riverine Biodiversity: Lessons from a Hypothetical Redesign of the Kruger National Park. *Biol. Conserv.* **2008**, *141*, 100–117. [CrossRef]
48. Nel, J.L.; Turak, E.; Linke, S.; Brown, C. Integration of Environmental Flow Assessment and Freshwater Conservation Planning: A New Era in Catchment Management. *Mar. Freshw. Res.* **2011**, *62*, 290–299. [CrossRef]
49. Nel, J.L.; Reyers, B.; Roux, D.J.; Cowling, R.M. Expanding Protected Areas beyond Their Terrestrial Comfort Zone: Identifying Spatial Options for River Conservation. *Biol. Conserv.* **2009**, *142*, 1605–1616. [CrossRef]
50. Leal, C.G.; Lennox, G.D.B.; Ferraz, S.F.; Ferreira, J.; Gardner, T.A.; Thomson, J.R.; Berenguer, E.; Lees, A.C.; Hughes, R.M.; Mac Nally, R.; et al. Integrated Terrestrial-Freshwater Planning Doubles Conservation of Tropical Aquatic Species. *Science* **2020**, *370*, 117–121. [CrossRef]
51. FBIS Freshwater Biodiversity Information System (FBIS). FBIS Version 3. Available online: https://freshwaterbiodiversity.org/ (accessed on 30 July 2023).
52. R Core Team. *R: A Language and Environment for Statistical Computing*; R Foundation for Statistical Computing: Vienna, Austria, 2023; Available online: http://www.r-project.org (accessed on 30 July 2023).
53. Esri Inc ArcGIS Pro (Version 3.0). Esri Inc. 2022. Available online: https://www.esri.com/en-us/arcgis/products/arcgis-pro/overview (accessed on 15 November 2023).
54. IUCN (International Union for Conservation of Nature). The IUCN Red List of Threatened Species (Spatial Data). Available online: https://www.iucnredlist.org/resources/spatial-data-download (accessed on 30 August 2023).
55. Robertson, M.P.; Visser, V.; Hui, C. Biogeo: An R Package for Assessing and Improving Data Quality of Occurrence Record Datasets. *Ecography* **2016**, *39*, 394–401. [CrossRef]
56. Beck, J.; Böller, M.; Erhardt, A.; Schwanghart, W. Spatial Bias in the GBIF Database and Its Effect on Modeling Species' Geographic Distributions. *Ecol. Inf.* **2014**, *19*, 10–15. [CrossRef]
57. Anderson, R.P.; Araújo, M.; Guisan, A.; Lobo, J.M.; Martínez-Meyer, E.; Peterson, A.T.; Soberón, J. *Final Report of the Task Group on GBIF Data Fitness for Use in Distribution Modelling Are Species Occurrence Data in Global Online Repositories Fit for Modeling Species Distributions? The Case of the Global Biodiversity Information Facility (GBIF) Authors (in Alphabetical Order)*; Global Biodiversity Information Facility: Copenhagen, Denmark, 2016.
58. Hughes, A.C.; Orr, M.C.; Ma, K.; Costello, M.J.; Waller, J.; Provoost, P.; Yang, Q.; Zhu, C.; Qiao, H. Sampling Biases Shape Our View of the Natural World. *Ecography* **2021**, *44*, 1259–1269. [CrossRef]
59. Daru, B.H.; Park, D.S.; Primack, R.B.; Willis, C.G.; Barrington, D.S.; Whitfeld, T.J.S.; Seidler, T.G.; Sweeney, P.W.; Foster, D.R.; Ellison, A.M.; et al. Widespread Sampling Biases in Herbaria Revealed from Large-Scale Digitization. *New Phytol.* **2018**, *217*, 939–955. [CrossRef]
60. Bowler, D.E.; Callaghan, C.T.; Bhandari, N.; Henle, K.; Benjamin Barth, M.; Koppitz, C.; Klenke, R.; Winter, M.; Jansen, F.; Bruelheide, H.; et al. Temporal Trends in the Spatial Bias of Species Occurrence Records. *Ecography* **2022**, *2022*, e06219. [CrossRef]
61. Hugo, S.; Altwegg, R. The Second Southern African Bird Atlas Project: Causes and Consequences of Geographical Sampling Bias. *Ecol. Evol.* **2017**, *7*, 6839–6849. [CrossRef]
62. Jiménez-Valverde, A.; Peña-Aguilera, P.; Barve, V.; Burguillo-Madrid, L. Photo-Sharing Platforms Key for Characterising Niche and Distribution in Poorly Studied Taxa. *Insect. Conserv. Divers.* **2019**, *12*, 389–403. [CrossRef]
63. Meyer, C.; Weigelt, P.; Kreft, H. Multidimensional Biases, Gaps and Uncertainties in Global Plant Occurrence Information. *Ecol. Lett.* **2016**, *19*, 992–1006. [CrossRef]
64. Rocha-Ortega, M.; Rodriguez, P.; Córdoba-Aguilar, A. Geographical, Temporal and Taxonomic Biases in Insect GBIF Data on Biodiversity and Extinction. *Ecol. Entomol.* **2021**, *46*, 718–728. [CrossRef]
65. Garcia-Rosello, E.; Gonzalez-Dacosta, J.; Guisande, C.; Lobo, J.M. GBIF Falls Short of Providing a Representative Picture of the Global Distribution of Insects. *Syst. Entomol.* **2023**, *48*, 489–497. [CrossRef]
66. Yang, W.; Ma, K.; Kreft, H. Geographical Sampling Bias in a Large Distributional Database and Its Effects on Species Richness-Environment Models. *J. Biogeogr.* **2013**, *40*, 1415–1426. [CrossRef]
67. Ballesteros-Mejia, L.; Kitching, I.J.; Jetz, W.; Nagel, P.; Beck, J. Mapping the Biodiversity of Tropical Insects: Species Richness and Inventory Completeness of African Sphingid Moths. *Glob. Ecol. Biogeogr.* **2013**, *22*, 586–595. [CrossRef]
68. Coro, G.; Magliozzi, C.; Vanden Berghe, E.; Bailly, N.; Ellenbroek, A.; Pagano, P. Estimating Absence Locations of Marine Species from Data of Scientific Surveys in OBIS. *Ecol. Model.* **2016**, *323*, 61–76. [CrossRef]
69. Kanno, Y.; Vokoun, J.C.; Dauwalter, D.C.; Hughes, R.M.; Herlihy, A.T.; Maret, T.R.; Patton, T.M. Influence of Rare Species on Electrofishing Distance When Estimating Species Richness of Stream and River Reaches. *Trans. Am. Fish. Soc.* **2009**, *138*, 1240–1251. [CrossRef]
70. Hughes, R.M.; Herlihy, A.T.; Peck, D.V. Sampling Efforts for Estimating Fish Species Richness in Western USA River Sites. *Limnologica* **2021**, *87*, 125859. [CrossRef]
71. Impson, D. Have Our Provincial Aquatic Scientists Become Critically Endangered?: Capacity Development-Feature. *Water Wheel* **2016**, *15*, 20–23.

72. Chakona, A.; Swartz, E.R.; Gouws, G. Evolutionary Drivers of Diversification and Distribution of a Southern Temperate Stream Fish Assemblage: Testing the Role of Historical Isolation and Spatial Range Expansion. *PLoS ONE* **2013**, *8*, e70953. [CrossRef]
73. Chakona, A.; Gouws, G.; Kadye, W.T.; Jordaan, M.S.; Swartz, E.R. Reconstruction of the Historical Distribution Ranges of Imperilled Stream Fishes from a Global Endemic Hotspot Based on Molecular Data: Implications for Conservation of Threatened Taxa. *Aquat. Conserv.* **2020**, *30*, 144–158. [CrossRef]
74. Chakona, G.; Swartz, E.R.; Chakona, A. Historical Abiotic Events or Human-aided Dispersal: Inferring the Evolutionary History of a Newly Discovered Galaxiid Fish. *Ecol. Evol.* **2015**, *5*, 1369–1380. [CrossRef]
75. Marr, S.M.; Ellender, B.R.; Woodford, D.J.; Alexander, M.E.; Wasserman, R.J.; Ivey, P.; Zengeya, T.; Weyl, O.L.F. Evaluating Invasion Risk for Freshwater Fishes in South Africa. *Bothalia-Afr. Biodivers. Conserv.* **2017**, *47*, a2177. [CrossRef]
76. Shelton, J.M.; Weyl, O.L.F.; Chakona, A.; Ellender, B.R.; Esler, K.J.; Impson, N.D.; Jordaan, M.S.; Marr, S.M.; Ngobela, T.; Paxton, B.R. Vulnerability of Cape Fold Ecoregion Freshwater Fishes to Climate Change and Other Human Impacts. *Aquat. Conserv.* **2018**, *28*, 68–77. [CrossRef]
77. Van Der Walt, J.A.; Weyl, O.L.F.; Woodford, D.J.; Radloff, F.G.T. Spatial Extent and Consequences of Black Bass (*Micropterus* spp.) Invasion in a Cape Floristic Region River Basin. *Aquat. Conserv.* **2016**, *26*, 736–748. [CrossRef]
78. Chakona, G.; Swartz, E.R.; Chakona, A. The Status and Distribution of a Newly Identified Endemic Galaxiid in the Eastern Cape Fold Ecoregion, of South Africa. *Aquat. Conserv.* **2018**, *28*, 55–67. [CrossRef]
79. Chakona, A.; Skelton, P.H. A Review of the Pseudobarbus Afer (Peters, 1864) Species Complex (Teleostei, Cyprinidae) in the Eastern Cape Fold Ecoregion of South Africa. *ZooKeys* **2017**, *140*, 109–140. [CrossRef] [PubMed]
80. Kambikambi, M.J.; Kadye, W.T.; Chakona, A. Allopatric Differentiation in the Enteromius Anoplus Complex in South Africa, with the Revalidation of Enteromius Cernuus and Enteromius Oraniensis, and Description of a New Species, Enteromius Mandelai (Teleostei: Cyprinidae). *J. Fish. Biol.* **2021**, *99*, 931–954. [CrossRef] [PubMed]
81. Chakona, A.; Gouws, G.; Kadye, W.T.; Mpopetsi, P.P.; Skelton, P.H. Probing Hidden Diversity to Enhance Conservation of the Endangered Narrow-Range Endemic Eastern Cape Rocky, Sandelia Bainsii (Castelnau 1861). *Koedoe Afr. Prot. Area Conserv. Sci.* **2020**, *62*, 1–6.
82. Kajee, M.; Henry, D.A.W.; Dallas, H.F.; Griffiths, C.L.; Pegg, J.; Van der Colff, D.; Impson, D.; Chakona, A.; Raimondo, D.C.; Job, N.M.; et al. How the Freshwater Biodiversity Information System (FBIS) Is Supporting National Freshwater Fish Conservation Decisions in South Africa. *Front. Environ. Sci.* **2023**, *11*, 303. [CrossRef]
83. Kleynhans, C.J. *Desktop Estimates of the Ecological Importance and Sensitivity Categories (EISC), Default Ecological Management Classes (DEMC), Present Ecological Status Categories (PESC), Present Attainable Ecological Management Classes (Present AEMC), and Best Attainab*; Department of Water Affairs and Forestry: Pretoria, South Africa, 2000.
84. Department of Forestry Fisheries and the Environment National Web Based Environmental Screening Tool. Available online: https://screening.environment.gov.za/screeningtool/#/pages/welcome (accessed on 22 August 2022).
85. Hara, M.; Muchapondwa, E.; Sara, J.; Weyl, O.; Tapela, B. *Inland Fisheries Contributions to Rural Livelihoods: An Assessment of Fisheries Potential, Market Value Chains and Governance Arrangements Report to the Water Research Commission*; Water Research Commission: Pretoria, South Africa, 2021; ISBN 978-0-6392-0240-2.
86. Britz, P.J.; Hara, M.M.; Weyl, O.; Tapela, B.N.; Rouhani, Q.A. *Scoping Study on the Development and Sustainable Utilisation of Inland Fisheries in South Africa. Volume 1: Research Report Report to the Water Research Commission*; Water Research Commission: Pretoria, South Africa, 2015.

 **fishes**

Article

Molecular Analysis of Two Endemic *Squalius* Species: Evidence for Intergeneric Introgression among Cyprinids and Conservation Issues

Damir Valić [1,*], Matej Kristan Mirković [2], Višnja Besendorfer [3] and Emin Teskeredžić [4]

[1] Laboratory for Biological Effects of Metals, Division for Marine and Environmental Research, Ruđer Bošković Institute, Bijenička 54, 10000 Zagreb, Croatia
[2] Department of Biology, Faculty of Science, University of Zagreb, Ravnice 48, 10000 Zagreb, Croatia; matej.k.mirkovic@gmail.com
[3] Division of Molecular Biology, Department of Biology, Faculty of Science, University of Zagreb, Horvatovac 102a, 10000 Zagreb, Croatia; visnja.besendorfer@biol.pmf.unizg.hr
[4] Laboratory for Aquaculture and Pathology of Aquatic Organisms, Division of Molecular Biology, Ruđer Bošković Institute, Bijenička 54, 10000 Zagreb, Croatia
* Correspondence: dvalic@irb.hr; Tel.: +385-1-4561-076

Abstract: Conservation of indigenous species, especially endemic ones, is of the utmost importance. Morphological determination of species is usually not sufficient; therefore, molecular phylogenetic analyses of the Illyrian chub, *Squalius illyricus*, and the Zrmanja chub, *Squalius zrmanjae*, from the Krka River were performed. For the genetic characterization of the mitochondrial gene cytochrome *b* and the non-coding nuclear region *Cyfun P*, 15 specimens from each species were subjected to analysis. The obtained sequences were aligned with similar ones from GenBank to determine the taxonomic and phylogenetic position of these species. The obtained molecular results imply that *S. zrmanjae* from the Krka River has a nuclear region that resembles Dalmatian rudd, *Scardinius dergle*. This result implies an introgression event and the transfer of genetic information between the two genera. The investigated species are on the IUCN Red List of Threatened Species, their biological data are scarce, and further investigation and protection are needed.

Keywords: Leuciscidae; cyt *b*; endemic species; *Squalius*

Key Contribution: We investigated two fishes endemic to the Adriatic basin—the Illyrian chub, *Squalius illyricus* and the Zrmanja chub, *Squalius zrmanjae*—both on the IUCN Red List of Threatened Species. They are protected by national law but their biological and molecular data are scarce, and data for their nuclear markers do not exist. The results obtained from molecular markers imply an introgression event and the transfer of genetic information between the two genera, *Squalius* and *Scardinius*, which have to be taken into consideration in the future conservation of these species.

Citation: Valić, D.; Mirković, M.K.; Besendorfer, V.; Teskeredžić, E. Molecular Analysis of Two Endemic *Squalius* Species: Evidence for Intergeneric Introgression among Cyprinids and Conservation Issues. *Fishes* **2024**, *9*, 4. https://doi.org/10.3390/fishes9010004

Academic Editors: Robert L. Vadas Jr. and Robert M. Hughes

Received: 17 November 2023
Revised: 14 December 2023
Accepted: 17 December 2023
Published: 20 December 2023

1. Introduction

The cyprinid genus *Squalius* includes around 28 species in Europe [1]. They are small- to large-sized fishes adapted to diverse habitats. Specimens can be found in streams, slow-flowing rivers, and even lakes. According to Kottelat and Freyhof [1], seven species can be found in Croatia: Chub *Squalius cephalus* (L. 1758); Illyrian chub *S. illyricus*, Heckel & Kner 1858; Makal dace *S. microlepis*, Heckel 1843; Cavedano chub *S. squalus* (Bonaparte, 1837); Neretva chub *S. svallize*, Heckel & Kner 1858; Livno masnica *S. tenellus*, Heckel 1843; and Zrmanja chub *S. zrmanjae*, Karaman 1928).

Squalius illyricus (Figure 1a) is on the IUCN List of Near-Threatened species (NT) (https://www.iucnredlist.org/species/61381/12469652, accessed on 10 February 2023) and is internationally protected by the Bern Convention. Moreover, it is an indigenous Croatian species, as well as endemic to the Adriatic Basin [2]. It inhabits a very small area—the

Cetina and Krka river basins— and it can be found in clean and swift karstic rivers and lakes. Biological data on this species are very scarce. *S. zrmanjae* (Figure 1b) is a Croatian Near-Threatened (NT) species (https://www.iucnredlist.org/species/60794/12399825, accessed on 10 February 2023) that can be found only in the Zrmanja and Krka catchment areas [2–4]. It is also indigenous to Croatia and endemic to the Adriatic basin [2]. Karstic streams are a natural habitat for this species, but the biological data are not sufficient. These two species are morphologically very similar and scale pigmentation is used as a distinguishing criterion, as are eye diameter and snout appearance. Therefore, molecular analysis will provide additional information in determination of the species.

Figure 1. Investigated species, (**a**)—*Squalius illyricus* and (**b**)—*Squalius zrmanjae*.

Research in the fields of taxonomy and phylogeny of fishes has been both intensive and extensive over the last three decades, using a wide range of different techniques. Molecular methods such as the sequencing of a specific gene or the entire genome have generally been used for these purposes. The Leuciscidae is the most investigated fish family in Europe, particularly the genera *Leuciscus*, *Chondrostoma*, *Scardinius* and *Rutilus* [5–11].

DNA sequences of mitochondrial genes, especially of cytochrome *b* (cyt *b*), are widely used to establish phylogenetic relationships between different organisms [12]. Cyt *b* is a highly conserved protein-encoding region that evolves slowly, is suitable for monitoring evolutionary processes, and is convenient for detecting differences between closely related species [13,14]. Mitochondrial markers represent only the maternal lineage and provide very little information about hybridization unless combined with information obtained from nuclear markers or geographic data [15].

Cyprinid fishes are known for frequent intergeneric hybridization and numerous different intrageneric and intergeneric hybrids have been described [16–21]. In the subfamily Leuciscinae, the nuclear region Cyfun P (Cyprinid formerly unknown nuclear Polymorphism) displays large intergeneric length variations caused by various deletion or insertion events. This nuclear region is useful for the detection of intergeneric hybridization [22,23]. Therefore, nuclear DNA is suitable for resolving relationships among higher taxonomic levels and detecting hybridization due to segregation of species-specific alleles. Due to the fact that nuclear genes have a low evolutionary rate and are unlikely to undergo mutational saturation, they provide important information complementary to mitochondrial genes.

As suggested by Vadas & Orth [24], the habitat use of fishes as well as their associations within the habitat can provide us with additional information about possible hybridization events between species occupying the same habitat. From this point of view, *S. illyricus* is a benthopelagic species that inhabits clean, fast karst rivers and lakes with a water temperature of 5–25 °C. *S. zrmanjae* is a rheophilic species that mainly inhabits karst streams but can also be found in lake areas. During sampling, these two species have been found at the same locations, potentially giving them the opportunity for hybridization, although their spawn timings have not been studied. Seeing hybridization as a viable mode of speciation [25], these two species have numerous ways and opportunities to develop all the possible different forms of this process. Being a Near-Threatened species with a very small distributional area, this is an extremely important issue, which is essential to the future conservation of these two species. Looking at the history of potential hybrids in fishes [25–28], it is evident that hybridization as a process is constantly present in nature, potentially as a possible mode of speciation and evolution, but it is also true that hybrids in fishes are difficult to detect and to determine. For certain species of fish that were supposed to be hybrids, it was found not to be. To prove this, it was necessary to apply modern techniques and methods that have been developed for years. Sequencing is a convenient tool for identifying differences that is especially applicable to small populations of native species that face extinction. However, determining whether a certain species is a hybrid or a valid species requires numerous other data.

The purpose of this study was to perform molecular analysis of specimens of *S. illyricus* and *S. zrmanjae* from the Krka River, due to their morphological resemblance, in order to genetically characterize these species. The aim was to obtain insight into the phylogenetic structure of the closely related species and to verify the taxonomic status of these two species, which are endemic to Croatia with a relatively small distribution area. Using mitochondrial and nuclear markers, we tried to assess the possibility of hybridization. The results obtained will provide additional information on these two endemic species and stimulate further research and the necessary measures to protect these Near-Threatened species.

2. Materials and Methods

The Krka River belongs to the Adriatic Sea basin and is 72.5 km long with a total slope of 224 m. The hydrological basin of the Krka River covers an area of approximately 2500 km^2. It is not considered polluted [29], and forms the basis of the Krka National Park (NP). About 20 species of fish inhabit the Krka River, more than half of which are endemic to the Croatian ichthyofauna [2,4,30].

Seven sampling sites (Figure 2, Table 1) were selected along the length of the Krka River. Sampling for the molecular analyses was carried out during the field trips from December 2006 to October 2008.

The fish were sampled with an electrofishing device (Hans Grassl, EL63 II GI, 5.0 KW, Honda GX270, 300/600 V max., 27/15A max.) according to the Croatian standard: HRN EN 14011:2005 Water quality—Sampling of fish with electricity. The captured fish were kept alive in a tank with aerated water until further processing. The fish were anesthetized in a separate tank with buffered MS-222 (Sigma-Aldrich, Taufkirchen, Germany). Vouchers (fixation with buffered 4% formaldehyde, deposited in 75% ethanol) were deposited at the Faculty of Science, University of Zagreb, Croatia. Identification of fishes was performed according to Kottelat and Freyhof [1]. First, biometric data were recorded, including total length, standard length and total mass, while the Fulton condition index was later calculated according to Rätz and Lloret [31]. The tip of the anal fin was cut off and preserved in 96% ethanol or stored at −80 °C until further DNA analyses in the laboratory.

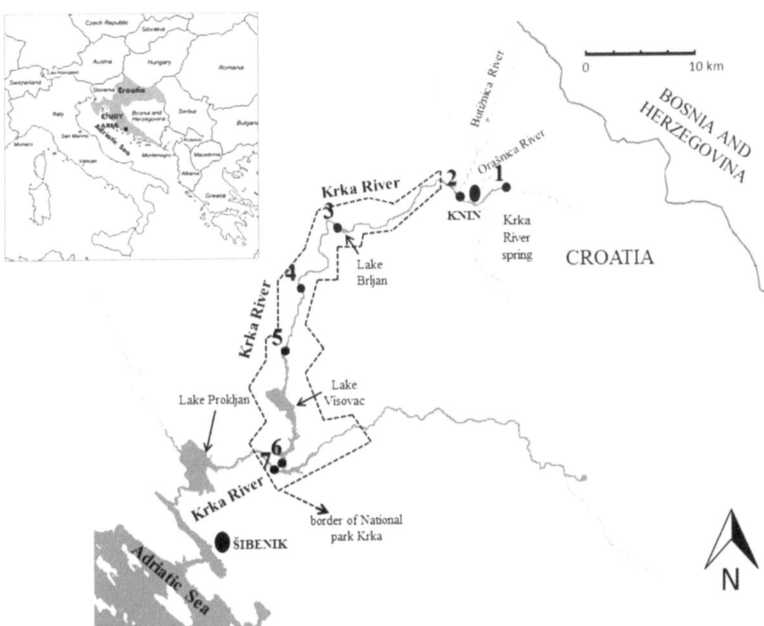

Figure 2. Study area marked with sampling sites along the length of the Krka River (1—Krka River spring; 2—Krka River upstream of the Butižnica River tributary; 3—Lake Brljan; 4—"Krka" monastery; 5—Roški Slap waterfall; 6—Lake Visovac; 7—Skradinski Buk waterfall).

Table 1. Sampling sites on the Krka River.

No.	Name of the Sampling Site	Coordinates
1.	Krka River spring	44°02.563′ N, 16°14.412′ E
2.	Krka River upstream of the Butižnica River tributary	44°02.295′ N, 16°10.347′ E
3.	Lake Brljan	44°00.343′ N, 16°02.444′ E
4.	"Krka" monastery	43°57.538′ N, 15°59.833′ E
5.	Roški Slap waterfall	43°54.140′ N, 15°58.815′ E
6.	Visovac Lake upstream from waterfall Skradinski Buk	43°48.349′ N, 15°58.693′ E
7.	Skradinski Buk waterfall	43°48.426′ N, 15°58.110′ E

A total of 237 specimens of *S. illyricus* and 267 of *S. zrmanjae* were caught in the Krka River during sampling, representing 18.96% and 21.38% of the total catch, respectively (Table S1). Altogether, 30 specimens were subjected to DNA analyses, 15 from each species. The partial cyt *b* sequence (1140 bp) was used for phylogenetic analysis. Newly obtained cyt *b* sequences were uploaded into the GenBank database (GenBank ID: JQ663535.1, JQ663536.1 and OR791603-OR791610). The nucleotide composition of the cyt *b* was in accordance with previous findings on closely related genera [11,32–37].

Total genomic DNA was extracted using the DNeasy Blood & Tissue Kit (Qiagen, Hilden, Germany) according to the spin-column protocol. The partial cyt *b* gene (1140 bp) was amplified in two overlapping fragments using the following primers: L15267 (5′-AATGACTTGAAGAACCACCGT-3′) and H16526 (5′-CTTTGGGAGYYRRGGGTGRGA-3′) [36]. The PCR reaction mixtures contained 4 U AmpliTaq DNA Polymerase (Applied Biosystems, Waltham, MA, USA), 1 × PCR buffer (without MgCl$_2$, Applied Biosystems, Waltham, MA, USA), 2.5 mM MgCl$_2$, 200 μM dNTPs (Sigma, Steinheim, Germany), 400 nM of each primer and approximately 5 μg/μL of DNA in a final volume of 100 μL. Reactions were subjected to the following thermocycling protocol: initial denaturation (94 °C: 3 min), 30 cycles (94 °C: 30 s; 55 °C: 30 s; 72 °C: 1 min) and final extension (72 °C: 7 min). PCR products were examined on a 1.7% agarose gel using electrophoresis and subsequently purified (QIAquick Gel Extraction Kit, Qiagen, Hilden, Germany). Sequencing was performed using the ABI PRISM® 3100 Avant Genetic Analyzer (Ruđer Bošković Institute DNA service) or at Macrogen Inc. in Seoul, Republic of Korea.

To test for putative hybridization and to support the cyt *b* results, the highly variable noncoding nuclear region, Cyfun P [23], was sequenced. Primers for this nuclear region have already been used for several species of fish [22]: Cyp_unFLP1F (5′-AAGTGGTGC-ATCGTGTTGTG-3′) and Cyp_unFLP1R (5′-CAGCCTGAACAATCAAAACAG-3′). The PCR reaction mixtures were identical to those used for cyt *b*, but the cycling protocol was somewhat different: initial denaturation (94 °C: 3 min), 35 cycles (94 °C: 15 s; 55 °C: 20 s; 72 °C: 45 s) and final extension (72 °C: 7 min). Purification and further sequencing of the PCR products were performed as described for cyt *b*.

The BLAST network service (http://www.ncbi.nlm.nih.gov, accessed on 13 November 2023) was used for sequence homology search. Multiple alignments were performed with CLUSTALW Ver. 1.6 [38] using the default parameters. Ambiguously aligned regions were determined using the program Gblocks 0.91b under less stringent parameters [39] and excluded from further analyses. Aligned sequences were imported into MEGA version 6 [40], where the phylogenetic relations of sequences in the datasets were analyzed using maximum likelihood (ML) [41]. The tench *Tinca tinca* (L. 1758) (Tincidae) was selected as the outgroup. Support for the nodes in the trees was estimated via bootstrapping (1000 bootstrap replicates in ML). The model for the ML analysis was selected using Modeltest 3.7 [42]. The Akaike Information Criterion (AIC) indicated TrN + I (Tamura Nei with invariant sites) for the cyt *b* and HKY + G (Hasegawa–Kishino–Yano with gamma distributed sites) for the Cyfun P. Initially, a large maximum likelihood tree was constructed, containing almost 150 total sequences from various species. The phylogenetic tree was then reduced to the current number of sequences by eliminating the sequences. Pruning was performed by hand, leaving at least one specimen of each original species and taking into consideration only the most unusual and distant sequences. To ensure the correctness of the pruned tree, a new maximum parsimony (MP) [43] tree containing all original sequences was constructed and compared with the pruned tree to ensure that no major differences occurred.

3. Results

S. illyricus and *S. zrmanjae* showed differences in nucleotide composition; A:T:G:C were 26.1:28.4:16.8:28.6 and 27.0:28.6:15.9:28.5, respectively. The sequence identity matrix showed 0.918 similarity between these two species from the Krka River.

Phylogenetic relation between *S. illyricus* and *S. zrmanjae* from the Krka River was established in the phylogenetic analysis, which included 44 sequences of closely related species obtained from GenBank, the majority of which originated from the work of Perea et al. [44]. Previously known and published sequences for these two species (GenBank ID: AJ251092.1, AJ251093.1, AJ251094.1, HM560183.1, HM560184.1, HM560213.1, MN166108.1 and MN166113.1) were also included in the analysis (Figure 3). Our *S. illyricus* sequences (GenBank ID: JQ663535.1 and OR791603-OR791607) grouped together with an *S. illyricus* (GenBank ID: MN166108.1) from the Krka River and other *S. illyricus* (GenBank ID: AJ251094.1, HM560183.1 and HM560184.1) haplotypes from the Cetina River (Croatia) with high bootstrap values. *S. zrmanjae* (GenBank ID: JQ663536.1 and OR791608-OR791610) grouped together with another *S. zrmanjae* (GenBank ID: AJ251092.1) from the Krka River (Croatia). The positions of investigated *S. illyricus* and *S. zrmanjae* from the Krka River in the phylogenetic tree indicate two divergent species.

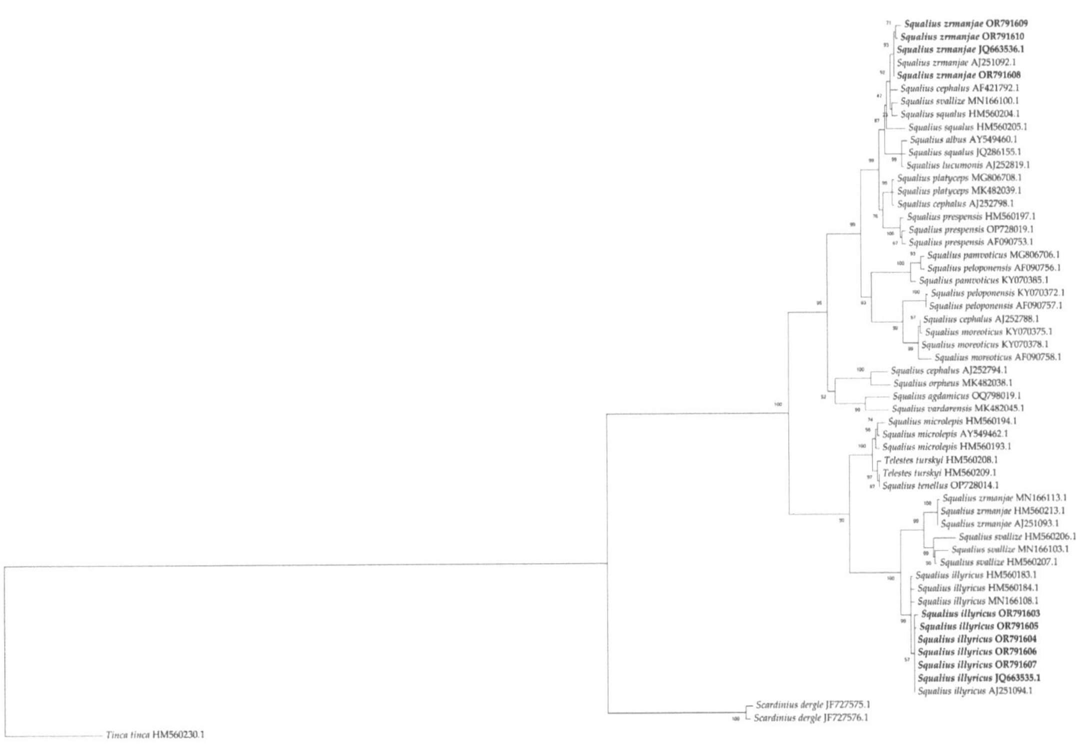

Figure 3. Phylogenetic relation of *Squalius illyricus* and *Squalius zrmanjae* (obtained sequences are shown in bold) based on mitochondrial cytochrome *b* (cyt *b*) nucleotide sequences. Bootstrap values (1000 replicates) are indicated by the line.

The same specimens subjected to cyt *b* analysis were used for characterization of the highly variable noncoding nuclear region, Cyfun P. The sequence identity matrix showed relatively low similarity between *S. illyricus* and *S. zrmanjae* from the Krka River (0.774), as base composition, A:T:G:C, was 39.8:28.4:16.2:15.5% and 35.0:31.2:16.9:16.9%, respectively. Further, the nucleotide sequence length of this marker was different in these two species, 394 and 343 bp, respectively. The Cyfun P sequences obtained were uploaded into the GenBank database (GenBank ID: JQ663537.1, JQ663538.1 and OR791611-OR791614).

Because hybridization of the species can be demonstrated with this marker, a phylogenetic analysis of the noncoding region was performed. The results show that the analyzed species were separated into two clusters supported by the high bootstrap values. *S. illyricus* was grouped with the *Squalius* species and *S. zrmanjae* with the *Scardinius* species (Figure 4). The sequence identity matrix showed a high similarity between *S. zrmanjae* and, the Dalmatian rudd, *Scardinius dergle* Heckel & Kner 1858 (GenBank ID: JF727578.1) from the Krka River (1.000), but also with the other sequences of *S. dergle* (GenBank ID: JF727577.1. and AY831428.1).

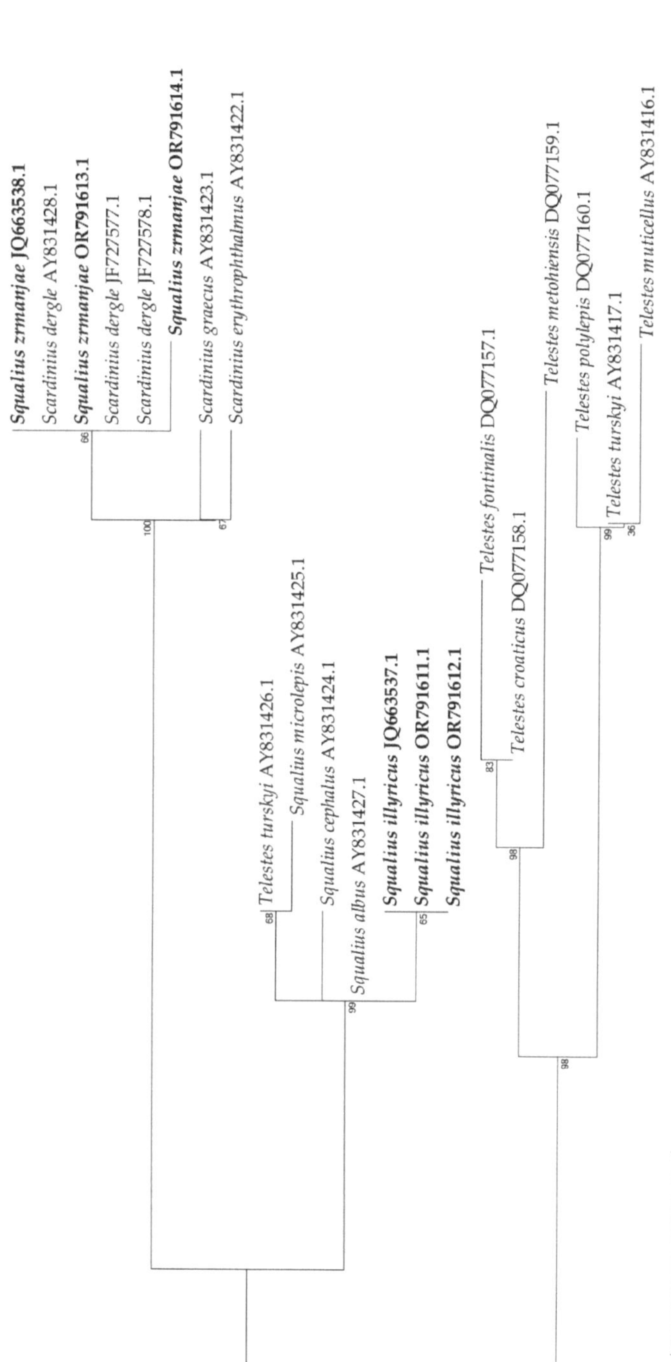

Figure 4. Phylogenetic position of *Squalius illyricus* and *Squalius zrmanjae* (obtained sequences are shown in bold) based on nuclear Cyfun P nucleotide sequences. Bootstrap values (1000 replicates) are indicated by the line.

4. Discussion

The analysis based on cyt *b* (Figure 3), showed that the *S. illyricus* haplotypes from the Krka River (GenBank ID: JQ663535.1 and OR791603–OR791607) were grouped in a cluster with the *S. illyricus* haplotypes (GenBank ID: AJ251094.1, HM560184.1, HM560183.1 and MN166108.1) from the Cetina River in Croatia. Because these two rivers are adjacent, this result was to be expected.

On the other hand, in the reconstructed phylogenetic tree there are two distinct, distant groups comprising the haplotypes of *S. zrmanjae*. In one group, which is closely related to the group of *S. illyricus*, there is one haplotype of *S. zrmanjae* from the Krka River (GenBank ID: MN166113.1) and two from the Zrmanja River, Croatia (GenBank ID: AJ251093.1 and HM560213), as well as three haplotypes of the closely related species *S. svallize* (GenBank ID: HM560206.1, HM560207.1 and MN166103.1) from Bosnia and Herzegovina. In the other group, which is distant from the previous one, our haplotypes of *S. zrmanjae* (GenBank ID: JQ663536.1 and OR791608–OR791610) are grouped together with another one from the Krka River (GenBank ID: AJ251092.1). The grouping of two haplotypes from the Krka River is to be expected, but it is curious that these two groups of *S. zrmanjae* are so far apart in the phylogenetic tree. The reason for this is probably that the haplotypes from the Krka and Zrmanja River do not belong to the same species. Although they are geographically close to each other (the Krka and Zrmanja river basins are connected through the karstic terrain), DNA sequences of fish from these two rivers comprise different species. It should be noted that the different grouping of haplotypes of *S. zrmanjae* probably presents two different and distinct species. This issue should be resolved with a larger number of samples of fishes from these two rivers.

The phylogenetic tree based on the highly variable noncoding nuclear region, Cyfun P, positions *S. illyricus* with other *Squalius* species and *S. zrmanjae* with *Scardinius* species (Figure 4). These results, together with the sequence identity matrix (high similarity between *S. zrmanjae* and *S. dergle* (GenBank ID: JF727578) from the Krka River—1.000), imply that our species, *S. zrmanjae*, in addition to obvious morphological characters of the genus *Squalius* (Table 2), had the nuclear region, Cyfun P, that resembles the genus *Scardinius*. "*S. zrmanjae*" could be a hybrid, and our results demonstrate an introgression event and transfer of genetic information between these two cyprinid genera. This would not be the first case of a hybrid specimen between *S. dergle* and a representative of the genus *Squalius*. In the work of Freyhof et al. [33], a hybrid specimen of "*S. dergle*" had a cyt *b* sequence similar to the sympatric species *S. tenellus*, although it did not show morphological characteristics of the genus *Squalius*. The authors demonstrated a transfer of genetic information between two related but distant cyprinid genera. The genera *Scardinius* and *Squalius* separated phylogenetically in the middle Miocene, approximately 10 Ma [45].

Table 2. Morphological characters of the genus *Squalius* and *Scardinius* [1,33].

Morphological Characters	*Scardinius*	*Squalius*
Body shape	Compressed	Cylindrical
Shape of the posterior anal-fin margin	Concave	Convex
Mouth position	Superior	Terminal
Pharyngeal teeth	Slightly serrated	Smooth
Branched anal rays	10	$8–9\frac{1}{2}$ *
Gill rakers	10–13	12–16 **
Lateral line scales	38–41	46–54 or 44–49*

* *S. illyricus* and *S. zrmanjae*; ** *S. tenellus*.

Hybridization between species involves mating between unrelated organisms regardless of the taxonomic status and in some cases may lead to gene transfer, a complex evolutionary process occurring in freshwater fishes [34]. Introgression occurs when hybrids backcross with one or both parental species [46]. For freshwater fish, cyprinids are known to exhibit higher rates of hybridization than other groups of fish, which is especially evident for Leuciscidae, for both interspecific introgressions [16,21,47,48] and intergeneric hybrids [33,49–52]. Of course, there are numerous cases of introgressive hybridization in other groups of fish [53–56], but at the beginning of the century, already, 62 different intra- and intergeneric hybrids were described for leuciscine species in the wild [19].

The geographic proximity of the investigated species explains the close phylogenetic relationship with the Italian species. The investigated species grouped together with the species from other countries in this region, which is in agreement with the hypothesis of peri-Mediterranean dispersal of freshwater fishes. This explanation concurs with our haplotypes of *S. zrmanjae* in the phylogenetic tree. The species related to our specimen occur in Albania, Greece, Italy, Slovenia and, of course, Croatia. Other haplotypes of *S. zrmanjae* from GenBank group together with our *S. illyricus* haplotypes. The majority of these haplotypes consist of native species from Croatia or the nearby countries of Bosnia and Herzegovina and Italy.

Research from Ketmaier et al. [57] supports the hypothesis of peri-Mediterranean dispersal of freshwater fishes during the "Lago Mare" phase of the Mediterranean Sea [58]. While this is accurate for the genus *Telestes*, the genus *Scardinius* had a different dispersal route. The intraspecific divergence of the genus *Scardinius* occurred between 1.35 ± 1 Ma [57]. Sea-level decline during the Messinian crisis probably increased the isolation of certain populations [59]. Future research on other groups of fish with a similar pattern of geographic dispersal can provide insight into the history and evolutionary relationships between species of fish, but also could clarify the events that led to the origin of new species and their dispersal in this region. The work of Sabolić et al. [60] on the morphological diversity and relationships of the populations of *S. dergle* and *S. plotizza* from Croatia and Bosnia and Herzegovina helps to clarify the taxonomic relationships and population status of one of the investigated genera in this work, but there is certainly a need for further molecular research on the phylogenetic relationships and taxonomy of these species.

Further work on the genus *Squalius* is needed, possibly with the introduction of additional markers (mitochondrial *COXI* and nuclear *RAG-1* and *S7* genes), as used by Buj et al. [35], and more samples from small tributaries should be included in this research. This is necessary because we found that the existing mitochondrial haplotypes of *S. zrmanjae* actually represent two different and distant species in the phylogenetic tree. Although further progress has been made on this topic [35], a similar finding has not been discovered. The authors were able to identify hybrid individuals among the Squalius species, confirming that hybridization is a widespread phenomenon in the Adriatic region. The haplotype from the Krka River is probably a hybrid whose mitochondrial genome originates from the species *S. dergle*, although morphological characteristics indicate a similarity with the genus *Squalius*. Buj et al. [35] found one sample from the Zrmanja River with *S. illyricus* mtDNA and the other from the Krka River with *S. zrmanjae* mtDNA. This shows that mtDNA intrageneric introgression and, in our case, intergeneric introgression are present.

Knowledge of the systematic relationships in the Croatian freshwater ichthyofauna is still incomplete, and further molecular analyses are required to solve taxonomic and systematic problems where morphology is not sufficient. There are 52 native species of fish in Croatian freshwaters [2] and intensive research is of great importance for their protection. If the probability of hybridization between endemic species of two different genera is relatively high, as we have found, then the importance of further molecular as well as morphological research of these species becomes even greater.

5. Conclusions

The aim of this study was to investigate both mitochondrial and nuclear markers of two native and Near-Threatened species. These species have very small distribution areas and phylogenetic analysis is of the utmost importance in addition to morphological analysis. Using the mitochondrial marker cyt *b*, we confirmed the previous phylogenetic position of the investigated species. However, the results of the nuclear marker indicated that *S. zrmanjae* may be a hybrid and that there may have been a transfer of genetic information between two cyprinid genera, *Squalius* and *Scardinius*. For conservation measures, the occurrence of hybridization presents a potential problem that needs to be resolved. Following the guidelines proposed by Allendorf et al. [28], further genetic investigations with additional markers (e.g., COI, RAG1) are needed to identify which type of hybridization occurred, define hybrid zones and accordingly consider which protection measures would be most efficient.

Supplementary Materials: The following supporting information can be downloaded at: https://www.mdpi.com/article/10.3390/fishes9010004/s1, The data in supplementary Table S1 are available from the corresponding author upon request.

Author Contributions: D.V., V.B. and E.T. developed the original research idea and designed the molecular analysis. D.V. and E.T. carried field sampling. D.V. performed molecular analysis, made analysis of aligned sequences and wrote the paper. M.K.M. performed analysis of aligned sequences. All authors have read and agreed to the published version of the manuscript.

Funding: The financial support by the Ministry of Science, Education and Sport of the Republic of Croatia (Project No. 098-0982934-2752) is acknowledged. This research was funded by the Public Institution Krka National Park (Grant No. 2962/8-2003).

Institutional Review Board Statement: The fish were returned to the wild after being held in an aerated water tank. No ethical review or approval was needed for this study.

Informed Consent Statement: Not applicable.

Data Availability Statement: Data available in a publicly accessible repository (GenBank ID: JQ663535.1-JQ663538.1 and OR791603-OR791614).

Acknowledgments: The authors are grateful to all laboratory members for their help during the field and laboratory work, and to Bruna Pleše for helping with the analysis of aligned sequences.

Conflicts of Interest: The authors declare no conflict of interest.

References

1. Kottelat, M.; Freyhof, J. *Handbook of European Freshwater Fishes*; Kottelat: Cornol, Switzerland; Freyhof: Berlin, Germany, 2007; ISBN 978-2-8399-0298-4.
2. Ćaleta, M.; Buj, I.; Mrakovčić, M.; Mustafić, P.; Zanella, D.; Marčić, Z.; Duplić, A.; Mihinjač, T.; Katavić, I. *Endemic Fishes of Croatia*; Agencija za zaštitu okoliša: Zagreb, Croatia, 2015; ISBN 978-953-7582-14-2.
3. Mrakovčić, M.; Brigić, A.; Buj, I.; Ćaleta, M.; Mustafić, P.; Zanella, D. *Red Book of Freshwater Fish of Croatia*; Ministry of Culture, State Institute for Nature Protection: Zagreb, Croatia, 2006; ISBN 953-7169-21-9.
4. Ćaleta, M.; Marčić, Z.; Buj, I.; Zanella, D.; Mustafić, P.; Duplić, A.; Horvatić, S. A Review of Extant Croatian Freshwater Fish and Lampreys—Annotated list and distribution. *Ribar. Croat. J. Fish.* **2019**, *77*, 137–234. [CrossRef]
5. Doadrio, J.; Carmona, J.A. Genetic divergence in Greek populations of the genus Leuciscus and its evolutionary biogeographical implications. *J. Fish. Biol.* **1998**, *53*, 591–613. [CrossRef]
6. Durand, J.D.; Bianco, P.G.; Laroche, J.; Gilles, A. Insight into the origin of the endemic Mediterranean ichthyofauna: Phylogeography of the *Chondrostoma* genus (Teleostei, Cyprinidae). *J. Hered.* **2003**, *94*, 315–328. [CrossRef] [PubMed]
7. Ketmaier, V.; Bianco, P.G.; Cobolli, M.; De Matthaeis, E. Genetic differentiation and biogeography in southern European populations of the genus *Scardinius* (Pisces, Cyprinidae) based on allozyme data. *Zool. Scr.* **2003**, *32*, 13–22. [CrossRef]
8. Milošević, D.; Winkler, K.A.; Marić, D.; Weiss, S. Genotypic and phenotypic evaluation of *Rutilus* spp. from Skadar, Ohrid and Prespa lakes supports revision of endemic as well as taxonomic status of several taxa. *J. Fish. Biol.* **2011**, *79*, 1094–1110. [CrossRef] [PubMed]
9. Valić, D.; Vardić Smrzlić, I.; Kapetanović, D.; Teskeredžić, Z.; Pleše, B.; Teseredžić, E. Identification, phylogenetic relationships and a new maximum size of two rudd populations (*Scardinius*, Cyprinidae) from the Adriatic Sea drainage, Croatia. *Biologia* **2013**, *68*, 539–545. [CrossRef]
10. Tan, M.; Armbruster, J.W. Phylogenetic classification of extant genera of fishes of the order Cypriniformes (Teleostei: Ostariophysi). *Zootaxa* **2018**, *4476*, 6–39. [CrossRef]
11. Schönhuth, S.; Vukić, J.; Šanda, R.; Yang, L.; Mayden, R.L. Phylogenetic relationships and classification of the Holarctic family Leuciscidae (Cypriniformes: Cyprinoidei). *Mol. Phylogenet. Evol.* **2018**, *127*, 781–799. [CrossRef]
12. Avise, J.C. *Molecular Markers, Natural History and Evolution*; Chapman & Hall: New York, NY, USA, 1994.

13. Zardoya, R.; Meyer, A. Phylogenetic performance of mitochondrial protein-coding genes in resolving relationships among vertebrates. *Mol. Biol. Evol.* **1996**, *13*, 933–942. [CrossRef]
14. Briolay, J.; Galtier, N.; Brito, R.M.; Bouvet, Y. Molecular phylogeny of Cyprinidae inferred from cytochrome b DNA sequences. *Mol. Phylogenet. Evol.* **1998**, *9*, 100–108. [CrossRef]
15. Mallet, J. Hybridization as an invasion of the genome. *Trends Ecol. Evol.* **2005**, *20*, 229–237. [CrossRef] [PubMed]
16. Durand, J.D.; Ünlü, E.; Doadrio, I.; Pipoyan, S.; Templeton, A.R. Origin, radiation, dispersion and allopatric hybridization in the chub *Leuciscus cephalus*. *Proc. Royal Soc. B* **2000**, *267*, 1687–1697. [CrossRef] [PubMed]
17. Scribner, K.T.; Page, K.S.; Bartron, M. Hybridization in freshwater fishes: A review of case studies and cytonuclear methods of biological inference. *Rev. Fish. Biol. Fish.* **2001**, *10*, 293–323. [CrossRef]
18. Salzburger, W.; Brandstätter, A.; Gilles, A.; Parson, W.; Hempel, M.; Sturmbauer, C.; Meyer, A. Phylogeography of the vairone (*Leuciscus souffia*, Risso 1826) in central Europe. *Mol. Ecol.* **2003**, *12*, 2371–2386. [CrossRef] [PubMed]
19. Yakovlev, V.N.; Slyn'ko, Y.V.; Grechanov, I.G.; Krysanov, E.Y. Distant hybridization in fish. *J. Ichthyol.* **2000**, *40*, 298–311.
20. Hayden, B.; Pulcini, D.; Kelly-Quinn, M.; O'Grady, M.; Caffrey, J.; McGrath, A.; Mariani, S. Hybridisation between two cyprinid fishes in a novel habitat: Genetics, morphology and life-history traits. *BMC Evol. Biol.* **2010**, *10*, 169. [CrossRef]
21. Ramoejane, M.; Gouws, G.; Swartz, E.R.; Sidlauskas, B.L.; Weyl, O.L.F. Molecular and morphological evidence reveals hybridisation between two endemic cyprinid fishes. *J. Fish. Biol.* **2020**, *96*, 1234–1250. [CrossRef]
22. Freyhof, J.; Lieckfeldt, D.; Bogutskaya, N.G.; Pitra, C.; Ludwig, A. Phylogenetic position of the Dalmatian genus *Phoxinellus* and description of the newly proposed genus *Delminichthys* (Teleostei: Cyprinidae). *Mol. Phylogenet. Evol.* **2006**, *38*, 416–425. [CrossRef]
23. Lieckfeldt, D.; Hett, A.K.; Ludwig, A.; Freyhof, J. Detection, characterization and utility of a new highly variable nuclear marker region in several species of cyprinid fishes (Cyprinidae). *Eur. J. Wildl. Res.* **2006**, *52*, 63–65. [CrossRef]
24. Vadas, R.L., Jr.; Orth, D.J. Species associations and habitat use of stream fishes: The effects of unaggregated-data analysis. *J. Freshw. Ecol.* **1997**, *12*, 27–37. [CrossRef]
25. Stauffer, J.R., Jr.; Hocutt, C.H.; Denoncourt, R.F. Status and distribution of the hybrid *Nocomis micropogon* X *Rhinichthys cataractae*, with a discussion of hybridization as a viable mode of vertebrate speciation. *Am. Midl. Nat.* **1979**, *101*, 355–365. [CrossRef]
26. Goodfellow, W.L., Jr.; Hocutt, C.H.; Morgan, R.P., II; Stauffer, J.R., Jr. Biochemical assessment of the taxonomic status of "*Rhinichthys bowersi*" (Pisces: Cyprinidae). *Copeia* **1984**, *3*, 652–659. [CrossRef]
27. Stauffer, J.R., Jr.; Hocutt, C.H.; Mayden, R.L. *Pararhinichthys*, a new monotypic genus of minnows (Teleostei: Cyprinidae) of hybrid origin from eastern North America. *Ichthyol. Explor. Freshw.* **1997**, *7*, 327–336.
28. Allendorf, F.W.; Leary, R.F.; Spruell, P.; Wenburg, J.P. The problems with hybrids: Setting conservation guidelines. *Trends Ecol. Evol.* **2001**, *16*, 613–622. [CrossRef]
29. Filipović Marijić, V.; Kapetanović, D.; Dragun, Z.; Valić, D.; Krasnići, N.; Redžović, Z.; Grgić, I.; Žunić, J.; Kružlicová, D.; Nemeček, P.; et al. Influence of technological and municipal wastewaters on vulnerable karst riverine system, Krka River in Croatia. *Environ. Sci. Pollut. Res.* **2018**, *25*, 4715–4727. [CrossRef] [PubMed]
30. Kerovec, M.; Marguš, D.; Mrakovčić, M.; Kučinić, M. A review of research into the fauna of the Krka National Park. In *Book of Papers, Symposium River Krka and National Park "Krka": Natural and Cultural Heritage, Protection and Sustainable Development*; Marguš, D., Ed.; Javna ustanova "Nacionalni park Krka": Šibenik, Croatia, 2005; pp. 49–61, ISBN 978-953-7406-01-6.
31. Rätz, H.-J.; Lloret, J. Variation in fish condition between Atlantic cod (*Gadus morhua*) stocks, the effect on their productivity and management implications. *Fish. Res.* **2003**, *60*, 369–380. [CrossRef]
32. Durand, J.D.; Persat, H.; Bouvet, Y. Phylogeography and postglacial dispersion of the chub (*Leuciscus cephalus*), in Europe. *Mol. Ecol.* **1999**, *8*, 989–997. [CrossRef]
33. Freyhof, J.; Lieckfeldt, D.; Pitra, C.; Ludwig, A. Molecules and morphology: Evidence for introgression of mitochondrial DNA in Dalmatian cyprinids. *Mol. Phylogenet Evol.* **2005**, *37*, 347–354. [CrossRef]
34. Perea, S.; Vukić, J.; Šanda, R.; Doadrio, I. Ancient mitochondrial capture as factor promoting mitonuclear discordance in freshwater fishes: A case study in the genus *Squalius* (Actinopterygii, Cyprinidae) in Greece. *PLoS ONE* **2016**, *11*, e0166292. [CrossRef]
35. Buj, I.; Marčić, Z.; Čavlović, K.; Ćaleta, M.; Tutman, P.; Zanella, D.; Duplić, A.; Raguž, L.; Ivić, L.; Horvatić, S.; et al. Multilocus phylogenetic analysis helps to untangle the taxonomic puzzle of chubs (genus *Squalius*: Cypriniformes: Actinopteri) in the Adriatic basin of Croatia and Bosnia and Herzegovina. *Zool. J. Linn. Soc.* **2020**, *189*, 953–974. [CrossRef]
36. Brito, R.M.; Briolay, J.; Galtier, N.; Bouvet, Y.; Coelho, M.M. Phylogenetic relationships within genus *Leuciscus* (Pisces, Cyprinidae) in Portuguese fresh waters, based on mitochondrial DNA cytochrome b sequences. *Mol. Phylogenet Evol.* **1997**, *8*, 435–442. [CrossRef] [PubMed]
37. Cantatore, P.; Roberti, M.; Pesole, G.; Ludovico, A.; Milella, F.; Gadaleta, M.N.; Saccone, C. Evolutionary analysis of cytochrome b sequences in some perciformes: Evidence for a slower rate of evolution than in mammals. *J. Mol. Evol.* **1994**, *39*, 589–597. [CrossRef] [PubMed]
38. Thompson, J.D.; Higgins, D.G.; Gibson, T.J. CLUSTAL W: Improving the sensitivity of progressive multiple sequence alignment through sequence weighting, position-specific gap penalties and weight matrix choice. *Nucleic Acids Res.* **1994**, *22*, 4673–4680. [CrossRef] [PubMed]
39. Castresana, J. Selection of conserved blocks from multiple alignments for their use in phylogenetic analysis. *Mol. Biol. Evol.* **2000**, *17*, 540–552. [CrossRef] [PubMed]

40. Tamura, K.; Stechter, G.; Peterson, D.; Filipski, A.; Kumar, S. MEGA6: Molecular Evolutionary Genetics Analysis Version 6.0. *Mol. Biol. Evol.* **2013**, *30*, 2725–2729. [CrossRef] [PubMed]
41. Guindon, S.; Gascuel, O. A simple, fast, and accurate algorithm to estimate large phylogenies by maximum likelihood. *Syst. Biol.* **2003**, *52*, 696–704. [CrossRef] [PubMed]
42. Posada, D.; Crandall, K.A. Modeltest: Testing the model of DNA substitution. *Bioinformatics* **1998**, *14*, 817–818. [CrossRef]
43. Nei, M.; Kumar, S. *Molecular Evolution and Phylogenetics*; Oxford University Press: New York, NY, USA, 2000; ISBN 9780195135855.
44. Perea, S.; Böhme, M.; Zupančič, P.; Freyhof, J.; Šanda, R.; Özuluğ, M.; Abdoli, A.; Doadrio, I. Phylogenetic relationships and biogeographical patterns in circum-Mediterranean subfamily Leuciscinae (Teleostei, Cyprinidae) inferred from both mitochondrial and nuclear data. *BMC Evol. Biol.* **2010**, *10*, 265. [CrossRef]
45. Zardoya, R.; Doadrio, I. Molecular evidence of the evolutionary and biogeographical patterns of European cyprinids. *J. Mol. Evol.* **1999**, *49*, 227–237. [CrossRef]
46. Rhymer, J.M.; Simberloff, D. Extinction by hybridization and introgression. *Annu. Rev. Ecol. Syst.* **1996**, *27*, 83–109. [CrossRef]
47. Sousa-Santos, C.; Collares-Pereira, M.J.; Almada, V.C. Evidence of extensive mitochondrial introgression with nearly complete substitution of the typical *Squalius pyrenaicus*-like mtDNA of the *Squalius alburnoides* complex (Cyprinidae) in an independent Iberian drainage. *J. Fish. Biol.* **2006**, *68*, 292–301. [CrossRef]
48. Tancioni, L.; Russo, T.; Cataudella, S.; Milana, V.; Hett, A.K.; Corsi, E.; Rossi, A.R. Testing species delimitations in four Italian sympatric leuciscine fishes in the Tiber River: A combined morphological and molecular approach. *PLoS ONE* **2013**, *8*, e60392. [CrossRef] [PubMed]
49. Ünver, B.; Erk'akan, F. A natural hybrid of *Leuciscus cephalus* (L.) and Chalcalburnus chalcoides (Güldenstädt) (Osteichthyes: Cyprinidae) from Lake Tödürge (Sivas, Turkey). *J. Fish. Biol.* **2005**, *66*, 899–910. [CrossRef]
50. Robalo, J.I.; Sousa-Santos, C.; Levy, A.; Almada, V.C. Molecular insights on the taxonomic position of the paternal ancestor of the *Squalius alburnoides* hybridogenetic complex. *Mol. Phylogenet. Evol.* **2006**, *97*, 143–149. [CrossRef] [PubMed]
51. Sousa-Santos, C.; Collares-Pereira, M.J.; Almada, V. Reading the history of a hybrid fish complex from its molecular record. *Mol. Phylogenet. Evol.* **2007**, *45*, 981–996. [CrossRef] [PubMed]
52. Aboim, M.A.; Mavárez, J.; Bernatchez, L.; Coelho, M.M. Introgressive hybridization between two Iberian endemic cyprinid fish: A comparison between two independent hybrid zones. *J. Evol. Biol.* **2010**, *23*, 817–828. [CrossRef] [PubMed]
53. Allen, B.E.; Anderson, M.L.; Mee, J.A.; Coombs, M.; Rogers, S.M. Role of genetic background in the introgressive hybridization of rainbow trout (*Oncorhynchus mykiss*) with westslope cutthroat trout (*O. clarkia lewisi*). *Conserv. Genet.* **2016**, *17*, 521–531. [CrossRef]
54. Magalhaes, I.S.; Ornelas-Garcia, C.P.; Leal-Cardin, M.; Ramirez, T.; Barluenga, M. Untangling the evolutionary history of a highly polymorphic species: Introgressive hybridization and high genetic structure in the desert cichlid fish *Herichthys minckleyi*. *Mol. Ecol.* **2015**, *24*, 4505–4520. [CrossRef]
55. Desvignes, T.; Le François, N.R.; Goetz, L.C.; Smith, S.S.; Shusdock, K.A.; Parker, S.K.; Postlethwait, J.H.; William Detrich, H., III. Intergeneric hybrids inform reproductive isolating barriers in the Antarctic icefish radiation. *Sci. Rep.* **2019**, *9*, 5989. [CrossRef]
56. Kwan, Y.-S.; Ko, M.-H.; Jeon, Y.-S.; Kim, H.-J.; Won, Y.-J. Bidirectional mitochondrial introgression between Korean cobitid fish mediated by hybridogenetic hybrids. *Ecol. Evol.* **2019**, *9*, 1244–1254. [CrossRef]
57. Ketmaier, V.; Bianco, P.G.; Cobolli, M.; Krivokapić, M.; Canigla, R.; De Matthaeis, E. Molecular phylogeny of two lineages of Leuciscinae cyprinids (*Telestes* and *Scardinius*) from the peri-Mediterranean area based on cytochrome b data. *Mol. Phylogenet. Evol.* **2004**, *32*, 1061–1071. [CrossRef] [PubMed]
58. Bianco, P.G. Potential role of the paleohistory of the Mediterranean and Paratethys basins on the early dispersal of Euro-Mediterranean freshwater fishes. *Ichthyol. Explor. Freshw.* **1990**, *1*, 167–184.
59. Doadrio, I. Origen y evolución de la ictiofauna continental Española. In *Atlas y Libro Rojo de los Peces Españoles*; Doadrio, I., Ed.; DGCONA-CSIC: Madrid, Spain, 2001; pp. 20–34, ISBN 8480143134.
60. Sabolić, M.; Buj, I.; Ćaleta, M.; Marčić, Z. Morphological diversity and relationships of *Scardinius dergle* and *Scardinius plotizza* populations (Actinopteri; Cypriniformes) from Croatia and Bosnia and Herzegovina. *J. Cent. Eur. Agric.* **2021**, *22*, 713–734. [CrossRef]

Article

Adaptation of the European Fish Index (EFI+) to Include the Alien Fish Pressure

Enric Aparicio [1,*], Carles Alcaraz [2], Rafel Rocaspana [3], Quim Pou-Rovira [4] and Emili García-Berthou [1]

[1]   GRECO, Institute of Aquatic Ecology, University of Girona, 17003 Girona, Spain; emili.garcia@udg.edu
[2]   IRTA Marine and Continental Waters, 43540 La Ràpita, Spain; carles.alcaraz@irta.cat
[3]   Gesna Estudis Ambientals, 25240 Linyola, Spain; rafel@gesna.net
[4]   Sorelló, Estudis al Medi Aquàtic, 17003 Girona, Spain; quim.pou@sorello.net
*   Correspondence: enric.aparicio@ictio.cat

Abstract: The European Fish Index EFI+ is the only fish-based multimetric index for the assessment of the ecological status of running waters that is validated and thus applicable across most countries of the European Union. Metrics of the index rely on several attributes of the species present in the fish assemblage, irrespective of their native/alien status. The abundance of alien fish, together with other anthropogenic impacts, is one of the most important threats to the conservation of native fish and ecosystem health and is also an indicator of degraded stream conditions. Therefore, to improve the performance of the EFI+ in regions with high incidence of alien species, the EFI+ was adapted to include alien fish pressure as a new metric that reflects the number of alien species as well as the proportional abundance of alien individuals. The application of the adapted index (A-EFI+) is illustrated with data from several Iberian Mediterranean basins and showed similar or stronger correlations than the original EFI+ with anthropogenic pressure (land-use variables and alterations in hydrology and river morphology) and with other regional fish indices. EFI+ has been invaluable to intercalibrate fish indices across Europe, and A-EFI+ is similar but explicitly includes alien pressure, thus helping to provide a more comprehensive assessment of ecosystem health and to communicate it to society.

Keywords: ecological status; index of biotic integrity; non-native species; water framework directive

Key Contribution: This study developed a modification of the EFI+ fish index to include alien fish pressure to improve its performance in assessing the biotic and ecological status of rivers with high incidence of alien species. The application of the adapted index is demonstrated using data from various Mediterranean basins in the Iberian region.

Citation: Aparicio, E.; Alcaraz, C.; Rocaspana, R.; Pou-Rovira, Q.; García-Berthou, E. Adaptation of the European Fish Index (EFI+) to Include the Alien Fish Pressure. *Fishes* **2024**, *9*, 13. https://doi.org/10.3390/fishes9010013

Academic Editors: Robert L. Vadas Jr. and Robert M. Hughes

Received: 6 December 2023
Revised: 24 December 2023
Accepted: 27 December 2023
Published: 29 December 2023

1. Introduction

The Water Framework Directive (WFD) requires all member countries of the European Union (EU) to assess the ecological status of running waters using biological indicators of several organism groups, including fish [1]. The WFD defines reference conditions (i.e., equivalent to high ecological status) as those water bodies with no or minor presence of anthropogenic changes in which all expected native species are present, populations are in good biological condition, and no alien species exist. Biotic indices facilitate rapid and cost-effective assessments of the environmental degradation of aquatic systems. They benefit from a standardized approach using a set of metrics or measures that represent various aspects of biological assemblage structure, function, or other measurable characteristics. This standardization facilitates consistent comparisons across different locations and time periods. By combining multiple metrics, the biotic indices provide an indication of the overall biological condition and can help identify and quantify the impacts of human-induced stress on aquatic communities at wide temporal and spatial scales [2]. In aquatic environments, fish are excellent ecological indicators due to their sensitivity to environmental changes and have several advantages as indicator organisms [3]. Fish are found in most of lotic ecosystems and are long-living organisms that reflect the cumulative effects of long-term anthropogenic stressors. Their high mobility allows them to use various habitats within river ecosystems, making them particularly sensitive to disturbances in river morphology and connectivity [4]. Fish-based indices have been used to assess the quality of river ecosystems since the 1980s, when the Index of Biotic Integrity (IBI) was first introduced [2].

The project FAME (Fish-based Assessment Method for the ecological status of European rivers) [5] was the first attempt to develop a pan-European fish index applicable in all the EU member states, resulting in the creation of the European Fish Index (EFI) [5,6]. The initial formulation of the EFI was primarily based on data collected in northern Europe. The index was subsequently improved by expanding the database with data from southern Europe, resulting in a new version called EFI+ [7]. EFI+ quantifies the deviation between the predicted fish assemblage (reference conditions) and the observed fish assemblage (sampling data), and is computed as the average of two metrics that vary with river type (salmonid or cyprinid). The river type is assessed automatically by the EFI+ software, based on physical parameters and proportion of salmonid species. The EFI+ model places each species in functional trait categories (guilds). The index for the cyprinid type uses two metrics based on species with rheophilic and lithophilic reproduction habitats, and the index for the salmonid type is composed of two metrics based on intolerant species to oxygen depletion and habitat degradation. The index value is calculated as the arithmetic mean of the two metrics scores. The EFI+ metrics rely on the whole fish assemblage, without any distinction between native or alien species. Therefore, the presence and abundance of alien species belonging to the guilds included in the metrics (i.e., rheophilic, lithophilic or intolerant) positively influences index scores.

The main pressures and impacts that affect surface waters in Europe are eutrophication, chemical pollution, water abstraction and hydromorphological alterations [8]. Alien species also constitute one of the most important threats to the conservation of native fish and ecosystem health and the impact may be as severe as that of other stressors [9,10]. The presence and abundance of alien species reflects biological pollution and causes disturbance to native species, mainly from predation and competition [11]. Higher pressure from alien species has been related to a greater loss of native species, reduced density and unbalanced size structures of native fish [12,13]. Furthermore, alien species are also an indicator of degraded conditions because their proliferation is facilitated with increasing eutrophication and the construction of dams with the subsequent reduction in seasonal flooding and stabilization of downstream flows [14,15]. Despite these negative implications, the inclusion of alien fish metrics in the ecological quality assessment of rivers has not been considered in the majority of WFD assessment methods [16]. Thus, only 5 of 25 (20%) fish assessment methods have an explicit metric for alien species (WISER "Water bodies in Europe: Integrative Systems to assess Ecological status and Recovery"; www.wiser.eu (accessed on 27 December 2023)).

The EFI+ has been proved effective in determining the ecological status of European rivers [6,17,18], but the absence of negative scoring when alien fish are present may be a serious shortcoming since an ideal indicator should be sensitive to all stressors and impacts [2]. Although including this type of impact may be unimportant in European regions with a low proportion of alien species (e.g., [19]), in regions where alien fish are widespread, the inclusion of a negative alien fish metric is considered crucial to properly assess ecosystem health [12,13]. Some European countries such as Poland have been aware of this limitation and have modified the index to solve it [20]. In contrast, Spain has chosen the EFI+ as the primary fish index for evaluating the ecological status required by the WFD, despite being a country with a high incidence of alien fish introductions and most basins having more alien than native species [9,21].

The fundamental issue with the EFI+ in relation with alien species lies in the fact that the index's development did not exclude reference to sites with alien species, nor did it employ metrics solely based on native species, which is currently recommended to develop multimetric indices [22]. Moreover, it is also recommended to include negative alien species metrics to improve WFD assessments [23]. As the usefulness of a global index at the European level is extremely valuable as a common metric for intercalibration among indices developed for smaller regional scales [17,24], one way to improve the index without losing its advantages is to include the alien fish pressure. Hence, the objectives of this study were to (1) adapt the EFI+ index to include the alien fish pressure to improve its performance in regions with a significant impact of alien fish and (2) illustrate its application using data from several Mediterranean basins of the Iberian Peninsula.

2. Materials and Methods

The modification of the EFI+ consisted of a weighted combination of original EFI+ metrics with a measure of alien fish pressure to produce an adapted version of the index (A-EFI+). We consider alien species those that occur outside their natural range and have been introduced to new areas by human activity, either intended or unintended. This encompasses species that, while native to a particular country, have been translocated and are now found outside their native range within the same country. Alien fish pressure should reflect the number of alien species as well as the relative abundance of alien individuals in relation to native fish. Thus, the alien fish pressure metric (AFP) was calculated as the average between the proportion of alien species and the proportion of alien individuals in the sample. The scores of the AFP metric range from 0 (absence of alien species) to 1 (all individuals belong to alien species). To calculate the A-EFI+, a third metric (i.e., AFP) is incorporated alongside the two metrics of the original index. This extra metric is given a one-third (33.3%) weight in the adjusted index; thus, each metric contributes equally to the final score. Therefore, A-EFI+ was calculated as follows:

$$\text{A-EFI+} = \text{EFI+} - \left(\frac{\text{AFP} \times \text{EFI+}}{3} \right) \tag{1}$$

where A-EFI+ is the adapted index, EFI+ is the original EFI+ index, and AFP is the alien fish pressure. When alien fish are present, the A-EFI+ scores are lower than the EFI+ scores, up to a maximum reduction of 33.3% of the original EFI+ score when all individuals are alien. The A-EFI+ ranges between zero and one, like the original EFI+.

The application of the modified index was illustrated using data collected within WFD monitoring programs and available from public databases from 344 sites of the Mediterranean slope of the Iberian Peninsula (230 sites in Catalonia and 114 sites in the Jucar River Basin District; latitudinal range of 38.2−42.8° N). This region was selected because it is severely affected by alien fish introductions [13,25,26]. Most of the streams have a typical Mediterranean hydrological regime, with dry summers and irregular precipitation in autumn and spring. Thus, flow regimes are highly variable, from temporary (seasonal flow) to perennial (continuous flow). A detailed description of the study area can be found elsewhere [13,25].

Fish data originated from electrofishing during low flow periods, following the CEN 14011 standard protocol [27]. A single upstream pass was made including all mesohabitat channel units present in the reach, with a minimum sampled length of 50 m or minimum area of 100 m^2. Fish were sampled between June and September in 2007, 2008 and 2009. The EFI+ was calculated with the software provided by the Spanish Ministry for Ecological Transition and the Demographic Challenge (https://www.miteco.gob.es (accessed on 27 December 2023)). Several alien species in this area positively score in the EFI+ metrics (i.e., are considered rheophilic, lithophilic or intolerant) such as *Phoxinus septimaniae* and *Phoxinus dragarum* (included as *Phoxinus phoxinus* in the EFI+ software), *Oncorhynchus mykiss*, *Ameiurus melas* and *Squalius cephalus*. Also, there are some translocated species originated from other Iberian basins, such as *Pseudochondrostoma polylepis*, *Luciobarbus graellsii* and *Squalius alburnoides*. Besides the EFI+ and A-EFI+, we calculated two other fish indices used for ecological monitoring in the study region: IBICAT2010 [28] and IBI-JUCAR [13]. The IBICAT2010 uses a set of metrics derived from the functional traits and characteristics of fish species, such as feeding guilds, habitat preferences, reproductive strategies and other life history traits. While it distinguishes between native and alien species in some metrics, it lacks a specific metric to negatively score the presence of alien species [28]. The IBI-JUCAR uses five metrics to evaluate the ecological health of streams based on the loss of native species, the presence of alien species, the abundance of native fish, the age (size) structure of native fish, and the presence of individuals with anomalies [13]. Therefore, this index includes a specific metric for alien species. Other data from biotic and abiotic indices widely applied for ecological monitoring in Spain were also gathered for the same sites and periods with fish data in order to be compared with the A-EFI+. The indices compared were the following: the Riparian vegetation quality index, QBR [29]; the Fluvial habitat index, IHF [30]; the Specific Pollution Sensitivity index based on diatoms, IPS [31], and a macroinvertebrate-based index, IBMWP [32]. At each sampling site, land uses and hydrological and morphological alterations were used as indicators of anthropogenic pressure [33]. The Corine Land Cover database (available at http://www.eea.europa.eu (accessed on 27 December 2023)) was used to quantify land-use variables. Land use was categorized as urbanized areas, including urban and industrial units (Artificial), agricultural areas (Agriculture) and forested/natural areas (Forest), and then the percentages of each category were calculated within the drainage basin upstream of the site. Data on alterations in hydrology (water abstraction and modified flow regimes) and morphology (presence of barriers, riverbank structures and physical channel modification) were compiled from the River Basin Management Plan reports for the years 2009–2015 in the basins studied. These data are derived from monitoring surveys for the identification of pressures and assessment of impacts within the characterization of water bodies. Data of stressor categories were grouped in two variables (Hydrology and Morphology) that measure on a discrete scale the pressure intensity at each sampling site. To study the performance and behavior of the A-EFI+, bivariate relationships among biological indices, land-use variables and hydrological and morphological alterations were analyzed using Spearman rank order correlation coefficients (r_s), which are adequate to describe monotonous relationships and do not assume bivariate normality or linearity. We also used multiple regression analyses to consider all the indicators of anthropogenic perturbation simultaneously. All statistical analyses were performed using R version 4.3.2 [34].

3. Results

A total of 39 species were recorded in the compiled dataset, of which 19 (48.7%) were non-native. Alien species were present in 187 (54.4%) of the 344 sampling sites, with similar proportions in the two groups of basins studied (55.2% in Catalonia and 52.6% in the Jucar River Basin District). The mean percentage of alien individuals at sites with presence of alien species was 63.2% (range 11.2–100%). Both alien metrics and its average (AFP, alien fish pressure) showed a positive correlation with artificial and agricultural land use, as well as with hydrological and morphological alterations, and were negatively correlated with percentage of forest land (Figure 1), thus supporting their relevance as metrics of stream degradation. Higher pressure of alien species was also related to a lower richness of rheophilic spawning species ($r_s = -0.675$) and lower density of lithophilic spawning species ($r_s = -0.495$) and thus to the EFI+ index (Figure 2).

After calculating the A-EFI+, the resulting scores were lower than EFI+ in 183 of the 187 sites with alien species. In the remaining four sites with alien fish, the EFI+ was already zero before calculating the A-EFI+ and could not be reduced (Figure 3). In these sites, the mean reduction in the score of the A-EFI+ with respect to the EFI+ was 18.6% (range: 3.1–33.3%). This led to a lowering in ecological status classes of 40.4% of these sites, mainly from Good to Moderate and from Moderate to Poor. The decreased proportion of the A-EFI+ score was higher as AFP increased, as expected (Figures 2 and 3).

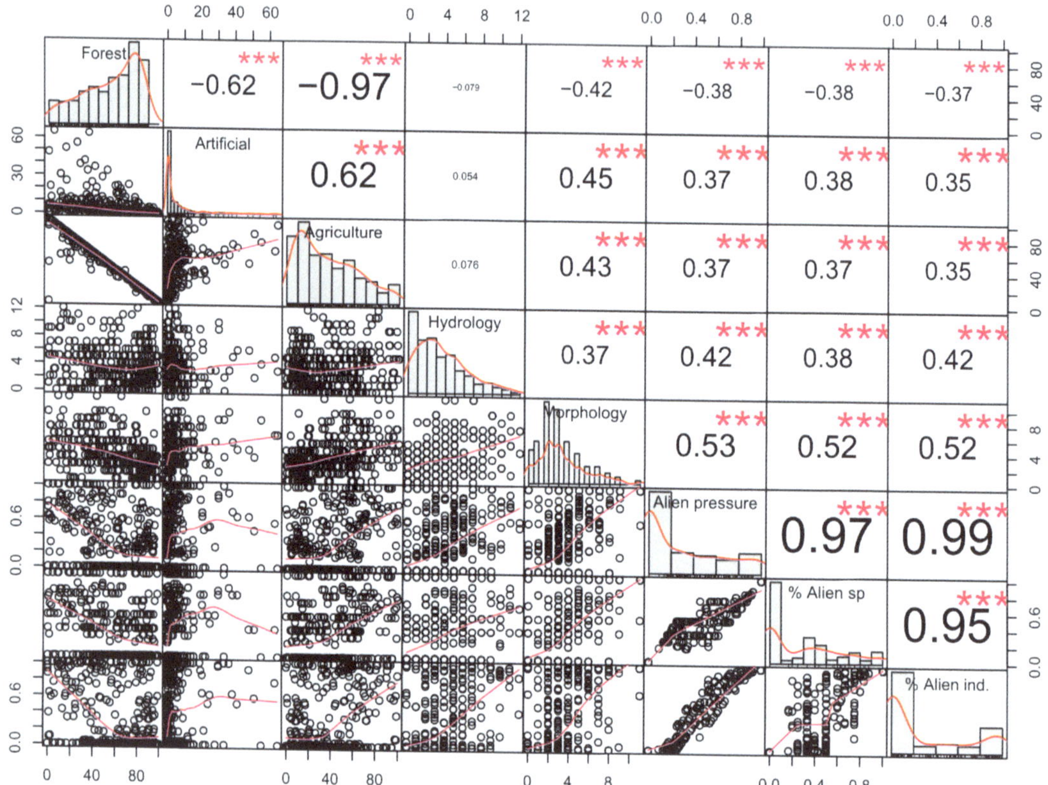

Figure 1. Relationship between fish alien metrics, land-use variables and hydrological and morphological alterations. The panels above the diagonal show the Spearman rank correlation coefficients with significance level (*** $p < 0.001$) and the panels below show the pairwise scatterplot with a smoothing curve (LOESS, red line). In the scatterplots, the *Y*-axis corresponds to the variable in the row diagonal and the *X*-axis to the column diagonal (e.g., the scatterplot on the bottom left has % alien individuals in the *Y*-axis and % forest cover in the *X*-axis).

The performance of the A-EFI+ was compared with several other biological indices and land-use variables by bivariate correlations and multiple regression. All fish indices (EFI+, A-EFI+, IBICAT and IBI-JUCAR) declined with increased catchment disturbance. They were negatively correlated with the percentage of artificial and agricultural land use, and positively correlated with the percentage of forest land (Figure 4). Compared to EFI+, the correlation of A-EFI+ with land use variables (Figure 4) was slightly higher, whereas the correlation with other biotic indices, such as the macroinvertebrate index (IBMWP) and the diatom index (IPS) or habitat index (IHF) was slightly lower (Figure 5). Hydrological and morphological alterations showed a negative correlation with both EFI+ and A-EFI+, with slightly better correlations for the adapted index (Figure 6). A multiple regression model suggested that the relationship with artificial land use was nonlinear but overall very similar for EFI+ and A-EFI+ (Table 1). Unsurprisingly, a regional fish index such as IBI-JUCAR performed slightly better (Table 1, Figures 4 and 5) but was quite correlated with those two indices (Figure 4).

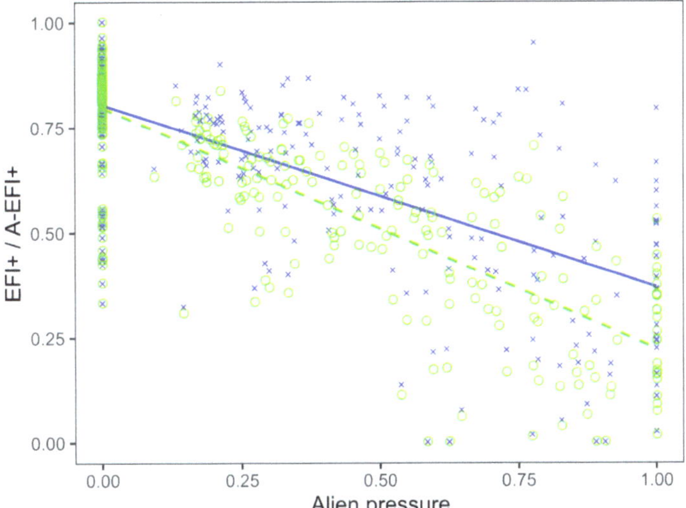

Figure 2. Relationship between the EFI+ and A-EFI+ indices (blue crosses and green circles, respectively) and the alien fish pressure (AFP) (average of % alien species and % alien individuals) in the study area. The simple regression lines are also shown (EFI+ = 0.803 − 0.435 AFP, R^2_{adj} = 0.465, $p < 0.001$; A-EFI+ = 0.795 − 0.575 AFP, R^2_{adj} = 0.683, $p < 0.001$).

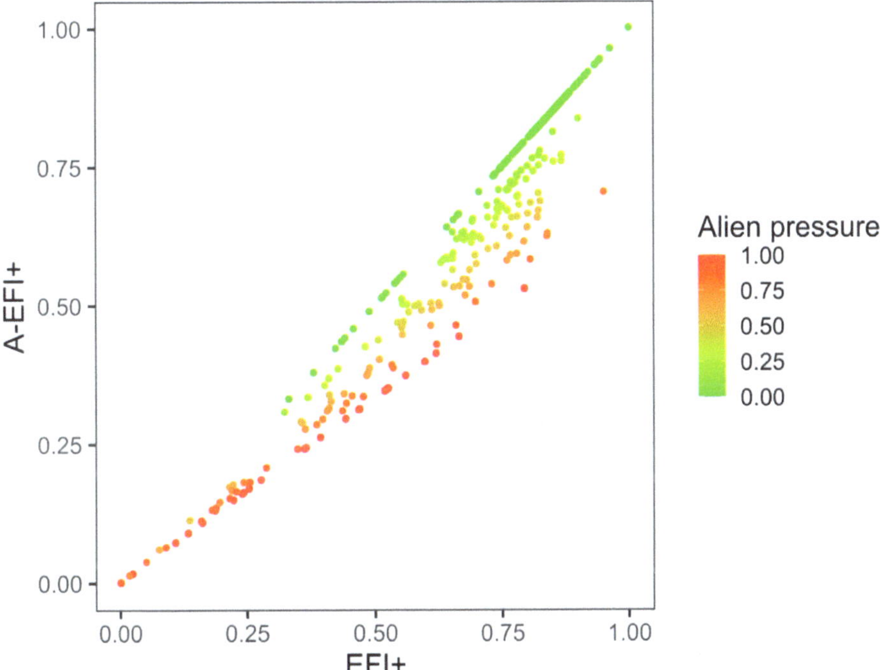

Figure 3. Relationship between the A-EFI+ and EFI+ indices and the effects of the alien fish pressure. The A-EFI+ index can be easily estimated from EFI+ with the following linear regression functions: A-EFI+ = −0.089 + 1.058 EFI+, R^2_{adj} = 0.940, $p < 0.001$; A-EFI+ = 0.1308 + 0.827 EFI+ − 0.215 AFP, R^2_{adj} = 0.991, $p < 0.001$.

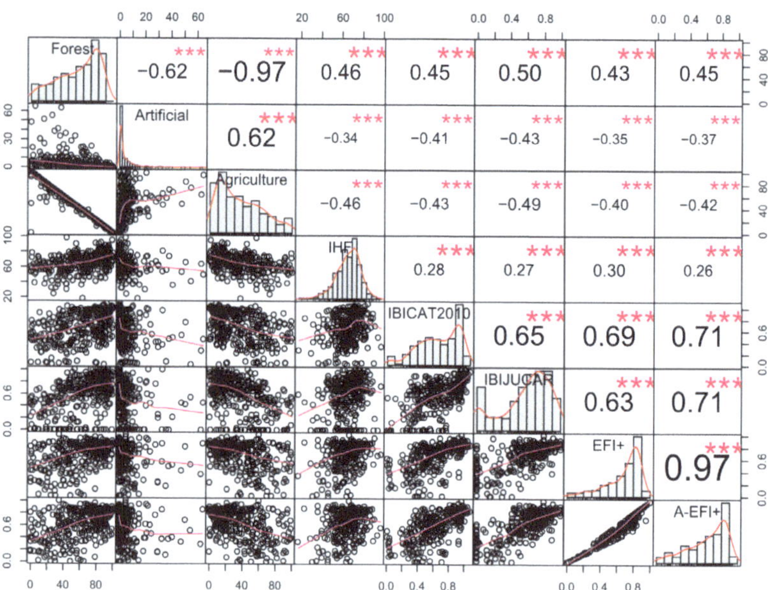

Figure 4. Relationship between fish indices and land-use variables. The panels above the diagonal show the Spearman rank correlation coefficients with significance level (*** $p < 0.001$) and the panels below the pairwise scatterplot with a smoothing curve (LOESS, red line). In the scatterplots, the Y-axis corresponds to the variable in the row diagonal and the X-axis to the column diagonal (e.g., the scatterplot on the bottom left has A-EFI+ in the Y-axis and % forest cover in the X-axis).

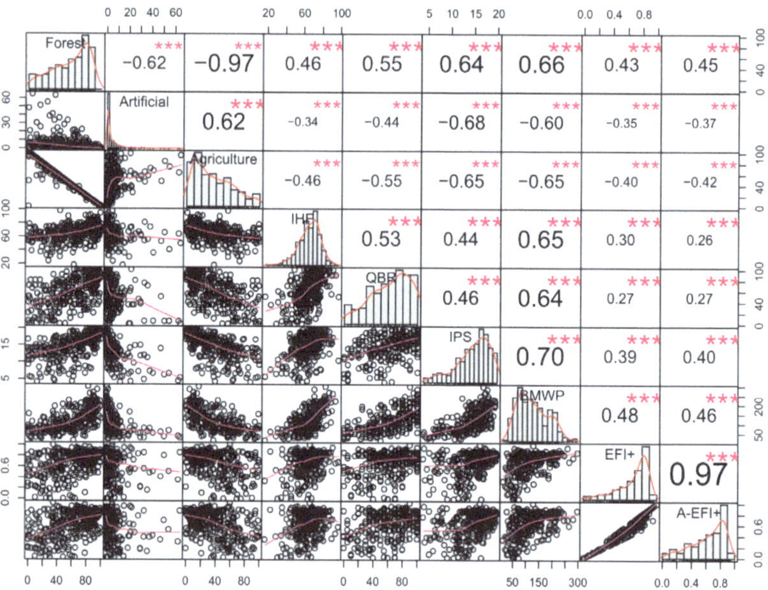

Figure 5. Relationship between biotic indices and land-use variables. The panels above the diagonal show the Spearman rank correlation coefficients with significance level (*** $p < 0.001$) and the panels below the pairwise scatterplot with a smoothing curve (LOESS, red line). In the scatterplots, the Y-axis corresponds to the variable in the row diagonal and the X-axis to the column diagonal (e.g., the scatterplot on the bottom left has A-EFI+ in the Y-axis and % forest cover in the X-axis).

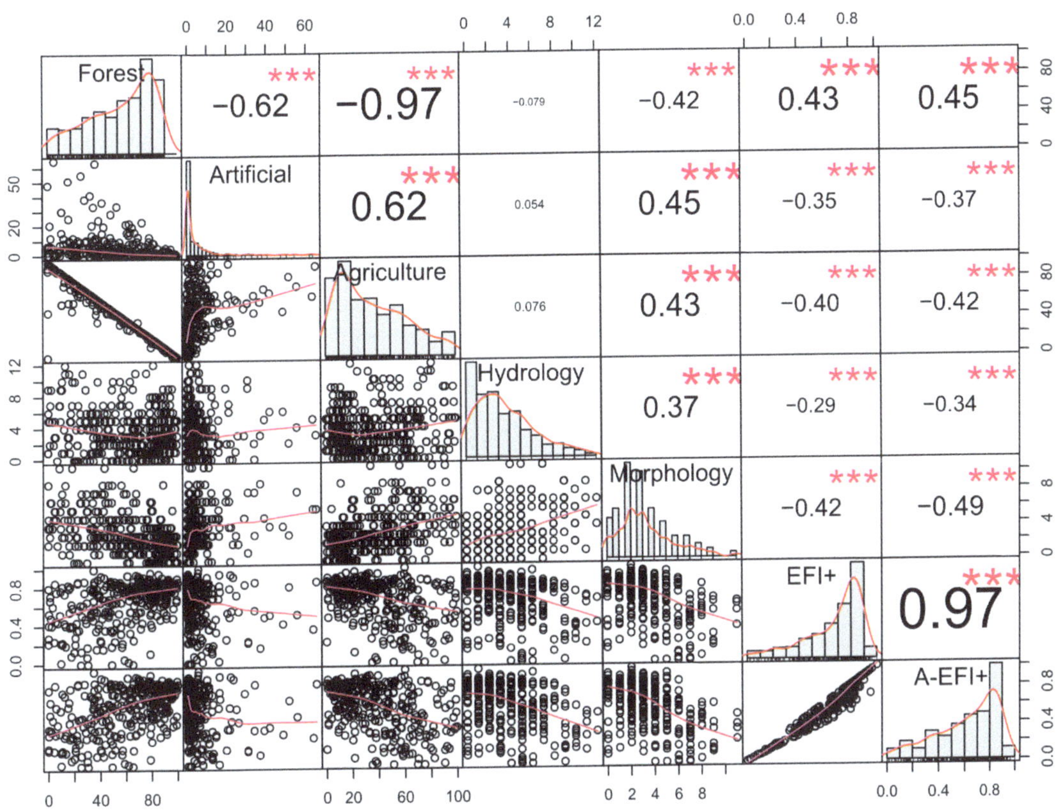

Figure 6. Relationship between land-use variables and hydrological and morphological alterations with EFI+ and A-EFI+ indices. The panels above the diagonal show the Spearman rank correlation coefficients with significance level (*** $p < 0.001$) and the panels below the pairwise scatterplot with a smoothing curve (LOESS, red line). In the scatterplots, the Y-axis corresponds to the variable in the row diagonal and the X-axis to the column diagonal (e.g., the scatterplot on the bottom left has A-EFI+ in the Y-axis and % forest cover in the X-axis).

Table 1. Multiple regression models of four fish indices with a habitat quality index (IHF) and % agricultural and artificial land uses. A quadratic component of artificial land use was also included because it was significant for most models and was supported by Akaike information criteria. * $p < 0.05$; ** $p < 0.01$; *** $p < 0.001$. All models were highly significant ($p \ll 0.001$).

	EFI+	A-EFI+	IBICAT2010	IBI-JUCAR
Intercept	0.5332 ***	0.5947 ***	0.7095 ***	0.7453 ***
IHF	0.0037 ***	0.0024 *	0.0015	0.0005
Agriculture	−0.0022 ***	−0.0029 ***	−0.0030 ***	−0.0044 ***
Artificial	−0.0071 *	−0.0099 **	−0.0082 *	−0.0117 **
Artificial^2	0.0001 *	0.0002 **	0.0001	0.0002 *
N	319	319	318	319
Residual standard error	0.1899	0.2079	0.2384	0.2489
R^2_{adj}	0.2193	0.2229	0.1849	0.2552

4. Discussion

The modification of the EFI+ proposed here is simple and easy to compute and adds an explicit consideration of alien fish pressure in the index, allowing for a wider assessment of stream health and more strictly following WFD guidelines. This modification is only relevant in areas highly impacted with alien species since the index remains unchanged when alien species are absent or reduces the score only slightly when the incidence of alien species is low. Therefore, its applicability should be more important in the western Mediterranean area because of the higher incidence of alien species than in Eastern or Northern Europe [35].

The positive relationships between the AFP (alien fish pressure) metric and agricultural and urban land uses, used here as a measure of anthropogenic pressures, is consistent with the view that alien fish species are generally more tolerant to environmental alterations than native ones [36] and therefore constitute good indicators of stream degradation [37]. Although the two original metrics of the EFI+ were negatively correlated with AFP and, therefore, some alien fish impact is already included in the index, the incorporation of the AFP metric into the EFI+ emphasizes the impact in sites where alien fish pressure is high, resulting in a more appropriate assessment of ecological integrity.

The A-EFI+ scores were highly consistent and proportional to the extent of alien species presence, showing a rapid decline in the quality status classes as the proportion of alien fish increased. Furthermore, the incorporation of the AFP metric into the EFI+ improved the index's performance as an indicator of degradation in comparison with the original version, as demonstrated by its stronger correlation with land-use pressures and hydrological and morphological alterations. The strength of the response to agricultural and artificial land-uses when the alien fish metric was included in the A-EFI+ did not decrease but rather increased. The A-EFI+ also showed higher correlation values than the original EFI+ with the two fish indices locally developed for the region (IBICAT and IBI-JUCAR) because they also consider explicitly alien species [13].

No important differences were found comparing the correlations of the EFI+ and A-EFI+ with biological indices for other organism groups (IBMWP, IPS) or abiotic indices (QBR, IHF), which suggests that the presence of alien species is not influenced by the ecological quality measured with these indices. Furthermore, the correlations of both EFI+ and A-EFI+ with diatom and macroinvertebrate indices were relatively low, as has been commonly reported [13,38]. This could be related due to the variable response of the different organism groups to degradation and is one of the reasons to consider multiple organism groups for stream health assessment [39]. For example, indices based on macroinvertebrates and diatoms often showed stronger responses to water quality parameters whereas fish, being more mobile organisms, appear to be more vulnerable to hydrological and habitat alterations [13,38].

Most EU member countries use fish indices locally developed for their respective territories to account for the specific characteristics of the fish assemblages. Local fish indices often provide a more accurate and precise evaluation of local conditions, but indices that are widely applicable across extensive regions are also desirable to improve the integration of results from various measurement methods and to enhance resource management [22]. Furthermore, national fish indices are required by the WFD for intercalibration to ensure consistency in ecological assessments throughout Europe [24,40]. The EFI+ is the only fish index currently available for use in most European countries (mainly those that provided data for the development of the index) and, therefore, an important application is to be used to convert to a common scale the assessments made with different local fish indices and allowing comparison [17]. The modification of the EFI+ proposed here does not affect the utility of this approach, since the transformation of the A-EFI+ to EFI+ and vice versa is simple and straightforward. Moreover, in regions with a significant presence of alien species, the A-EFI+ should provide a more comprehensive tool to assess and communicate the ecosystem health or rivers.

5. Conclusions

The presence of alien species poses a significant threat to the preservation of freshwater native diversity and should be considered an indicator of degraded stream conditions and anthropogenic pressures on aquatic ecosystems. Consequently, any biotic index must be able to effectively assess this type of impact. The EFI+ index lacks a specific metric to weight the presence of alien fish. The adapted version of the EFI+ presented here includes a negative metric for alien species and therefore can assess the extent of their presence and their potential impact on native fish communities. Furthermore, including a metric for alien species improves the accuracy and relevance of the index for a more comprehensive assessment of ecological status, helping to identify areas where conservation efforts should be focused and where management actions are needed to mitigate the negative impacts of alien species. The A-EFI+ is most suitable for assessing ecological status in areas with a high proportion and abundance of alien fish, such as some basins in the Iberian Peninsula, which also host many local endemic species, many of which are severely threatened.

Author Contributions: E.A., C.A. and E.G.-B. participated in the conceptualization of the study, performed the data analysis and drafted the manuscript. R.R. and Q.P.-R. helped to draft the manuscript by contributing their expertise. All authors have read and agreed to the published version of the manuscript.

Funding: Financial support was provided by the Spanish Ministry of Science, Innovation and Universities (MCIN/AEI/10.13039/501100011033) and the European Union (NextGenerationEU/PRTR) through projects PID2019-103936GB-C21, TED2021-129889B-I00 and RED2022-134338-T.

Institutional Review Board Statement: All the data come from previous research or other databases. In this case, ethical approval is not needed for this article.

Data Availability Statement: The data presented in this study are openly available in FigShare data repository at DOI: 10.6084/m9.figshare.24665571 (accessed on 27 December 2023).

Acknowledgments: We thank the Catalan Water Agency (ACA) and the "Confederación Hidrográfica del Júcar" (CHJ) for the data provided for this work. CA also acknowledges the support from the CERCA Programme (Generalitat de Catalunya).

Conflicts of Interest: The authors declare no conflicts of interest.

References

1. European Union Directive 2000/60/EC of the European Parliament and the Council of 23 October 2000 Establishing a Framework for Community Action in the Field of Water Policy. *Off. J. Eur. Communities* **2000**, *43*, L327.
2. Karr, J.R. Assessment of biotic integrity using fish communities. *Fisheries* **1981**, *6*, 21–27. [CrossRef]
3. Fausch, K.D.; Lyons, J.; Karr, J.R.; Angermeier, P.L. Fish communities as indicators of environmental degradation. In Proceedings of the American Fisheries Society Symposium, Pittsburgh, PA, USA, 26–30 August 1990; AFS Publications: Bethesda, Maryland, 1990; Volume 8, pp. 123–144.
4. Fullerton, A.H.; Burnett, K.M.; Steel, E.A.; Flitcroft, R.L.; Pess, G.R.; Feist, B.E.; Torgersen, C.E.; Miller, D.J.; Sanderson, B.l. Hydrological connectivity for riverine fish: Measurement challenges and research opportunities. *Freshw. Biol.* **2010**, *55*, 2215–2237. [CrossRef]
5. *FAME Consortium Manual for the Application of the European Fish Index—EFI. A Fish-Based Method to Assess the Ecological Status of European Rivers in Support of the Water Framework Directive*; BOKU: Viena, Austria, 2004.
6. Pont, D.; Hugueny, B.; Rogers, C. Development of a fish-based index for the assessment of river health in Europe: The European Fish Index. *Fish. Manag. Ecol.* **2007**, *14*, 427–439. [CrossRef]
7. *EFI+ Consortium Manual for the Application of the New European Fish Index-EFI+. A Fish-Based Method to Assess the Ecological Status of European Running Waters in Support of the Water Framework Directive*; BOKU: Viena, Austria, 2009.
8. European Environment Agency (EEA). *European Waters—Assessment of Status and Pressures*; European Environment Agency: Luxembourg, 2018.
9. Clavero, M.; García-Berthou, E. Homogenization dynamics and introduction routes of invasive freshwater fish in the iberian peninsula. *Ecol. Appl.* **2006**, *16*, 2313–2324. [CrossRef] [PubMed]
10. Hermoso, V.; Clavero, M.; Blanco-Garrido, F.; Prenda, J. Invasive species and habitat degradation in Iberian streams: An analysis of their role in freshwater fish diversity Loss. *Ecol. Appl.* **2011**, *21*, 175–188. [CrossRef] [PubMed]
11. Ribeiro, F.; Leunda, P.M. Non-native fish impacts on Mediterranean freshwater ecosystems: Current knowledge and research needs. *Fish. Manag. Ecol.* **2012**, *19*, 142–156. [CrossRef]

12. Ramos-Merchante, A.; Prenda, J. The ecological and conservation status of the Guadalquivir River Basin (s Spain) through the application of a fish-based multimetric index. *Ecol. Indic.* **2018**, *84*, 45–59. [CrossRef]

13. Aparicio, E.; Carmona-Catot, G.; Moyle, P.B.; García-Berthou, E. Development and evaluation of a fish-based index to assess biological integrity of Mediterranean streams. *Aquat. Conserv. Mar. Freshw. Ecosyst.* **2011**, *21*, 324–337. [CrossRef]

14. Aparicio, E.; Vargas, M.J.; Olmo, J.M.; De Sostoa, A. Decline of native freshwater fishes in a Mediterranean watershed on the Iberian Peninsula: A quantitative assessment. *Environ. Biol. Fishes* **2000**, *59*, 11–19. [CrossRef]

15. Radinger, J.; Alcaraz-Hernández, J.D.; García-Berthou, E. Environmental filtering governs the spatial distribution of alien fishes in a large, human-impacted Mediterranean river. *Divers. Distrib.* **2019**, *25*, 701–714. [CrossRef]

16. Boon, P.J.; Clarke, S.A.; Copp, G.H. Alien species and the EU Water Framework Directive: A comparative assessment of European approaches. *Biol. Invasions* **2020**, *22*, 1497–1512. [CrossRef]

17. Segurado, P.; Caiola, N.; Pont, D.; Oliveira, J.M.; Delaigue, O.; Ferreira, M.T. Comparability of fish-based ecological quality assessments for geographically distinct Iberian regions. *Sci. Total Environ.* **2014**, *476–477*, 785–794. [CrossRef] [PubMed]

18. Almeida, D.; Alcaraz-Hernández, J.D.; Merciai, R.; Benejam, L.; García-Berthou, E. Relationship of fish indices with sampling effort and land use change in a large Mediterranean river. *Sci. Total Environ.* **2017**, *605–606*, 1055–1063. [CrossRef] [PubMed]

19. Zogaris, S.; Tachos, V.; Economou, A.N.; Chatzinikolaou, Y.; Koutsikos, N.; Schmutz, S. A model-based fish bioassessment index for Eastern Mediterranean rivers: Application in a biogeographically diverse area. *Sci. Total Environ.* **2018**, *622–623*, 676–689. [CrossRef] [PubMed]

20. Adamczyk, M.; Prus, P.; Buras, P.; Wiśniewolski, W.; Ligięza, J.; Szlakowski, J.; Borzęcka, I.; Parasiewicz, P. Development of a new tool for fish-based river ecological status assessment in Poland (EFI+IBI_PL). *Acta Ichthyol. Piscat.* **2017**, *47*, 173–184. [CrossRef]

21. Muñoz-Mas, R.; García-Berthou, E. Alien animal introductions in Iberian inland waters: An update and analysis. *Sci. Total Environ.* **2020**, *703*, 134505. [CrossRef]

22. Vadas, R.L.; Hughes, R.M.; Bae, Y.J.; Baek, M.J.; Gonzáles, O.C.B.; Callisto, M.; Carvalho, D.R.D.; Chen, K.; Ferreira, M.T.; Fierro, P.; et al. Assemblage-based biomonitoring of freshwater ecosystem health via multimetric indices: A critical review and suggestions for improving their applicability. *Water Biol. Secur.* **2022**, *1*, 100054. [CrossRef]

23. Filipe, A.F.; Feio, M.J.; Garcia-Raventós, A.; Ramião, J.P.; Pace, G.; Martins, F.M.; Magalhães, M.F. The European Water Framework Directive facing current challenges: Recommendations for a more efficient biological assessment of inland surface waters. *Inland Waters* **2019**, *9*, 95–103. [CrossRef]

24. Birk, S.; Willby, N.J.; Kelly, M.G.; Bonne, W.; Borja, A.; Poikane, S.; Van De Bund, W. Intercalibrating classifications of ecological status: Europe's quest for common management objectives for aquatic ecosystems. *Sci. Total Environ.* **2013**, *454–455*, 490–499. [CrossRef]

25. Maceda-Veiga, A.; Monleon-Getino, A.; Caiola, N.; Casals, F.; De Sostoa, A. Changes in fish assemblages in catchments in north-eastern Spain: Biodiversity, conservation status and introduced species. *Freshw. Biol.* **2010**, *55*, 1734–1746. [CrossRef]

26. Alcaraz, C.; Carmona-Catot, G.; Risueño, P.; Perea, S.; Pérez, C.; Doadrio, I.; Aparicio, E. Assessing population status of *Parachondrostoma arrigonis* (Steindachner, 1866), threats and conservation perspectives. *Environ. Biol. Fishes* **2015**, *98*, 443–455. [CrossRef]

27. *EN 14011:2003*; Water QUALITY—Sampling of Fish with Electricity. European Standard. European Committee for Standardization: Brussels, Belgium, 2003.

28. Sostoa, A.D.; Caiola, N.; Casals, F.; García-Berthou, E.; Alcaraz, C.; Benejam, L.; Maceda-Veiga, A.; Solà, C.; Munné, A. *Ajust de l'Índex d'Integritat Biòtica (IBICAT) Basat en l'ús Dels Peixos com a Indicadors de la Qualitat Ambiental als Rius de Catalunya*; Agència Catalana de l'Aigua, Departament de Medi Ambient i Habitatge, Generalitat de Catalunya: Barcelona, Spain, 2010; p. 187.

29. Munné, A.; Prat, N.; Solà, C.; Bonada, N.; Rieradevall, M. A simple field method for assessing the ecological quality of riparian habitat in rivers and streams: QBR index. *Aquat. Conserv. Mar. Freshw. Ecosyst.* **2003**, *13*, 147–163. [CrossRef]

30. Pardo, I. El hábitat de los ríos mediterráneos. diseño de un Índice de diversidad de hábitat. *Limnetica* **2002**, *21*, 115–133. [CrossRef]

31. CEMAGREF. *Etude Des Méthodes Biologiques Quantitatives D'appréciation de La Qualité Des Eaux*; Agence de l'eau Rhône-Méditerranée-Corse: Lyon, France, 1982.

32. Alba-Tercedor, J. Caracterización del estado ecológico de ríos mediterráneos Ibéricos mediante el Índice IBMWP (Antes BMWP'). *Limnetica* **2002**, *21*, 175–185. [CrossRef]

33. Martínez-Fernández, V.; Solana-Gutiérrez, J.; García De Jalón, D.; Alonso, C. Sign, Strength and shape of stream fish-based metric responses to geo-climatic and human pressure gradients. *Ecol. Indic.* **2019**, *104*, 86–95. [CrossRef]

34. R Core Team, R. *A Language and Environment for Statistical Computing*; R Foundation for Statistical Computing: Vienna, Austria, 2019.

35. Magliozzi, C.; Tsiamis, K.; Vigiak, O.; Deriu, I.; Gervasini, E.; Cardoso, A.C. Assessing invasive alien species in European catchments: Distribution and impacts. *Sci. Total Environ.* **2020**, *732*, 138677. [CrossRef] [PubMed]

36. Maceda-Veiga, A.; De Sostoa, A. Observational evidence of the sensitivity of some fish species to environmental stressors in Mediterranean rivers. *Ecol. Indic.* **2011**, *11*, 311–317. [CrossRef]

37. Ferreira, T.; Oliveira, J.; Caiola, N.; De Sostoa, A.; Casals, F.; Cortes, R.; Economou, A.; Zogaris, S.; Garcia-Jalon, D.; Ilhéu, M.; et al. Ecological traits of fish assemblages from Mediterranean Europe and their responses to human disturbance. *Fish. Manag. Ecol.* **2007**, *14*, 473–481. [CrossRef]

38. Benejam, L.; Aparicio, E.; Vargas, M.J.; Vila-Gispert, A.; García-Berthou, E. Assessing fish metrics and biotic indices in a Mediterranean stream: Effects of uncertain native status of fish. *Hydrobiologia* **2008**, *603*, 197–210. [CrossRef]
39. Hering, D.; Johnson, R.K.; Kramm, S.; Schmutz, S.; Szoszkiewicz, K.; Verdonschot, P.F.M. Assessment of European streams with diatoms, macrophytes, macroinvertebrates and fish: A comparative metric-based analysis of organism response to stress. *Freshw. Biol.* **2006**, *51*, 1757–1785. [CrossRef]
40. *Guidance Document on the Intercalibration Process 2008–2011. Guidance Document No. 14. Common Implementation Strategy for the Water Framework Directive (2000/60/EC)*; European Commission: Luxembourg, 2011. [CrossRef]

 *fishes*

Article

Sicklefin Chub (*Macrhybopsis meeki*) and Sturgeon Chub (*M. gelida*) Temporal and Spatial Patterns from Extant Population Monitoring and Habitat Data Spanning 23 Years

Mark L. Wildhaber [1,*], Benjamin M. West [1], Kendell R. Bennett [1], Jack H. May [1], Janice L. Albers [2] and Nicholas S. Green [3]

[1] Columbia Environmental Research Center, U.S. Geological Survey (USGS), 4200 New Haven Road, Columbia, MO 65201, USA; bwest@usgs.gov (B.M.W.); krbennett@usgs.gov (K.R.B.); jhmay@usgs.gov (J.H.M.)
[2] Upper Midwest Environmental Science Center, U.S. Geological Survey (USGS), 2630 Fanta Reed Road, La Crosse, WI 54603, USA; jalbers@usgs.gov
[3] Department of Ecology, Evolution, and Organismal Biology, Kennesaw State University, 370 Paulding Ave NW, Kennesaw, GA 30144, USA; ngreen62@kennesaw.edu
* Correspondence: mwildhaber@usgs.gov

Citation: Wildhaber, M.L.; West, B.M.; Bennett, K.R.; May, J.H.; Albers, J.L.; Green, N.S. Sicklefin Chub (*Macrhybopsis meeki*) and Sturgeon Chub (*M. gelida*) Temporal and Spatial Patterns from Extant Population Monitoring and Habitat Data Spanning 23 Years. *Fishes* **2024**, *9*, 43. https://doi.org/10.3390/fishes9020043

Academic Editors: Robert L. Vadas, Jr. and Robert M. Hughes

Received: 30 September 2023
Revised: 9 January 2024
Accepted: 11 January 2024
Published: 23 January 2024

Abstract: Sicklefin (*Macrhybopsis meeki*) and sturgeon chub (*M. gelida*) historically occurred throughout the Missouri River (MR), in some tributaries, and Mississippi River downstream of the MR. They have been species of U.S. state-level conservation concern and U.S. Endangered Species Act listing candidates since the 1990s. We applied analytical approaches from occupancy modeling to correlation to monitoring data spanning 23 years to assess relationships between occupancy and time, space, environmental factors, habitat, and other species. Sicklefin chub occupancy appeared higher in the early to mid-2000s and mid-to-late 2010s. A potential decline in occupancy occurred for sturgeon chub in the mid-to-late 2010s. Spatially, chub occupancy was depressed for 159 to 438 km downstream of MR dams. Among macrohabitats, inside bends had relatively high occupancy for both species; secondary connected channels had relatively high values for sturgeon chub. Co-occurrence was likely between sicklefin and sturgeon chub and between chubs and shovelnose sturgeon (*Scaphirhybchus platorybchus*) and channel catfish (*Ictalurus punctatus*). The observed co-occurrence of chubs and pallid sturgeon (*Scaphirhynchus albus*; PS) was potentially higher than expected for adult PS. For juvenile PS, co-occurrence was lower than expected in the Lower MR and potentially higher than expected in the Upper MR, warranting future research. Results from this research suggest management for the improvement of sicklefin and sturgeon chub populations may benefit other MR fish populations.

Keywords: sicklefin chub; sturgeon chub; *Macrhybopsis*; occupancy model; Missouri River; channel catfish; shovelnose sturgeon; pallid sturgeon; predators

Key Contribution: Based on monitoring data spanning 23 years (1996 to 2018), we found sicklefin chub occupancy to be higher in the early to mid-2000s and mid-to-late 2010s, in contrast to a decline for sturgeon chub in mid-to-late 2010s. We also found lower site occupancy by chubs below dams, with inside bend macrohabitat having the highest chub occupancy for both chub species and a very strong co-occurrence between sicklefin and sturgeon chub and between chubs and shovelnose sturgeon and channel catfish, suggesting management for the improvement of sicklefin and sturgeon chub populations may benefit other Missouri River fish populations.

1. Introduction

Sicklefin chub (*Macrhybopsis meeki*) and sturgeon chub (*M. gelida*) have historical ranges throughout the mainstem Missouri River (MR) and in selected tributaries and the mainstem Mississippi River downstream from the confluence with the MR, where they are an important part of the benthic fish community [1,2]. However, analysis of annual fish monitoring surveys has indicated declining population trends in these chubs prior to 2010 [3]. Sicklefin chub is on state-level conservation concern lists in Montana, South Dakota, Iowa, Kentucky, North Dakota, Nebraska, Kansas, Missouri, and Tennessee. Sturgeon chub is listed as a conservation concern in these states, as well as in Wyoming and Illinois. Both species have been candidate species for listing under the United States Endangered Species Act (ESA) since the 1980s and 1990s [4]. In 2001, the 12-month finding for sicklefin and sturgeon chub found they did not warrant listing under the ESA but documented losses of over 46% and 45% of their historical habitat in the MR basin, respectively [4]. These results, along with the U.S. Fish and Wildlife Service (USFWS) 12-month finding in 2001 [4], motivated USFWS to produce an assessment in 2023 investigating whether sicklefin and sturgeon chub warranted listing as threatened or endangered. Following the 2016 petition to list these chub species under ESA [4] and the apparent declines of both chub species in recent years, USFWS reassessed the status of both species and did not find definitive evidence for listing in 2023 [5]. Despite the 2023 decision not to list these species, the variation in patterns presented in the 2023 assessment highlight the need to further investigate sicklefin and sturgeon chub trends spatially, temporally, and relative to abiotic and biotic factors to help determine whether they are prone to further declines.

Sicklefin and sturgeon chub are components of the diet of adult federally listed endangered pallid sturgeon (*Scaphirhynchus albus*). Pallid sturgeon consume chubs as a large part of their diet [6]. Given that over 175,000 pallid sturgeon have been stocked into the MR between 1992 and 2017 [7], it is possible that increased predation on chubs may have reduced sicklefin and sturgeon chub populations. Additionally, pallid sturgeon diet and dependency on fish prey were also identified as a top priority in a 2004 assessment of research needs for pallid sturgeon research on the Missouri River [8]. Chubs are also consumed by walleye (*Sander vitreus*), sauger (*Sander canadensis*), and a variety of piscivorous fishes, birds, and mammals [9].

Past activities associated with river or fish species management may have either degraded or improved conditions for sicklefin and sturgeon chub. The U.S. Army Corps of Engineers (USACE) Missouri River Recovery Program's current channel reconfiguration efforts are focused on interception-rearing complexes for juvenile pallid sturgeon, i.e., modifying hydrologic conditions to promote free-floating pallid sturgeon juveniles' arrival in suitable nursery habitat [10]. By 2014, USACE channel reconfiguration through dike notching, top-width widening, and side channel addition activities have modified the majority of dikes in the MR to gain shallow water habitat credit for pallid sturgeon recovery. Current research is limited as to how these activities affect MR sicklefin and sturgeon chub populations. However, Ridenour et al. [11,12] reported a loss of chub nursery habitat due to dike notching activities.

In support of USFWS management decisions associated with sicklefin and sturgeon chub, there is a need for comprehensive, long-term data analyses of both chub populations. The only such data to date have been collected on the MR and Kansas River (KR). Two large data sources for sicklefin and sturgeon chub and a wide range of other species are the MR Benthic Fishes Study (MRBFS) [13] and the Pallid Sturgeon Population Assessment Project (PSPAP) [3]. Since the MR represents a major portion of the historical range of both species [14], these data can provide valuable insight into their management. Previous studies using these data have documented declines in sicklefin and sturgeon chub populations [3,15], along with other anecdotal observations from practitioners.

Occupancy models predict species occurrences while accounting for imperfect detection, i.e., the fact that species can be undetected where they are present [16]. When assessing habitat associations for a species, occupancy models can reduce the bias compared to models that use relative abundance data without accounting for imperfect detection [17,18]. In addition to their ability to reduce bias compared to other methods, occupancy models can be useful for rare or hard-to-detect species, allowing researchers to evaluate occurrence trends [19]. Sicklefin and sturgeon chub are difficult to detect; a detection-focused occupancy modeling research study in the Lower MR found detection probabilities in benthic trawls ranging from approximately 0.05 to 0.48 [20]. In terms of rarity, other research suggests that these species are in low densities in the MR, with only 18.46% and 5.41% of benthic trawl deployments catching sicklefin and sturgeon chub, respectively, in the Lower MR [3]. In the context of other fish species of conservation concern, occupancy modeling has been used to investigate the effects of water development on imperiled fish species in the deserts of the Southwestern United States and to assess the validity of managing these species as one unit [21]. One group of occupancy models, multispecies occupancy models, can be used to examine interactions among species that affect co-occurrence [22,23]. For example, two-species occupancy models were created for threatened eastern sand darter (*Ammocrypta pellucida*) with the goal of informing reintroduction efforts based on interaction with other species [24]. Because occupancy and detection rates are often related to abundance [25], determining factors related to the occupancy of sicklefin and sturgeon chubs could also provide insight into factors related to population expansion and contraction and changes in overall population size over space and time.

Because of their potential rarity and the relative importance of sicklefin and sturgeon chub as prey for the endangered pallid sturgeon, there is a need to develop a better understanding of the relationship between system conditions and sicklefin and sturgeon chub populations, and how these conditions may help quantify the associated risk of chub declines. The goal of this study is to better understand how the occurrence patterns of sicklefin and sturgeon chub in the MR are related to abiotic and biotic factors, with the expectation that this information would inform USFWS efforts in assessing the current status of these two chub species, as well as future conservation of these species. To do so, this study used PSPAP and MRBFS data and multiple analytical approaches to (1) describe temporal, spatial, and environmental factors associated with the occupancy patterns of sicklefin and sturgeon chub, (2) assess the co-occurrence of the two chub species, (3) assess the relationship between annual site occupancy coefficients and river discharge, and (4) investigate the co-occurrence patterns of these chub species with other important fish species found in the MR benthic fish community using either three-species occupancy models or comparisons of observed co-occurrence.

2. Materials and Methods

2.1. Extant Data and Study Area

To inform models for sicklefin and sturgeon chub, we used extant data from two projects, the Missouri River Benthic Fishes Study (MRBFS) and the Pallid Sturgeon Population Assessment Project (PSPAP). The MRBFS was conducted from 1996 to 1998, sampling benthic fishes and environmental variables in the mainstem of the MR and lower portion of the Yellowstone River [13,26]. The PSPAP began in the fall of 2003 and sampled fish species in the Missouri River from Fort Peck Dam to the confluence with the Mississippi River [3,27]. The PSPAP was divided into two seasons: fish community season from 1 July to 31 October, focused on sampling small, juvenile pallid sturgeon and the fish community, and sturgeon season, when colder water temperatures allowed for the use of gill nets to catch larger juvenile and adult pallid sturgeon with a decreased chance for mortality from stress. For our analyses of PSPAP data, we used data from the fish community season because the sampling was consistent and used types of gear aimed at catching non-pallid sturgeon species [28]. We only used PSPAP data collected between 2003 and 2018, due to large changes in protocol in 2019 that eliminated the fish community season [10].

Both projects used stratified, random sampling of locations and divided sampling locations into macrohabitats including main-channel crossover, outside bend, inside bend, tributary mouth, secondary connected channel, and secondary non-connected channel. Both projects collected the following water conditions: depth, velocity, temperature, and turbidity. For more information on MRBFS and PSPAP, see Appendix A, associated MRBFS data release [29], PSPAP Standard Operation Procedure [27], and previous studies incorporating these data sets [3,13].

For our analyses, the study area encompassed two regions of the MR, the Upper MR (UMR) and the Lower MR (LMR), and the Lower Kansas River (Figure 1). For the purposes of this study, we defined the UMR as the 8-digit Hydrologic Unit Code (HUC 8) subbasins [30] containing mainstem MR between Fort Peck Dam (MR kilometer 2850) and Lake Sakakawea (MR kilometer 2546). We defined the LMR as the HUC 8 subbasins containing the mainstem MR just upstream of Gavins Point Dam (MR kilometer 1416) to the MR confluence with the Mississippi River (MR kilometer 0). These sections were the only MR sections used in this study because they were sampled in the PSPAP at least 75% of the years between 2003 and 2018; tributaries other than the Kansas River were not part of the standard sampling protocol [10,27]. Reservoirs were also excluded because the focus was on riverine species. Importantly, dams along the mainstem MR, such as Fort Peck and Gavins Point dams, are hydro-electric dams.

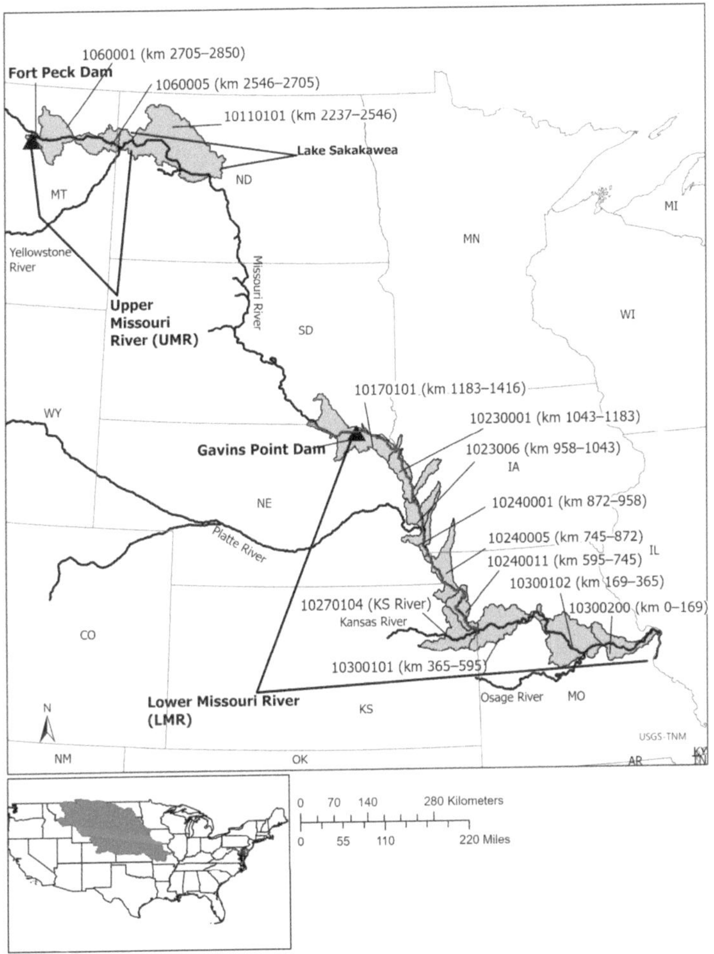

Figure 1. Study area map showing HUC 8 subbasins along the Upper Missouri River (UMR), Lower Missouri River (LMR), and Lower Kansas River, the locations of Gavins Point and Fort Peck Dams (filled black triangles), and the Missouri River Basin (thick, medium gray outline in large map). The small map on the lower left shows the entire Missouri River Basin, in gray. Geospatial data for HUC 8 subbasins and the Missouri River Basin are from the U.S. Geological Survey (USGS) National Hydrography Dataset, and base map is from USGS.

2.2. Benthic Fish Species of Interest

Although sicklefin and sturgeon chub were the primary species of interest in this study, we also wanted to investigate potential interactions of these two chub species with other species in the MR benthic fish community. Channel catfish (*Ictalurus punctatus*) was included in this study because of its recreational value [31] and to investigate the hypothesis of high co-occurrence of chubs and channel catfish. Channel catfish are found in similar riverine macrohabitats as sicklefin and sturgeon chub [13,32]. If co-occurrence exists, it may suggest that management actions that improve sicklefin and sturgeon chub populations could also improve channel catfish status and vice versa.

Shovelnose sturgeon (*Scaphirhynchus platorynchus*) was included in this study because of its potential for competition for habitat and food resources with sicklefin and sturgeon chub [33] and piscivorous species such as pallid sturgeon, sauger, and walleye. Pallid sturgeon were chosen for their potential roles as competitors to chubs in their juvenile stage and predators of chubs as adults [34,35]. Sauger and walleye were chosen due to their role as potential chub predators since they are known to prey on benthic fishes of similar size to chubs [36].

2.3. Data Processing

Prior to modeling, we used several data filtering and processing steps. A summary of these steps is provided here; more detailed information is contained in Appendix B. The purpose of these steps was to ensure that the data used to fit our occupancy models (1) fell within standard sampling protocols for the PSPAP and the MRBFS; (2) included primary gear type that caught sicklefin and sturgeon chub; (3) were properly structured for modeling; (4) met model assumptions; and (5) allowed for use of PSPAP and MRBFS data together.

Except for analyses detailed in Section 2.4.3 "Observed co-occurrence", we exclusively used benthic trawl samples because no other gear utilized caught sicklefin or sturgeon chub consistently. In fact, of all sites (i.e., macrohabitat within a river bend) where either a sicklefin or sturgeon chub was detected, >95% of chub samples had at least one detection of that species in a benthic trawl. The PSPAP and MRBFS used different benthic trawls; the PSPAP used a 4.8 m-wide otter trawl (OT16; [27]), and the MRBFS used a 2 m bottom trawl (BT; [26,29]). To structure data for an occupancy model with spatially replicated visits [37], the full definition of a site was a macrohabitat within a river bend sampled during a particular year. Each river bend was only visited once during the fish community season within a year. A visit was a pull of a benthic trawl. Only sites with multiple visits were retained, as multiple visits are required for occupancy modeling [38]. Additionally, only macrohabitats sampled by both the MRBFS and PSPAP, i.e., main-channel crossover, outside bend, inside bend, tributary mouth, secondary connected channel, and secondary non-connected channel, were used. All environmental data were averaged across visits to the level of the site. As a way of examining spatial patterns with a categorical variable, each site was assigned to a subbasin designated by a U.S. Geological Survey-standardized 8-digit Hydrologic Unit Code [30] (HUC 8, Figure 1).

Due to the shorter duration of MRBFS compared to the PSPAP, MRBFS data were only analyzed in combination with PSPAP-FC data. Spatially, MRBFS data were limited to standard sampling areas of the PSPAP [27]. When the PSPAP-FC and MRBFS data were used in combination, Kansas River data were excluded from the PSPAP data due to this river not being sampled by the MRBFS. In the combined MRBFS+PSPAP-FC data set, there were 12,160 trawls and totals of 6618 sicklefin and 3360 sturgeon chub individuals collected. For the raw detection rate by trawl, 17.7% and 12.4% of trawls collected at least one sicklefin or sturgeon chub, respectively. For the raw detection rate by site, that is, macrohabitat, 24.7% and 29.2% of sites had at least one sicklefin or sturgeon chub detected, respectively. The size range of sicklefin and sturgeon chub collected was 11–177 mm and 16–121 mm, respectively. The low detection rates for sicklefin and sturgeon chub was one reason for using occupancy modeling, and not relative abundance, in this study. Final data sets at the end of processing were PSPAP–fish community season (PSPAP-FC; n = 3631 sites, v = 11,500 visits across sites) and MRBFS+PSPAP-FC (n = 3926, v = 12,160).

2.4. Statistical Analyses

2.4.1. Occupancy Models

To address our primary goals, we developed three multispecies, single-season occupancy models. The first was a two-species occupancy model to examine sicklefin and sturgeon chub occupancy and co-occurrence relationships to temporal, spatial, and environmental factors. This model was fit using the MRBFS+PSPAP-FC data set because this data set contained the longest span of data with the most sites. The two-species model had fully parameterized detection and occupancy components and a co-occurrence component parameterized with continuous water condition covariates. The remaining models were three-species occupancy models of occupancy and co-occurrence of focal chub species and either channel catfish or shovelnose sturgeon. Three-species models were fit using only the PSPAP-FC data because the BT in the MRBFS was inefficient at capturing channel catfish and shovelnose sturgeon. At MRBFS sites where a channel catfish or shovelnose sturgeon had been captured by at least one type of gear, 62.6% of channel catfish sites had channel catfish caught in BT, and 39.6% shovelnose sturgeon sites caught shovelnose sturgeon in BT. Corresponding rates were much higher for the OT16 in the PSPAP (90.7% for channel catfish and 87.2% for shovelnose sturgeon; see Appendix B). Three-species models had fully parameterized detection components and intercept-only occupancy and co-occurrence components. The primary purpose of the three-species models was to examine the co-occurrence of each chub species with either channel catfish or shovelnose sturgeon; they were also used to estimate the overall occupancy and co-occurrence of sicklefin and sturgeon chub more easily due to their simpler model structure. For all models, we used a multispecies occupancy modeling framework [23] within statistical software (R version 4.2.2; "occuMulti" function of "unmarked" package version 1.2.5 [39,40]). Given the structure of extant data, we followed the spatially replicated visit definition used by previous fishery research [20,37], not traditionally repeated visits [38].

The detection component of occupancy models was used to calculate a detection probability, that is, probability that a species is detected during a visit to a site where that species is present [38]. For the type of multispecies occupancy models we used, detection probabilities were calculated independently for each species, and these probabilities were modeled as functions of covariates [23]. Across all models and species, the detection component had trawling distance, depth, and velocity as covariates (Table 1). Trawling distance (i.e., level of effort) is an important factor in detecting most fish species using trawls, including focal chub species [3,13,41]. Additionally, depth and velocity can affect the efficiency of benthic trawls [42,43], but they were collected for only a fraction of visits (i.e., trawl deployments) at a site (i.e., macrohabitat). Consequently, site-level depth and velocity means across visits were used for the detection component of models. For the two-species models, which used MRBFS+PSPAP-FC data, we added a covariate for project because each monitoring program used different types of benthic trawls (i.e., BT versus OT16) and predicted detection probabilities for each program separately, using median continuous covariate values specific to each project. For the detection component of each three-species model, we report predicted occupancy probability for the median value of each continuous covariate. Detection probability estimates were made with 10,000 bootstrap samples to estimate 95% confidence intervals (default "predict" methods for the detection component of multispecies occupancy models in "unmarked" version 1.2.5 [23,40]).

Table 1. Variables included in detection, occupancy, and co-occurrence components of the two- and three-species occupancy models of benthic fishes sampled using benthic trawls in the Missouri River basin. Subheadings, in italics, are variable types, either *"Catergorical"* or *"Continuous."* * = single species occupancy component of model; ** = parameters were averaged across an entire site and not necessarily recorded for every trawl deployment. The number of categories for each categorical variable is listed in parentheses.

Model Type	Detection	Occupancy *	Co-Occurrence
Two-species	*Categorical*	*Categorical*	*Categorical*
	Project (2)	Sampling year (19)	(none)
		HUC 8 subbasin (12)	
		Macrohabitat (5)	
	Continuous	*Continuous*	*Continuous*
	Trawl distance	Water temp. **	Water temp. **
	Water depth **	Water depth **	Water depth **
	Water velocity **	Water velocity **	Water velocity **
		Turbidity	Turbidity
Three-species	*Continuous*	*Continuous*	*Continuous*
	Trawl distance	(none)	(none)
	Water depth **		
	Water velocity **		

For the two-species model, the occupancy component for individual species was modeled using the categorical factors of year, HUC 8 subbasin, and macrohabitat (Table 1). Year and subbasin provided means of assessing large-scale temporal and spatial patterns of occupancy, and macrohabitat is known to be related to the occurrence of our focal species in the Missouri River [13]. We assessed differences among levels of different categorical variables by comparing confidence interval overlap among coefficients; levels where 95% confidence intervals of coefficients did not overlap were considered significantly different. The categorical variable levels used as a reference condition were included in these comparisons as a value of 0. Continuous covariates used in the individual species occupancy component and co-occurrence components of the two-species models were four measures of water conditions: depth, velocity, turbidity, and temperature (Table 1). These variables were the only site-level environmental variables consistently collected by the PSPAP [3]; they were also among the environmental variables collected by the MRBFS [13]. Depth, velocity, turbidity, and temperature have all been shown to affect the presence or spawning of sicklefin chub, sturgeon chub, and similar species [13,15,44]. As temperature and depth are related to the co-occurrence patterns of other fish species [45]; we hypothesized that these factors—along with velocity and turbidity—could also affect the co-occurrence of focal chub species. The categorical factors of years, HUC 8 subbasins, and macrohabitats were not included in the co-occurrence component of these models because their use made it impossible to estimate uncertainty for all factor levels. This was due to limited information as a result of smaller sample sizes on observed co-occurrence for years, HUC 8 subbasins, and macrohabitats.

In the three-species models, we sought to describe differences in sicklefin or sturgeon chub occupancy given the presence or absence of each other, shovelnose sturgeon, or channel catfish. We also used these models to report an overall occupancy probability for each species independent of the other species. Because we were not interested in the factors underlying overall occupancy or co-occurrence for these purposes, we set both occupancy and co-occurrence components of the three-species models as intercept-only (Table 1). The co-occurrence components of the three-species models only included second-order interactions between species because we only wanted to compare pairs of species. To obtain overall occupancy probabilities, we calculated marginal occupancy probabilities. To calculate differences in sicklefin or sturgeon chub occupancy given the presence or absence of other chub species, shovelnose sturgeon, or channel catfish, we calculated conditional occupancy for each chub species in each model based on the conditions of presence and absence of one of each of the other species in the model. For marginal and conditional occupancy probabilities, we used 10,000 bootstrap samples to estimate 95% confidence intervals (default "predict" methods for occupancy models in "unmarked" version 1.2.5 [23,40]). For conditional occupancy probabilities, we checked for overlap of 95% confidence intervals between each pair of estimates, where a pair consisted of occupancy probability conditional on presence and occupancy probability conditional on absence for each possible combination of primary chub species (sicklefin or sturgeon chub), secondary species used as the condition (the other chub species, shovelnose sturgeon, or channel catfish), and model (shovelnose sturgeon or channel catfish). We compared predicted conditional occupancy probabilities instead of coefficients because, in preliminary analyses, we found that significant differences in co-occurrence coefficients did not always translate into differences in predicted occupancy.

Single-season occupancy models, including specific multispecies models we used, have five assumptions. Given the limitations imposed by the data sets and specific model outputs required by our research objectives, relaxing some of these assumptions was sometimes necessary. The assumptions are as follows: (1) closure: occupancy status at each site does not change over the sampling period; (2) occupancy probability is constant or modeled as a function of covariates; (3) there is no unmodeled heterogeneity in detection probability; (4) independence: detections of a species and detection histories are independent among sites and visits; and (5) identification: species are correctly identified such that there are no false positives [23,38].

To address assumption 1, closure, we filtered the PSPAP component of each data set to include only visits conducted ≤7 days apart at the same site. In cases where >7 days elapsed between samples, the first 7-day period with at least two samples was retained per site. These criteria resulted in the exclusion of 85 (~0.007%) possible visits from the PSPAP-FC data set. Multi-day sampling was rare and >98% of sites had all visits for each site conducted on the same day for each data set. We address assumption 2 in the two-species models by modeling occupancy probability as a function of habitat covariates. In the three-species model, assumption 2 is not explicitly addressed using habitat covariates, but we treat co-occurrence like a covariate for the purpose of this assumption. We address assumption 3 across all models by modeling detection using covariates known to affect trawl capture probability (i.e., trawl distance, water depth, water velocity; [42]). For assumption 4, we used multiple levels of data processing to maintain independence throughout data sets (Section 2.3). Though three main-channel macrohabitats of MRBFS (i.e., inside bend, outside bend, and channel crossover) were considered statistically non-independent in the stratified, random design of Wildhaber et al. [13], independence for the purpose of occupancy modeling was considered met given the size of each MR bend; this made sites and visits, i.e., trawl samples, spatially far enough apart to minimize potential for sampling of the same fish. Assumption 5 was addressed by MRBFS and PSPAP protocols requiring crews to be trained in fish identification and to send unknown specimens to experts [28].

To assess the predictive ability of occupancy models, we calculated the area under the receiver operating characteristic (ROC) curve (AUC; [46]) with statistical software ("roc" function of the "pROC" R package version 1.18.4, [47]). Predicted values for ROC curves were the products of (predicted cumulative detection probability across visits to each site) × (predicted occupancy probability at that site). These products were calculated at each site for each species included in a model, and each combination of species × site was used in the predicted data set for the ROC curve. Mean predictions for visit-level detection probabilities and site-level occupancy probabilities were calculated directly from data used from each specific model ("predict" methods for multispecies occupancy models in "unmarked" version 1.2.5 [39,40]). Cumulative detection probabilities were calculated as $1 - \prod_{i=1}^{n}(1 - p_i)$, where i was a visit, n was the total number of visits to a site, and p_i was the predicted detection probability for visit i. For the AUC analysis, known site-level occurrence of co-occurring species was used to inform occupancy predictions for each primary species, i.e., the species for which occupancy probability was being calculated. If a co-occurring species was detected at a site during at least one visit, the predicted occupancy of the primary species was calculated as conditional upon the presence of the detected species [23]. If a potentially co-occurring species was not detected, no prediction condition was applied for non-detected species.

In terms of interpreting AUC values, AUC > 0.5 indicates the model makes predictions better than random chance [46]. AUC = 1.0 means that all sites with (occupancy probability) × (detection probability) ≥ 0.5 had observations of a given species and that all sites with (occupancy probability) × (detection probability) < 0.5 had no observations of a given species [46]. One potential way of qualitatively assessing AUC can be found in Hosmer et al. [48].

2.4.2. Post Hoc Missouri River Discharge Analyses

Previous research indicates that the abundance and distribution of focal chub species in the MR may be linked to variability in annual river discharge [12]. We wanted to conduct exploratory analyses to further examine potential patterns of focal chub species site occupancy related to annual mean MR discharge across a time scale spanning multiple decades. Annual hourly mean MR discharge, in m^3/s, was calculated for each sampling year using data from USGS hydrological stations near Wolf Point, Montana (Site ID 06177000), and Hermann, Missouri (Site ID 06934500), representing the UMR and LMR, respectively [49]. Wolf Point had the only station between Fort Peck Reservoir and Lake Sakakawea with continuous discharge records during the study period, whereas Hermann was the downstream-most station on the MR with continuous discharge records during the study period. To calculate annual hourly mean discharge for each station, discharge was averaged hierarchically by hour, then day, and then sampling year, that is, November 1 of one calendar year to October 31 of the following year. For all analyses involving the UMR (i.e., Wolf Point station), sampling year 2011 was excluded because its annual hourly mean discharge of 798 m^3/s was an extreme outlier; the mean ± 1 standard deviation across 19 sampling years for Wolf Point station was 279 ± 150 m^3/s.

For the response variable of occupancy-related exploratory discharge analyses, we used yearly occupancy coefficients for each focal chub species from the two-species occupancy model. We were unable to include discharge as a covariate within occupancy models for several reasons. First, there were few stations along the MR that measured discharge for the duration of the study, making it difficult to assign discharge values to individual sites. Second, biological processes likely influenced by discharge, such as chub spawning and recruitment, generally happen on an annual scale [1,2]. Third, annual discharge is related to location along the river (i.e., HUC 8) and year, creating multicollinearity concerns. Fourth, because we averaged hourly discharge across entire years, there would be pseudo-replication of discharge observations within each year if incorporated into a site-scale occupancy model.

We conducted eight linear regression models for occupancy coefficients versus discharge. Models were separated by species (2; sicklefin or sturgeon chub), MR section (2, UMR or LMR), and relative year of discharge (2; current year or prior year). We analyzed prior-year discharge in addition to current-year discharge separately because both current sampling year and prior sampling year could result in possible lag effects of discharge that have been observed for previous fishes [50]. We only examined the current-year and prior-year discharge because most chubs in the Missouri River are less than 2 years old [1,2,51]. Linear regressions were conducted in statistical software ("lm" function of the "stats" package in R version 4.2.2 [39]).

2.4.3. Observed Co-Occurrence and Simulations

In addition to channel catfish and shovelnose sturgeon, we were also interested in focal chub species co-occurrence patterns related to uncommon secondary fish of interest (USFI), namely pallid sturgeon, walleye, and sauger. All three of these species were selected mainly due to their potential predation on focal chub species. Walleye and sauger are also important sportfishes [36], whereas pallid sturgeon is a federally listed, endangered species in the United States [35] that preys on focal chub and other small fish to reach adulthood and survive as adults [6,34,35,52–54]. In our comparisons, juvenile pallid sturgeon were considered as a separate USFI from adults because they are less piscivorous and instead as potential competitors for sicklefin and sturgeon chub [33]. The cutoff we used for juvenile pallid sturgeon was total length < 500 mm; pallid sturgeon ≥ 500 mm are almost exclusively piscivorous [6,34,35,52–54].

These larger USFI are generally mobile in the riverine environment and are likely to violate the closure assumptions of occupancy models [35,54,55]. Additionally, the implementation of the PSPAP resulted in few synchronized deployments of gear types that effectively detected both focal chub species and USFI. Because these issues precluded the use of occupancy modeling, we could not account for the imperfect detection of co-occurrence of focal chubs and USFI. Alternatively, we used Monte Carlo methods, detailed below, to compare proportional rates of observed co-occurrence to predicted rates of observed co-occurrence based on random chance. The primary purpose of these comparisons was to inform hypotheses to test in the future using a different sampling and analytical design.

The subset of data used for comparison between focal chub species and USFI detections started with only PSPAP-FC data, due to the unique presence of bends in that data set with some concurrent or nearly concurrent sampling using multiple types of actively pulled gear. OT16 data in PSPAP-FC were supplemented with captures from trammel nets 38.1 m in length (TN). Due to its larger size compared to the OT16 [27], the TN was able to detect USFI missed by OT16. Passive and bait-based gears types were not considered due to a lack of comparability to the actively pulled OT16. The mobility and seasonal migration of adult pallid sturgeon [35,52–54,56], walleye [55], and sauger [55] created multiple issues that required further consolidation and filtering of the TN-supplemented PSPAP-FC data set. Due to the potential for movement of USFI across macrohabitats within a river bend, detection/non-detection of focal chub species and USFI were consolidated to bend level. To capture coarse spatial variation in patterns of observed co-occurrence, this bend-level data set was portioned into three regions: (1) PSPAP-FC bends in the Upper Missouri River (UMR), i.e., upstream of Lake Sakakawea; (2) PSPAP-FC bends in the Lower Missouri River (LMR), i.e., downstream of Gavin's Point Dam; and (3) all bends in the filtered PSPAP-FC data set, i.e., in both the UMR and LMR.

To predict observed co-occurrence (OC) rates based on random chance, we calculated an expected probability of OC based on random chance as

$$\frac{C \times U}{\left(n_{bend}\right)^2} \tag{1}$$

where C was the number of bends where a given focal chub species was detected, U was the number of bends where a given USFI was detected, and n_{bend} was the region-specific (i.e., UMR, LMR, or combined UMR + LMR) total number of bends in the filtered PSPAP-FC data set. These expected probabilities of OC were calculated for each combination of focal chub species (2), USFI (4), and MR region (3) for 24 total test combinations. For each test combination, we conducted 100,000 Bernoulli-distribution-based Monte Carlo simulations using statistical software ("rbinom" function with size = 1 in the "stats" package of R version 4.2.2). For each simulation, the number of observations was the number of bends in a regional data set, and the probability of OC for each observation was the expected probability of OC based on random chance. The predicted OC rate for a simulation was calculated by dividing the number of bends with predicted OC by the number of bends in that simulation's region.

To compare the actual versus predicted OC rate for each test combination, we conducted unidirectional tests that calculated the proportion of simulated predictions either greater than or less than the actual OC rate. Greater than tests (prediction < actual) were used if the actual OC rate was less than the expected probability of OC for a test combination; otherwise, a less than test (prediction > actual) was used. Following the language of [57], we note proportions of simulated predictions meeting test criteria (q) at thresholds of $q \leq 0.10$, $q \leq 0.05$, $q \leq 0.01$, and $q \leq 0.001$ as weak evidence, moderate evidence, strong evidence, and very strong evidence of non-random actual OC, respectively.

3. Results
3.1. Two-Species Occupancy Models
3.1.1. Occupancy and Co-Occurrence

In the fully parameterized two-species model, occupancy probabilities varied between years and species, reflecting both increasing and decreasing chub spatial distribution and, possibly, abundance (Figure 2). For sicklefin chub, 2003 and 2005 had higher occupancy coefficient estimates than at least two other years. In addition, 1996, 2008, and 2011 had estimates lower than at least two other years (Figure 2). For sturgeon chub, 1996, 2005, 2006, 2012, and 2013 had higher occupancy coefficient estimates than at least two other years; 2016 and 2018 had lower estimates than at least two other years (Figure 2).

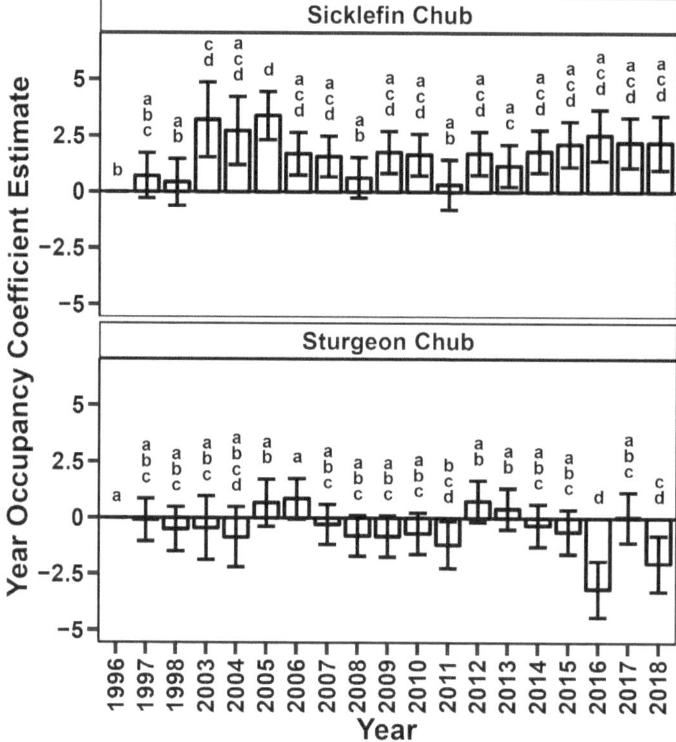

Figure 2. Mean occupancy coefficient estimates with 95% confidence intervals of year from multispecies occupancy model for sicklefin (*Macrhybopsis meeki*) (**top**) and sturgeon (*M. gelida*) (**bottom**) chub. Significant differences between means are indicated by compact letter display groups above each bar. All data were collected with benthic trawls. Year 1996 was used as the reference year (i.e., set to zero); however, note that each species has a separate occupancy intercept, so plots of different species cannot be directly compared.

For sicklefin chub HUC 8 occupancy coefficients, the UMR HUC 8 subbasins between the Redwater River confluence and Lake Sakakawea (2237 to 2705 km) were greater than the UMR HUC 8 just below Fort Peck Dam (river km 2705 to 2850), the reference condition (i.e., set to zero). The occupancy coefficient for the HUC 8 containing Gavins Point Dam was not different from that of the reference HUC 8; occupancy was similar for the four HUC 8 subbasins starting just upstream of Gavins Point Dam and ending 86 km downstream of the Platte River confluence (872 to 1416 km; Figures 1 and 3). Occupancy for the five HUC 8 subbasins from river km 0 to 872 was higher than the reference and the HUC 8 containing Gavins Point Dam.

Sturgeon chub HUC 8 subbasin occupancy patterns had similar overall trends as sicklefin chub such as relative decreases below hydro-electric dams, but patterns were less clearly defined (Figure 3). Sturgeon chub occupancy coefficients were (i) greater than the reference HUC 8 just below Fort Peck Dam (2705 to 2850 km) in both UMR HUC 8 subbasins between the Redwater River confluence and Lake Sakakawea (2237 to 2705 km) but (ii) less than the reference for LMR HUC 8 subbasins containing and just below Gavins Point Dam (1183 to 1416 km; Figures 1 and 3). The UMR reference HUC 8 subbasin immediately below Fort Peck Dam (2705 to 2850 km) was similar to several of the LMR HUC 8 subbasins downstream of those containing and just below Gavins Point Dam (Figure 3).

Sicklefin chub occupancy coefficients were significantly greater for inside bend macrohabitat compared to channel crossover, outside bend, and secondary connected channel macrohabitats, with the latter having the lowest mean occupancy coefficient (Figure 4). Sturgeon chub had a higher occupancy coefficient for inside bend and secondary connected channel compared to channel crossover macrohabitats (Figure 4).

Focal chub occupancy differed with environmental variables. Sicklefin chub occupancy was not significantly related to any water conditions included in the two-species model; coefficients were -0.066 (-0.260, 0.132) for depth, -0.055 (-0.856, 0.745) for water velocity, 0.001 (0.000, 0.002) for turbidity, and -0.039 (-0.112, 0.034) for water temperature. In contrast, sturgeon chub occupancy had a marginal negative relationship to depth -0.244 (-0.489, 0.001), was positively related to water velocity 0.883 (0.046, 1.720), was not significantly related to turbidity 0.000 (-0.002, 0.002), and was negatively related to water temperature -0.088, (-0.158, -0.018).

In the two-species model, the mean co-occurrence intercept was positive with a 95% confidence interval excluding 0. In the two-species model, there was no statistically significant relationship between environmental covariates assessed and the co-occurrence of sicklefin and sturgeon chub; mean coefficient estimates with 95% confidence intervals were 0.138 (-0.191, 0.467) for depth, -0.173 (-1.342, 0.997) for water velocity, -0.0008 (-0.003, 0.002) for turbidity, and -0.036 (-0.127, 0.054) for water temperature.

3.1.2. Model Performance and Detection

The AUC for the two-species model was 0.862, corresponding to what Hosmer et al. [48] calls "excellent discrimination." In species-specific detection components of the two-species model, the detection probability of sicklefin chub increased with trawl distance and water depth (Figure 5). The detection probability of sturgeon chub increased with trawl distance, decreased with increasing depth, and increased with velocity (Figure 5). Our expected finding of increased detection probability with increased trawl distance (effort) for both focal chub species, is comparable to results for fish species of similar size and habitats in rivers of Iowa, USA [41]. For the categorical effect of project, detection coefficients, as mean (95% confidence interval), for the PSPAP project were -0.382 (-0.164, -0.600) for sicklefin chub and -0.592 (-0.372, -0.812) for sturgeon chub relative to the MRBFS observations. Median covariate values used to predict detection probability were 150 m trawl distance, 2.50 m water depth, and 0.646 m/s water velocity for MRBFS data; for PSPAP data, they were 161 m trawl distance, 2.05 m water depth, and 0.381 m/s water velocity. Using the two-species model and its coefficients, mean estimates of visit-level detection probabilities based on these median values were 0.38 (0.36, 0.41) for sicklefin chub in the PSPAP and 0.48 (0.43, 0.52) in the MRBFS, while 0.26 (0.23, 0.29) and 0.38 (0.33, 0.42) for sturgeon chub, respectively. Regarding environmental effects on detection probability, previous occupancy modeling of the focal chubs found large-scale spatial heterogeneity in detection probability for the Lower MR [20]; our results regarding water conditions complement those findings by informing the effects of local water conditions on detection probability. The positive relationship of depth with detection probability for sicklefin chub differs from previously observed detection/non-detection data for this species, where there was not a significant relationship between the number of zero catches and depth [3]. For sturgeon chub, the negative relationship of depth and positive relationship of velocity with detection may signal abundance effects, that is, more sturgeon chub at shallow and higher velocity sites. In the current study, sturgeon chub occupancy probability was marginally negatively related to depth and positively related to velocity. As occupancy can be correlated with abundance [21], this supports previous research showing non-zero catches and abundance of sturgeon chub being negatively correlated with depth [3]. Though less clear than the relationships between detection and trawl distance, these environmental effects on detection still provide new insight into water conditions related to the capture of these chub species.

3.2. Three-Species Occupancy Models

3.2.1. Occupancy and Co-Occurrence

In both channel catfish and shovelnose sturgeon models, mean predicted marginal occupancy probabilities (with 95% confidence intervals) for individual chub species at the macrohabitat scale—across the entire study area and time period—were 0.42 (0.40, 0.45) for sicklefin chub and 0.41 (0.39, 0.44) for sturgeon chub.

For the three-species models, co-occurrence intercepts for sicklefin chub × sturgeon chub were positive with 95% confidence intervals excluding zero [58]. Predicted marginal occupancy probabilities were significantly higher for sicklefin chub when sturgeon chub were present, and vice versa (Figure 6). Predicted occupancy probability was, on average, 2.8× higher for sicklefin chub in the presence of sturgeon chub, compared to occupancy probability in the absence of sturgeon chub; it was, on average, 2.9× higher for sturgeon chub in the presence of sicklefin chub.

Sicklefin chub and sturgeon chub had higher occupancy probabilities when channel catfish and shovelnose sturgeon were present (Figure 6). The sicklefin chub occupancy probability was, on average, 3.5× higher in the presence of channel catfish as compared to their absence; this rate was 1.7× higher for sturgeon chub. The sicklefin chub occupancy probability, was, on average, 3.1× higher in the presence of shovelnose sturgeon as compared to their absence; this rate was 1.4× higher for sturgeon chub.

3.2.2. AUC and Detection

The AUC was 0.757 for the three-species model with channel catfish and 0.769 with shovelnose sturgeon, a decrease of 0.10 to 0.105 in AUC from the two-species models; these values correspond to what Hosmer et al. [48] call "acceptable discrimination." The detection probability of channel catfish increased with water depth and decreased with water velocity (Figure 6). The detection probability of shovelnose sturgeon increased with both trawl distance and water depth (Figure 6).

For the three-species models, median covariate values used to predict detection probability were 159 m for trawl distance, 2.06 m for water depth, and 0.372 m/s for water velocity. Mean estimates of visit-level detection probabilities (with 95% confidence intervals) using median detection covariate values were 0.36 (0.34, 0.39) for sicklefin chub and 0.21 (0.19, 0.23) for sturgeon chub in both the channel catfish and shovelnose sturgeon models. This pattern of detection is similar to that found by [20], including higher detection probabilities for sicklefin chub than sturgeon chub in the MR. Channel catfish model detection probability was 0.45 (0.41, 0.48) for channel catfish. Shovelnose sturgeon model detection probabilities were 0.37 (0.34, 0.41) for shovelnose sturgeon.

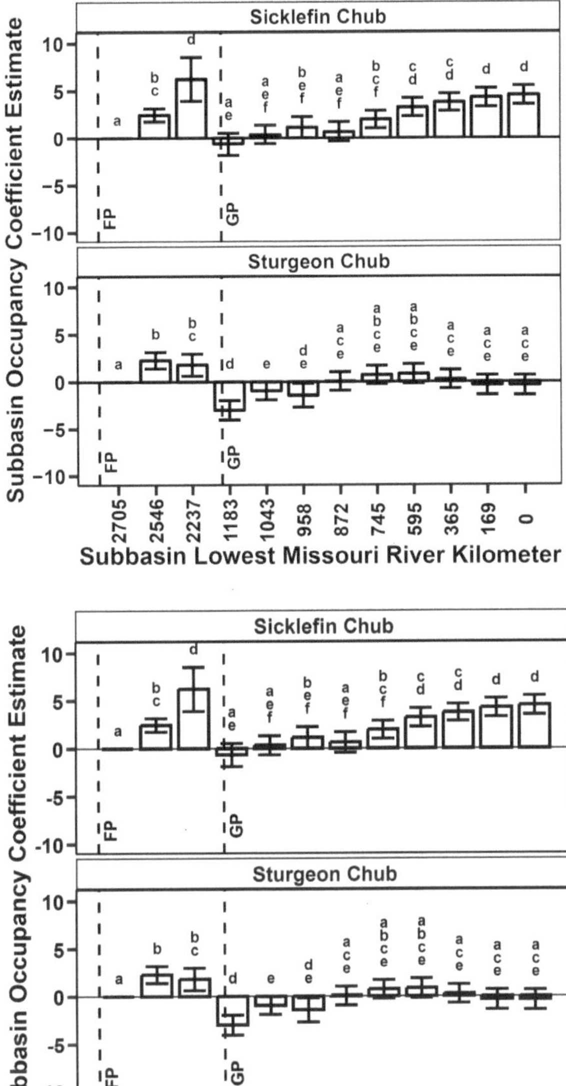

Figure 3. Mean occupancy coefficient estimates from multispecies occupancy model by National Hydrologic Database 8-digit Hydrologic Unit Code (HUC 8) subbassins for sicklefin (*Macrhybopsis meeki*) (**top**) and sturgeon (*M. gelida*) (**bottom**) chub, defined here by lowest HUC 8 Missouri River kilometer (MR km). All data were collected with benthic trawls. Significant differences between means are indicated by compact letter display groups above each bar. Vertical dashed lines represent locations of major dams. Fort Peck Dam (FP) is located at MR km 2850; Gavins Point Dam (GP) is located at MR km 1305 as indicated by vertical dashed lines. The HUC 8 corresponding to MR km 2705 was used as the reference condition (i.e., set to zero); however, note that each species has a separate occupancy intercept, so plots of different species cannot be directly compared.

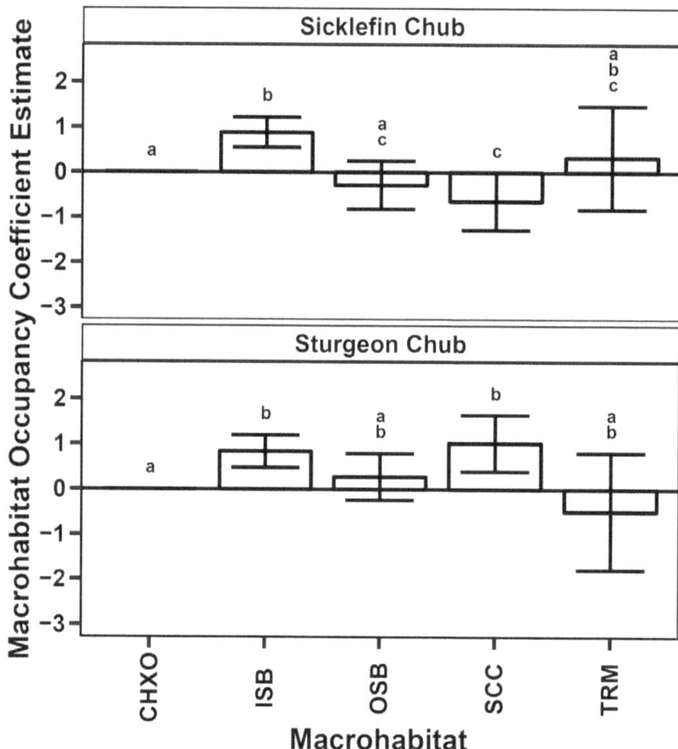

Figure 4. Mean occupancy coefficient estimate from multispecies occupancy model by macrohabitat type (channel crossover (CHXO), inside bend (ISB), outside bend (OSB), secondary connected channel (SCC), and tributary mouth (TRM)) for sicklefin (*Macrhybopsis meeki*) (**top**) and sturgeon (*M. gelida*) (**bottom**) chub. Significant differences between means are indicated by compact letter display groups above each bar. All data were collected with benthic trawls. CHXO was used as the reference condition (i.e., set to zero); however, note that each species has a separate occupancy intercept, so plots of different species cannot be directly compared.

3.3. Missouri River Discharge Relationships

When comparing annual discharge to year coefficients for occupancy, we found negative relationships between current-year discharge and yearly occupancy coefficients for sicklefin chub in both the UMR and LMR; there was also a negative relationship between sicklefin chub occupancy coefficients and LMR's previous-year discharge (Figure 7). There was no evidence of a relationship between the yearly occupancy coefficient and UMR discharge for the prior year for sicklefin chub (Figure 7). For sturgeon chub, there was marginal evidence for a negative relationship between current-year discharge in the LMR and yearly occupancy coefficients (Figure 7). There was no other evidence for relationships between mean annual discharge and sturgeon chub annual occupancy coefficients (Figure 7).

3.4. Observed Co-Occurrence and Simulations

Prior to additional filtering of the PSPAP-FC data set to account for the inclusion of trammel net (TN) captures, there were 474 PSPAP-FC river bends in the UMR, 1164 in the LMR, and 1638 bends in the entire MR. Filtering of PSPAP-FC bends to include only bends where TN sampling took place within the same week as OT16 sampling reduced the number of bends to 252 (53% of original bends) for the UMR, 201 (17%) for the LMR, and 453 (28%) for the entire MR.

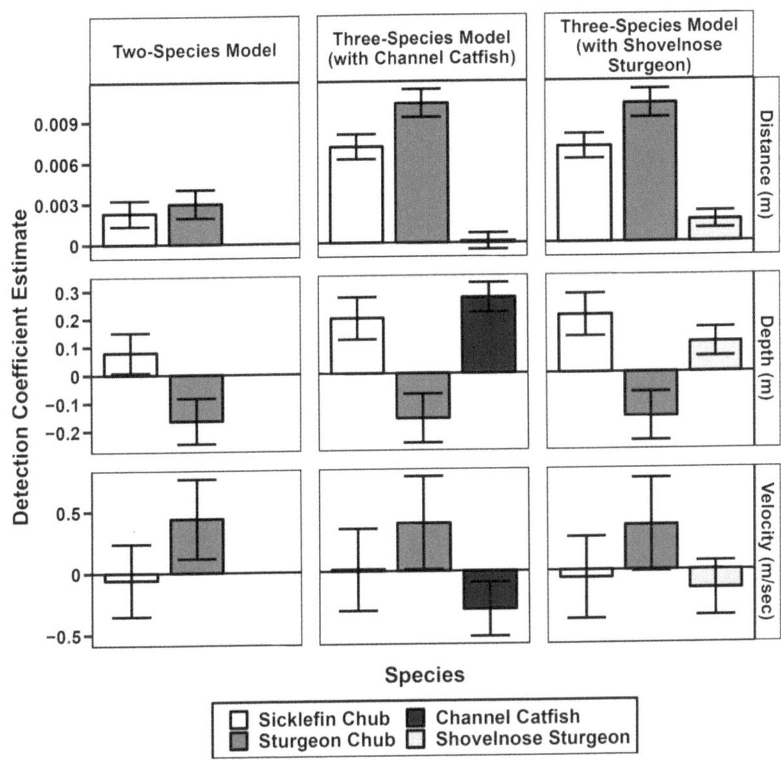

Figure 5. Mean detection coefficients with 95% confidence intervals of three continuous covariates for four Missouri River Basin benthic fish species across three multispecies occupancy models. Coefficients are on a logistic scale. For column headings, all models include sicklefin chub (*Macrhybopsis meeki*) and sturgeon chub (*M. gelida*); three-species models also include species in parentheses. All data were collected with benthic trawls.

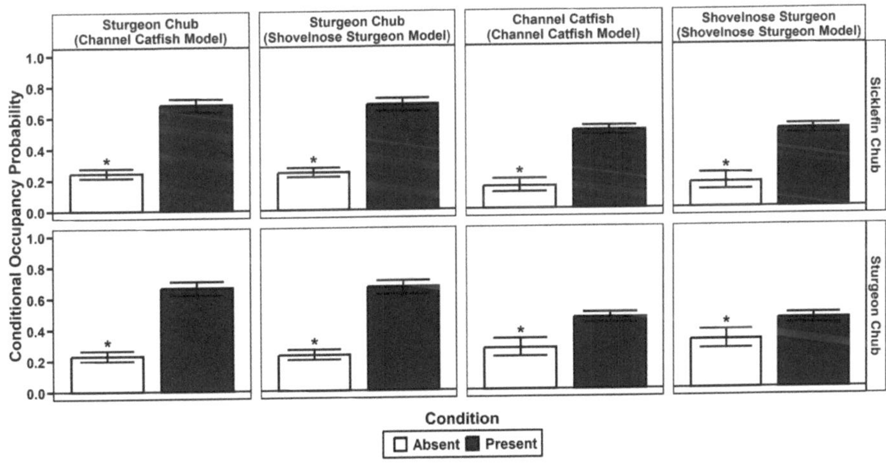

Figure 6. Conditional occupancy probability for sicklefin (*Macrhybopsis meeki*) (**top**) and sturgeon (*M. gelida*) (**bottom**) chub in three species models. All data were collected with benthic trawls. Significant differences between means are indicated by an asterisk (*) above the absent condition bar.

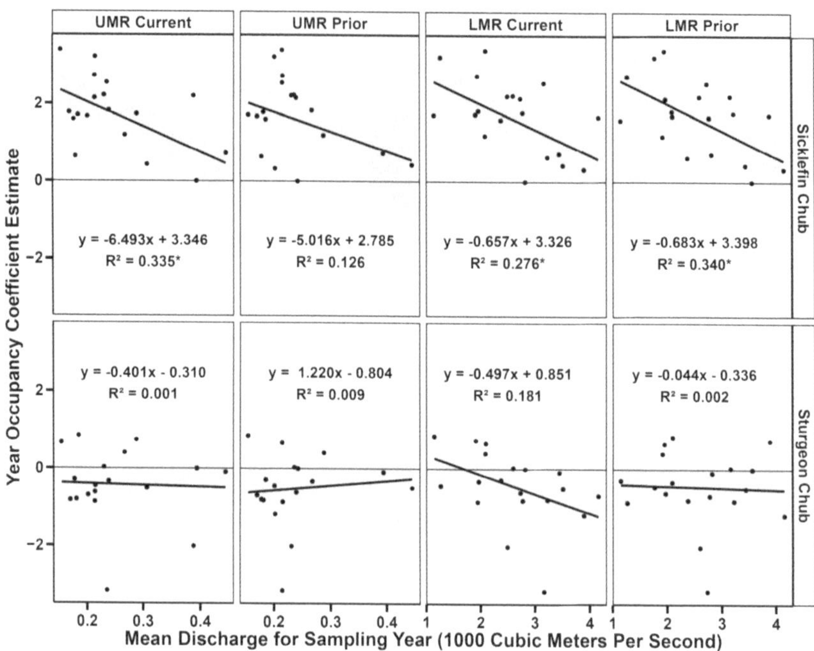

Figure 7. Linear regressions of annual mean discharge in Upper (UMR) and Lower Missouri River (LMR) and mean yearly occupancy coefficient estimates for sicklefin (*Macrhybopsis meeki*) and sturgeon (*M. gelida*) chub. "Current" represents discharge data from the same year as the occupancy coefficient. "Prior" represents discharge data from the year prior to the occupancy coefficient. Shaded area represents a 95% confidence interval.

Co-occurrence rates with piscivorous fishes differed between the chub species. The observed co-occurrence of sicklefin chub and adult pallid sturgeon was 44% higher than expected over the entire MR, 62% higher in the UMR, and 24% higher in the LMR (Figure 8). For sicklefin chub and juvenile pallid sturgeon, observed co-occurrence was 2% less than expected for the entire river and 66% less and 10% higher for the LMR and UMR, respectively. For sicklefin chub and sauger, observed co-occurrence was 2% lower than expected throughout the entire river, 23% lower for the LMR, and 7% higher for the UMR. For sicklefin chub and walleye, observed co-occurrence was 24% lower than expected for the entire river, and 41% higher than expected for the UMR. In the LMR, there was no observed co-occurrence between walleye and sicklefin chub (100% less than expected).

The observed co-occurrence of sturgeon chub and adult pallid sturgeon was 16% higher than expected for the entire river, 3% higher than expected for the LMR, and 20% higher than expected for the UMR (Figure 8). Observed co-occurrence between sturgeon chub and juvenile pallid sturgeon was 30% higher than expected for the entire river, 10% higher than expected for the UMR, and 68% lower than expected for the LMR. For sturgeon chub and sauger, observed co-occurrence was 24% higher than expected for the entire river, 1% higher than expected for the UMR, and 3% higher than expected for the LMR. Observed co-occurrence between sturgeon chub and walleye was 15% lower than expected for the entire river, 85% lower than expected for the LMR, and 4% higher than expected for the UMR.

For adult pallid sturgeon, there was weak evidence for higher-than-random observed co-occurrence (OC) with sicklefin chub in the UMR, and when the UMR and LMR were lumped (Figure 8). There was no evidence of different-from-random OC of adult pallid sturgeon and sturgeon chub for any MR region (Figure 8). Actual OC rates for adult pallid sturgeon and individual focal chub species were 0.04 and 0.01–0.03 across chub species in the UMR and LMR, respectively.

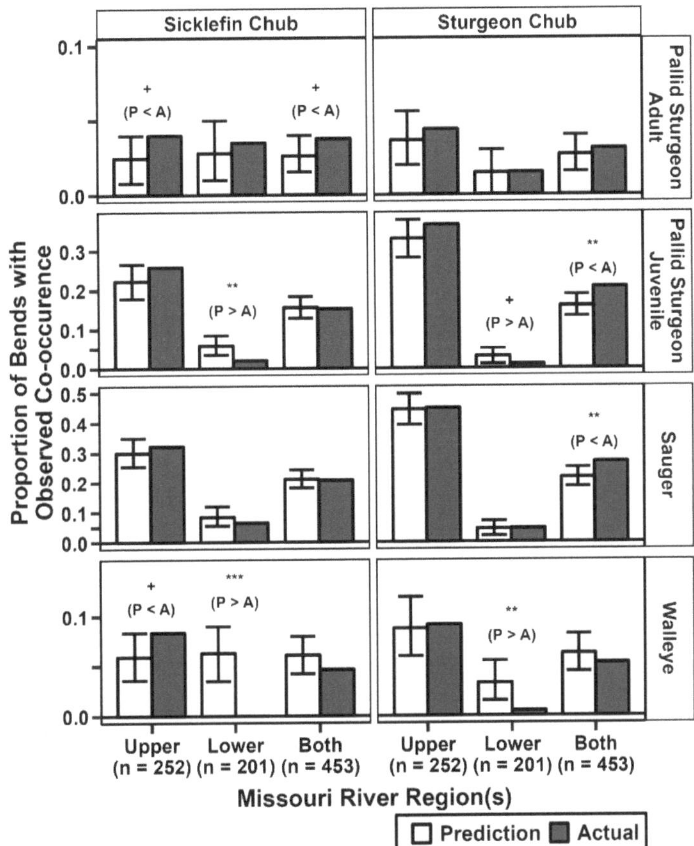

Figure 8. Comparisons of proportions of bends with known and predicted observed co-occurrence of sicklefin (*Macrhybopsis meeki*) and sturgeon (*M. gelida*) chub with pallid sturgeon (*Scaphirhynchus albus*) adults, pallid sturgeon juveniles, sauger (*Sander canadensis*), and walleye (*Sander vitreus*) as detected by otter trawl and trammel net. Labels on x-axes indicate if sampling occurred in the Upper Missouri River ("Upper"), Lower Missouri River ("Lower"), or a combination of the two ("Both"). Numbers in parentheses are bends sampled by trammel nets within a week of an otter trawl sample. Predictions were calculated based on the assumption of random chance of observed co-occurrence using 100,000 Bernoulli-distribution-based Monte Carlo simulations; error bars are 0.05 and 0.95 quantiles. Actual observed co-occurrences are directly from field sampling data. Proportion of simulations where prediction > actual (P > A) or proportion of simulations where prediction < actual (P < A). Symbols above equations indicate proportion of simulations meeting test criteria; no symbol or equation is >0.10, + is ≤0.10 but >0.05, ** is ≤0.01, and *** is ≤0.001.

For juvenile pallid sturgeon, there was strong evidence for lower-than-random OC with sicklefin chub in the LMR only (Figure 8). For juvenile pallid sturgeon and sturgeon chub, there was weak evidence of lower-than-random OC in the LMR and strong evidence for higher-than-random OC when the UMR and LMR were lumped (Figure 8). Actual OC rates for juvenile pallid sturgeon and individual focal chub species were 0.22–0.33 and 0.01–0.02 across chub species in the UMR and LMR, respectively.

There was no evidence of different-from-random OC of sauger and sicklefin chub for any MR region (Figure 8). For sauger and sturgeon chub, there was strong evidence of higher-than-random OC only when the UMR and LMR were lumped (Figure 8). Actual OC rates for sauger and individual focal chub species were 0.32–0.45 and 0.04–0.06 across chub species in the UMR and LMR, respectively.

For walleye and sicklefin chub, there was weak evidence for higher-than-random OC in the UMR and very strong evidence of lower-than-random OC in the LMR (Figure 8). For walleye and sturgeon chub, there was strong evidence of lower-than-random OC in the LMR (Figure 8). Actual OC rates for walleye and individual focal chub species were 0.08–0.09 and <0.01 across chub species in the UMR and LMR, respectively.

4. Discussion

This study examined the occupancy of sicklefin and sturgeon chub within the Missouri River. Overall, they appear to differ somewhat in the habitats they occupy and have shown differing occupancy patterns over time. For example, sicklefin chub occupancy was related to annual discharge; unlike sturgeon chub, sicklefin chub occupancy was higher at the inside bend habitat and lower at the secondary connected channel habitat, whereas sturgeon chub occupancy was equally high in both habitats. Lastly, sturgeon chub occupancy stayed fairly constant over the 23-year study period, with a general downward trend at the end of that period. Sicklefin chub had many more fluctuations in occupancy that, though partially recovering, never returned to its highest occupancy levels in the early 2000s. Even though this study found some major differences between sicklefin and sturgeon chubs, there were also general commonalities. When averaged across all sampling years and sites within the study area, the mean occupancy probabilities of sicklefin and sturgeon chubs were almost identical. Both chub species had lower occupancy rates below dams. Co-occurrence was strong for the two chubs with each other, channel catfish, and shovelnose sturgeon. The observed co-occurrence of chubs and pallid sturgeon was potentially higher than expected for adult sturgeon; for juvenile pallid sturgeon, it was lower than expected in the Lower MR and likely higher than expected in the Upper MR. Co-occurrence with the two chub species was mostly less than expected with walleye and sauger. As with many phenotypically similar species, habitats occupied by sicklefin and sturgeon chubs share some features, but the two species show differing historical patterns and habitat associations, suggesting individual species management may be needed.

4.1. Occupancy Patterns

From a temporal perspective, sicklefin chub occupancy tended to be higher in the early to mid-2000s and mid-to-late 2010s. For sturgeon chub, there was much less variation, aside from a potential decline in the mid-to-late 2010s. Due to the sampling design of the PSPAP, where the same sites were not sampled every year, these changes in occupancy coefficients represent changes in the overall proportion of occupied sites per year rather than the extirpation or colonization of specific sites and, potentially, changes in population size [25]. This potential decline in occupancy matches with results reported by [15], based on sampling 60 km of the MR just above the confluence with the Yellowstone River (YR) downstream toward Lake Sakakawea, which found the catch rate of sturgeon chub in the UMR both from age 0 to 1 and over 1 year old was negatively correlated with passing years from 2004 to 2016. We found sicklefin chub annual occupancy coefficients were more variable and showed a pattern of higher occupancy rates around 2003–2005 and 2014–2016 for sicklefin chub, compared to other years. This result is also supported by [15], which found no significant correlation between catch rate and year for sicklefin chub in the UMR but noted lower catch rates in 2010 and 2012. Data from a longer time period, as well as information on potential population drivers such as food availability and patterns from other small fish species, would increase temporal predictive ability for the species. Previous research based on PSPAP data [3] documented a pattern of decline in UMR sicklefin chub relative abundance starting in 2003 through 2010, which coincided with increased pallid sturgeon stocking in the UMR [7]. This pattern of decreased sicklefin chub relative abundance with increased stocking of pallid sturgeon, along with one for sturgeon chub, in the UMR was also demonstrated by [15]. In our analysis, this pattern is seen in our annual occupancy coefficients for sicklefin chub, with lower occupancy in the late 2000s into the early 2010s followed by what appears to be a partial recovery in the mid-to-late 2010s despite continued stocking of pallid sturgeon.

If we consider occupancy patterns as a potential index of population size, the current study results suggest there was an increase in MR sicklefin chub populations from 2011 to 2014. This increased occupancy appears to be somewhat less than the levels observed in the early 2000s and was followed by lower occupancy levels from 2015 to 2018. Our results suggest that sturgeon chub populations have been at relatively constant levels, similar to those identified as a concern by USFWS in 2000, but declining more recently.

From a large-scale spatial perspective, aside from potential dam effects, occupancy between the UMR and LMR was often similar for sicklefin chub, supporting the results of [3,13]. Spatial occupancy patterns for sturgeon chub relative to dam locations were less consistent, but there was a general pattern of higher occupancy coefficients in the UMR compared to the reference HUC 8 just below Fort Peck Dam. Based on our analyses, sturgeon chub occupancy appears to have been stable in the LMR at least since the late 1990s until the late 2010s, when occupancy began to drop. These patterns follow [3,13], where many more sturgeon chub were collected in the UMR and YR compared to the LMR. One difference between areas upstream and downstream of Gavins Point Dam is the increased channelization of the river segments downstream of the dam [59], a potential driver of sturgeon chub occupancy rates that could be a subject for future research. Sturgeon chub were also never detected in the Kansas River, despite being reported as commonly caught there prior to 1953 [60].

Both focal chub species had relatively high occupancy coefficients in the MR away from major mainstem MR dams compared to HUC 8 subbasins immediately below Fort Peck and Gavins Point Dams, providing evidence for an apparent effect of dams on occupancy. Relatively low occupancy downstream of Gavins Point Dam continued for $\approx$438 river km after the dam for sicklefin chub and $\approx$311 river km for sturgeon chub. Relatively low occupancy downstream of Fort Peck Dam continued for $\approx$159 river km for both sicklefin and sturgeon chub. Dieterman and Galat [61] found that sturgeon chub populations were vulnerable to habitat fragmentation and needed $\approx$300 continuous river kilometers for the persistence of metapopulations, so low occupancy rates in the LMR could represent local extinction of populations cut off from a population source. Our results regarding dams are reflected by prior relative-abundance-based analyses of the MRBFS and PSPAP data sets. For the MRBFS, capture rates of both chubs in river segments immediately downstream of Fort Peck and Gavins Point Dams were too low to effectively model relative abundance when that data set was considered by itself [13]. Like the occupancy coefficients in current analyses, Wildhaber et al. [3] used PSPAP data from 2003 to 2010 to find a lower relative abundance of focal chubs immediately downstream of both dams. From 1996 to 1998 during the MRBFS, both chub species were collected, primarily, in river segments above Fort Peck Lake (MR segments above Fort Peck Lake were not sampled during PSPAP) and between Fort Peck Dam and Lake Sakakawea [13]; collection numbers were too low for both species just below Fort Peck Dam to model. Therefore, the MRBFS results presented by [13] are similar to our occupancy results in that sicklefin chub numbers tended to be higher in the LMR than for sturgeon chub whose numbers were much higher in the UMR, whereas both species had low numbers just below Fort Peck and Gavins Point dams. Again, one possible reason for the lower numbers of sturgeon chub in the LMR may be due to its channelization making it a much deeper, faster river than it was historically [59]. The pattern of lower occupancy or relative abundance below dams for these chub species has been described elsewhere [1,2,61] for sicklefin chub, with authors predicting the highest presence for sicklefin chub at distances exceeding 301 km below an impoundment. For the MRBFS, this same pattern of limited chub collections below LMR dams is supported by the fact that the only LMR segments where chub species were collected enough for occupancy modeling were the last two segments from below the

Grand River to the Mississippi River confluence for sicklefin chub [12]. This observation is supported by the fact that previous research on the presence/absence of sicklefin in the MR and YR was possible only at the segment scale [61], similar in scale to the HUC 8 component of our model. Additionally, resampling by USFWS in 1994 of historic main channel and secondary connected channel sites sampled by [62] on the MR, limited to Missouri, collected only 93 sicklefin and 26 sturgeon chub above the Grand River based on 38 and 25 ft seine deployments covering a total of 14,120.6 m^2 [63], with neither species collected in 1997 at two of the same sites [64].

This study, as with previous research [1–3,13,65], described sicklefin and sturgeon chub as being found in main-channel, moderate-to-high-flow macrohabitats. Sicklefin chub have, at least in the past several decades, been found primarily in the mainstem MR, whereas sturgeon chub have been found both in the MR main channel and its tributaries [1–3,13,65]. In this study, inside-bend, main-channel macrohabitat tended to have relatively high occupancy coefficients for both species. Interestingly, secondary connected channel and inside bend had similar occupancy coefficients for sturgeon chub, whereas secondary connected channel was less occupied by sicklefin chub compared to other macrohabitats. This divergent pattern of secondary connected channel habitat occupancy may provide niche partitioning between sicklefin and sturgeon chub. Current analyses show similar results as the MRBFS [13] in that main and secondary connected channels tended to have the highest occupancy for sturgeon chub. During the PSPAP sampling, macrohabitats were sampled within a bend, so original comparisons possible among macrohabitats were nested within a bend [3]. Macrohabitat usage patterns found previously [3] indicate that sturgeon chub tended to be found in bends with additional macrohabitat types beyond the main channel (i.e., inside bends, outside bends, and channel crossover). However, the relative abundance of sicklefin and sturgeon chub where found tended also to be higher in bends with secondary connected channels, without additional macrohabitats in the UMR, and lower for sicklefin chub in the LMR [3]. Within a bend type, which particular flowing macrohabitats present tended to have the highest relative abundance varied between species and between the UMR and LMR [3]. These differences in macrohabitat usage between species suggest that sicklefin chub are more frequently found in main-channel macrohabitats than sturgeon chub. In the laboratory [33], further potential evidence for niche partitioning among our study species comes from the fact that sicklefin chub selection of mud or sand habitat was unaffected by the presence of food, whereas juvenile pallid, shovelnose, and pallid/shovelnose hybrid sturgeon selected habitat with food. The selection for mud over sand was also greater for juvenile pallid sturgeon compared to sicklefin and shoal chub (*Macrhybopsis hyostoma*) and pallid/shovelnose hybrid sturgeon [33].

Water conditions can be related to the occupancy coefficients of sturgeon chub but not sicklefin chub. In this study, sturgeon chub were found more often in cooler, faster water. Previous research using various subcomponents of data that we also used provides a mixed picture for these two chub species, with studies differing in what water conditions are considered best for their presence or abundance. Because the focus of our study is general patterns across much of the range of these species, we only report patterns from other literature as opposed to specific values, as these values may not be applicable to the scale of our study. Sicklefin chub have been found more often in cooler water (MRBFS and PSPAP, [3,13]), but work MR 8 km upstream and 48.3 km downstream of the confluence with YR and 24 km of the lower YR [66] found them in warmer waters. Wildhaber et al. [3] found that sturgeon chub occurred most often in cooler water in the UMR. Sampling the same area as [59], with the addition of an inter-reservoir segment below Lake Sakakawea, sicklefin chub were reported to occur more often in slower water [44] where [66] reported faster water use; similar patterns have been reported for sturgeon chub [44,66]. The MR inter-reservoir segments lack water conditions conducive to chub survival, as we and [44] found neither chub in the inter-reservoir segments below Lake Sakakawea. Sicklefin and sturgeon chub were found to occur more often in deeper water in the UMR [3], but sturgeon chub were found to occur more often in shallower water in the LMR [3]. When analyzing the presence/absence of sicklefin chub with river segment (i.e., 48.3 to 191.6 km; similar in size to an HUC 8) as the site, sicklefin chub were found to occur more often in turbid waters [44]. Flow constancy and the percentage of annual flow occurring in August [61] also seem to affect sturgeon chub, which occur more often in faster water in the UMR and LMR [3]. Sturgeon chub occur more often in either clearer [3,66] or turbid waters [44,61].

Further research is needed to help clarify the relationship of sicklefin and sturgeon chub occupancy to water conditions. Whereas previous researchers came to some different conclusions compared to ours and others, the spatial scale model of their studies [61] and their spatial extents [15,44,66] differ greatly from ours and cover shorter time spans. Our model included both large and small spatial scale components examined across the previous studies. For example, Dieterman and Galat [61] may have included some additional MR segments above Fort Peck Dam and segments above the Grand River that were not included in our study, but their analyses were done at the scale of segment, similar to our HUC 8 subbasins, which account for the geographic differences along the MR. In our case, by including the HUC 8 subbasins in our model to account for geographic differences, any additional significance of temperature, depth, velocity, and turbidity in the model is after accounting for the geographical differences and represent range-wide, general patterns, so our analyses provide a wider-spanning picture as to temporal and spatial, long-term patterns of occupancy for sicklefin and sturgeon chub in the MR than previously available. Given the strong predictive ability of the occupancy component of the two-species occupancy model, as demonstrated by a high AUC value of 0.862, our broad-scale patterns related to water conditions could provide valuable information needed to make management decisions related to sicklefin and sturgeon chub, even at the local scale of macrohabitat.

4.2. Possible Discharge Patterns

Occupancy coefficient patterns relative to annual discharge suggest that discharge could be a contributing factor to changes seen in sicklefin chub over time. The annual pattern of occupancy coefficients for sicklefin chub appears to be negatively related to prior- and current-year discharge in the LMR and current-year discharge in the UMR. These results suggest higher annual discharges may be detrimental to sicklefin chub populations in the LMR. It is possible that this negative relationship is due to either the reduction in already reduced shallow spawning habitats or the loss of sicklefin chubs from the system by being swept into the Mississippi River by higher discharge in the more channelized LMR [1,33]. Relationships to river discharge have been shown for spawning in other cypriniform fishes such as grass carp (*Ctenopharyngodon idella*; [67] and references therein) and blue suckers (*Cycleptus elongatus*, [68]) and recruitment for other small, benthic fish species such as the Neosho madtom (*Noturus placidus*, [50]). Our results provide some additional factors that may result in niche partitioning between these two chub species, but determining this requires further investigation. Notably, our results suggest that sicklefin chub recruitment is related to discharge rate, with the greatest site occupancy rate increases in years with lower discharge; this follows the previously mentioned hypothesis that sicklefin chubs may be included with other fish species where recruitment benefits from low water discharge years [61]. Further research into the relationship between discharge and sicklefin and sturgeon chub population changes could help to determine if regulating discharge could be a possible management tool for maintaining and potentially increasing their populations and potentially improving prey resources for pallid sturgeon.

4.3. Co-Occurrence across Analyses

The co-occurrence of sicklefin and sturgeon chub with shovelnose sturgeon and channel catfish provides potential insight for species relationships. Sicklefin and sturgeon chub had a high rate of co-occurrence, as demonstrated from predictions derived from three-species occupancy models. However, when examining these patterns more closely in the two-species model, co-occurrence was not correlated with any of the water condition variables we tested. These results suggest that sicklefin and sturgeon chub co-occurrence is more a result of similar habitat selection at macrohabitat and HUC 8 scales, as opposed to a finer scale of selection not measured in these monitoring studies, where habitat partitioning may be occurring between these species. Sicklefin and sturgeon chub occupancy was higher in the presence of each other and channel catfish and shovelnose sturgeon. The association between the two chub species and shovelnose sturgeon indicates their selection for the same habitat and ability to co-exist therein. This could lead to potential competitive interactions for habitat and food resources if shared resources become limiting [33]; this may also be the case for channel catfish that are considered habitat generalists. However, our strong co-occurrence patterns among these four species suggest that management for the improvement of sicklefin and sturgeon chub populations may benefit shovelnose sturgeon and channel catfish populations.

Regarding bend-level observed co-occurrence, we found significantly higher-than-random co-occurrence of sturgeon chub with juvenile pallid sturgeon and sauger only when both MR regions were combined. This lack of within-region deviations from randomness is likely due to the fact that the occurrence of all three species is greater in the UMR compared to the LMR. Though there was weak or no statistical evidence of differences, sicklefin and sturgeon chub actual observed co-occurrence with adult pallid sturgeon was always at least slightly higher than predicted observed co-occurrence. Weak evidence for the higher-than-random co-occurrence of adult pallid sturgeon and walleye with sicklefin chub in the UMR suggests that an occupancy-model-based study specifically designed for rare species (e.g., [19]) could further explore co-occurrence relationships between adult pallid sturgeon and walleye with sicklefin chub and other potential prey species. Overall, these results will be important to resource managers if the potential for competition between these two chub species and juvenile pallid sturgeon, as well as shovelnose sturgeon, observed in laboratory studies [19] occurs in the natural environment. There is also the potential that adult pallid sturgeon, as chub predators, may have at least part of their demonstrated population declines [3,15] attributable to the loss of prey resources, e.g., chubs. The stocking of predatory fish, a common fisheries management tool, can cause declines in prey fish populations ([67] and references therein). Again, these combined results support the hypothesis that improving conditions for sicklefin and sturgeon chub conservation and recovery would be expected to improve conditions for shovelnose sturgeon, pallid sturgeon, and channel catfish in the MR indirectly through increased habitat availability and associated food resources and directly for adult pallid sturgeon via increased prey resources.

Regarding observed co-occurrence levels, there are some other potential directions for further study for adult and juvenile pallid sturgeon, sauger, and walleye. Sicklefin and sturgeon chub appear to co-occur with adult pallid sturgeon at higher-than-predicted levels, with a similar pattern for juvenile pallid sturgeon in the UMR. Support for these patterns based on these simulations was low to marginal, possibly due to low sample size, warranting future research. In the channelized LMR, there was evidence suggesting lower-than-expected co-occurrence of juvenile pallid sturgeon and chubs, but sample sizes were low. This result is nevertheless supported by mesocosm habitat selection studies [33], where selection for mud over sand was greater for juvenile pallid sturgeon compared to sicklefin and shoal chub and juvenile pallid/shovelnose hybrid sturgeon. As with shovelnose sturgeon and channel catfish, the association between the two chub species and juvenile pallid sturgeon does not refute previous observations that they may be potential competitors for habitat and food resources [33]. For adult pallid sturgeon, the association with two chub species may reflect its almost exclusive consumption of fish, including sicklefin and sturgeon chub, to attain historically observed adult sizes [6,34,35,53,54]. Data for examining the co-occurrence of sicklefin and sturgeon chub with pallid sturgeon were limited. However, if the patterns we observed could be validated through a more directed study, consistent association patterns observed among these species groups would also support the idea that management for the improvement of sicklefin and sturgeon chub populations could affect both juvenile and adult pallid sturgeon populations. For walleye and sauger, there was no consistent association with sicklefin and sturgeon chub based on limited data available for the MR. Therefore, we were unable to provide evidence supporting the hypothesis that walleye and sauger are important predators of sicklefin and sturgeon chub in the MR; this is also true for previous work by Dieterman [69].

Co-occurrence patterns observed in this study also illustrate the potential for further study to assess if some areas of the MR could support a greater number of benthic fish species such that sturgeon and sicklefin chub, shovelnose and pallid sturgeon, channel catfish, sauger, and walleye would be more likely to be found there. A primary area of study could be in the UMR where, on top of strong co-occurrence between both chub species, shovelnose sturgeon, and channel catfish, the proportion of bends with the observed co-occurrence of each chub species and adult and juvenile pallid sturgeon, sauger, and walleye was often higher than predicted given random distributions. If there are areas that support more benthic fish species than others, it would make sense for these predator species to concentrate in these areas. Interestingly, this pattern of increased association was reversed for both chub species and pallid sturgeon juveniles in the LMR. One potential focus for future research could be to assess potential factors that may drive these differences between the UMR and the LMR. If resources are more limited in general below Gavins Point Dam, then competition for food, appropriate spawning and egg-laying sites, or another resource could limit the potential of juvenile pallid sturgeon co-occurrence with these two chub species within a bend. Adult pallid sturgeon and sicklefin chub had a higher-than-predicted rate of co-occurrence in both the UMR and the river as a whole. This indicates that further study into the relationships between these species could be beneficial. Pallid sturgeon, as a large mobile species, can move freely between bends, so they may be selecting them based on the availability of prey or another environmental variable.

4.4. Other Potential Future Research

Overall, our research hints at several avenues for future research to further increase our understanding of sicklefin chub, sturgeon chub, and the Missouri River benthic fish assemblage as a whole. Studies designed specifically targeting chub species could select spatial and temporal sampling methods that are designed around species' life histories. This could allow the construction of multi-season occupancy models that include local extinction and colonization rates, providing a more complete picture of chub populations [38]. Along with this, developing multispecies occupancy models for rare species to further explore the relationship between pallid sturgeon and chub occurrence would enhance our understanding of the ecological requirements of both species. Additional models that include other small benthic species that are potential competitors as potential co-occurrent species would also help to elucidate factors influencing where the species occur in the MR. Studies including environmental variables (e.g., aquatic invertebrate abundance, discharge, distance to impoundment, and physical features of the river bed) other than those currently available would also allow researchers to better determine how environmental factors influence occupancy, co-occurrence, differences between upper and lower river segments, and sites potential for fish biodiversity. We hope that our research might provide a starting point for future research, could enhance our understanding of the MR, and benefit the conservation of its biotic community.

5. Conclusions

Using a variety of analytical approaches from occupancy modeling to simple correlation analyses, we elucidate relationships among time, space, environmental factors, and habitat and the occurrence and co-occurrence of key benthic fishes found in the MR. This work provides important information on how sampling conditions of the MR may affect detection probabilities and thus insight into additional considerations for future sampling gear and efforts. We also provide hypotheses to be considered in future research related to management associated with MR fish populations based on the most extensive data currently available. Such information is critical to informing continued efforts to modify the habitat of the LMR to improve conditions for chubs and potentially endangered pallid sturgeon. Given the long history of habitat loss and population declines, sicklefin and sturgeon chub should continue to be of interest for the foreseeable future. Future studies focusing solely on chubs could select sample sites of appropriate size and resample each site frequently enough to construct multi-season occupancy models with extinction and colonization rates for each site. This would help monitor changes in chub populations and any loss or gain in the range of sicklefin and sturgeon chubs.

Author Contributions: M.L.W. conceptualized this study. M.L.W. and B.M.W. developed the methodology. M.L.W., B.M.W. and J.H.M. conducted formal data analysis. M.L.W., B.M.W., K.R.B., J.H.M., J.L.A. and N.S.G. assisted in investigation. M.L.W. provided resources. M.L.W. and K.R.B. provided data curation. M.L.W., B.M.W., K.R.B. and J.H.M. wrote the original manuscript draft; all authors contributed to producing the final manuscript draft. M.L.W., B.M.W., K.R.B. and J.H.M. visualized the data. M.L.W. acquired funding. M.L.W. provided study administration and supervision. All authors have read and agreed to the published version of the manuscript.

Funding: The research was funded by the U.S. Fish and Wildlife Service (IA 4500146736) and with in-kind support from U.S. Geological Survey Columbia Environmental Science Center.

Institutional Review Board Statement: All data used in this study were from extant monitoring programs; this study did not involve the capture, observation, or use of additional animals.

Informed Consent Statement: Not applicable.

Data Availability Statement: Raw monitoring data for Missouri River Benthic Fishes are available as a U.S. Geological Survey (USGS) Data Release [29]. Occupancy model coefficients and observed co-occurrence simulation initial values and results are also available as a USGS Data Release [58]. The U.S. Army Corps of Engineers manages the Pallid Sturgeon Population Assessment Project and its associated raw data; inquiries can be directed to Tim Welker at tim.l.welker@usace.army.mil.

Acknowledgments: We would like to thank the U.S. Geological Survey Columbia Environmental Research Center staff who helped with manuscript analytical aspects, review, and editing: K.K. Ditter, Z.D. Beaman, and A.P. Moore, as well as U.S Army Core of Engineers, U.S. Bureau of Reclamation, U.S. Environmental Protection Agency, U.S. Fish and Wildlife Service, Kansas Department of Wildlife and Parks, Iowa Department of Natural Resources, Missouri Department of Conservation, Montana Department of Fish, Wildlife and Parks, Nebraska Game and Parks Commission, North Dakota Game and Fish Department, South Dakota Department of Game, Fish and Parks, The Wildlife Management Institute, The University of Missouri, Kansas State, University, Iowa State University, South Dakota State, University, Montana State University, and The University of Idaho. Any use of trade, firm, or product names is for descriptive purposes only and does not imply endorsement by the U.S. Government.

Conflicts of Interest: The authors declare no conflicts of interest.

Appendix A. Data Sources

Starting in 1996, the Missouri River (MR) Benthic Fishes Study (MRBFS) documented annual patterns, habitat, and water quality associations of 21 benthic fish species based on relative abundance, including sicklefin and sturgeon chub [13]. Sampling occurred in the unimpounded mainstem of the MR from above Fort Peck Reservoir, Montana, to its confluence with the Mississippi River and the last 48 km of the Yellowstone River (i.e., Lower Yellowstone River; Figure 1; [13]). For three years, 1996 to 1998, segments of these rivers were sampled annually between July and September using a stratified random design [13]. Sampling was stratified over six different macrohabitats: main-channel crossovers, outside bends, inside bends, tributary mouths, and connected and secondary non-connected channels. These divisions led here and in the PSPAP data set to a hierarchical nested sampling design of (a) river segments/HUC 8 sections; (b) river bends (primary sampling unit); (c) macrohabitat types; (d) specific sampling sites/trawling locations within macrohabitat within river bends. This design required the use of five types of gear for collecting fishes: drifting trammel net, electrofishing, stationary gill nets, bag seine, and bottom trawl. Fish species were identified by boat crews, and 461 voucher specimens were sent to an expert ichthyologist for verification [26]. Along with fish samples, information about the riverine environment was collected: water temperature, velocity, depth, conductivity, turbidity, and the proportions of sand, silt, and gravel within the riverbed were measured.

Starting in the fall of 2003, the Pallid Sturgeon Population Assessment Program (PSPAP) was initiated based on the MRBFS design [3,27]. The PSPAP encompassed the MR from Fort Peck Dam, Montana, at river km 2851 downstream to the confluence of the MR and Mississippi River near St. Louis, Missouri, at river km 0 and the lower 32.2 km of the KR ([3]; Figure 1). Sampling years ran from 1 November of the prior year through 31 October of the named year (e.g., sampling year 2004 was from 1 November 2003 to 31 October 2004). The PSPAP used 14 different types of gear (i.e., gill nets, modified gill nets, trammel nets, otter trawls, mini-fyke nets, trot lines, push trawls, beam trawls, larval fish drift nets, hoop nets, bag seines, set lines/bank lines, and fishing/angling) to sample fish species in the river. For example, sampling year 2004 was from 1 November 2003 to 31 October 2004. To minimize risk to pallid sturgeon, each PSPAP sampling year was divided into two seasons. The sturgeon sampling season (ST) occurred from autumn, when water temperature was less than 12.8 °C (usually October or November), to June 30 of the sampling year. These lower water temperatures allowed for the use of gill nets to safely catch pallid sturgeon; most non-gill net gear types, e.g., otter trawls, were deployed during the March through June portion of sturgeon season [27]. Fish community sampling season (FC) occurred from July 1 to October 31 of a sampling year; this season did not use gill nets and instead utilized a variety of other gear types to sample young-of-the-year pallid sturgeon and other species that make up the benthic fish community. Throughout our analyses, we only used FC data from the PSPAP because the sampling protocols for that season were consistent and used types of gear directed at capturing non-pallid sturgeon species. For the PSPAP, sampling sites consisted of bends of the river randomly selected from within each of 13 defined segments. These segments were defined based on a variety of hydrologic variables and differed slightly from the segments of the MRBFS. Fish were identified to species by the boat crews, or if field identification was not possible, specimens were preserved for laboratory identification [28]. Along with these samples, the environmental variables of water temperature, velocity, depth, turbidity, and substrate composition were measured. Lastly, although substrate characteristics were collected, inconsistencies in data collection prevented substrate from being included in any analyses [3].

Appendix B. Data Filtering and Processing

The PSPAP and MRBFS data were filtered and partitioned to reduce the risk of biased samples, ensuring only data that followed standard sampling protocols were used, which allowed for the creation of a data set that included both MRBFS and PSPAP data. Because neither the MRBFS nor PSPAP was designed to inform occupancy modeling (though see [20]), multiple steps were required to make the data appropriate for this type of model. Occupancy modeling and its data requirements were discussed further in Section 2.4.1. "Occupancy Models" of the main manuscript, but the general data structure required for these models are binary detection/non-detection data collected across sites, where each site is visited multiple times [38]. Two key occupancy model assumptions that affect data processing included (1) the closed occupancy state of a site—that is, the species does not immigrate to or emigrate from the site within a single season—and (2) occupancy and detection between sites and visits are independent [38]. For all data sets, we defined a site as a unique combination of river bend, macrohabitat, and year. An individual visit to a site was defined as one benthic trawl deployment there; we assumed that trawl deployments were spatially independent. This definition of visits allowed us to use spatial replicates as visits for occupancy models, as was done by Kelly et al. [37]. Details of these processed data sets and the steps used to arrive at them are explained below.

For all extant PSPAP data, we used random gear deployments from randomly sampled river bends, as this was the standard study protocol [3,27]; this filter contributed to ensuring independence between samples. We only used PSPAP samples collected using the standard gear that accounted for the majority of sicklefin and sturgeon chub captures, the 4.8 m-wide otter trawl (OT16).

We limited our data to trawls with lengths between 75 m and 300 m, the standard OT16 distance range for PSPAP samples. Changes in the PSPAP protocol implemented in 2019 drastically reduced the spatial and temporal extent of OT16 sampling and eliminated the fish community season [10]. These changes made data collected after 2018 incomparable to those collected during or before that. Therefore, we only used PSPAP data collected prior to sampling year 2019.

This PSPAP-FC data set included data collected from 1 July to 31 October of the sampling year [3,27]. We defined visits to a site using spatial separation within a bend. The design of the PSPAP sampling protocol included revisiting some of the same bends during the two different sampling seasons within a sampling year. We used data from only one sampling to ensure independence between sites.

The support for using OT16 data for our study is provided by the fact that the great majority of sicklefin and sturgeon chub, channel catfish, and shovelnose sturgeon collections came from OT16 deployments. Across all pre-filtered PSPAP data, samples collected using OT16 accounted for >91% and >98% of standard gear captures of sicklefin and sturgeon chub, respectively. Of all PSPAP sites sampled with OT16 where sicklefin chub were detected in any gear, 98.5% of these sites had sicklefin chub detected in OT16; this was 99.1% for sturgeon chub. For our secondary species of all PSPAP sites sampled with OT16 where channel catfish were detected in any gear, 90.7% of these sites had channel catfish detected in OT16. Among all OT16-sampled PSPAP sites where shovelnose sturgeon were detected, this percentage was 87.2% for shovelnose sturgeon.

The MRBFS data set was initially filtered in two stages: a first to make the data conform to standard protocols for a single sampling gear, and a second to create a combined data set with the PSPAP-FC data, named the "MRBFS+PSPAP-FC" data set. This combination was used due to the comparatively small size of the MRBFS data set by itself, 759 sites after the first filtering stage compared to the >3300 sites of each PSPAP data set. Prior to combining, we filtered the MRBFS to include only samples collected with standard gear that accounted for the majority of sicklefin and sturgeon chub captures, the 2 m bottom trawl (BT). The OT16 was not deployed during the MRBFS, and the BT was not used during the PSPAP. To make the data comparable to the PSPAP data, we limited our data to trawling distances of 75–300 m; the standard BT distance range for MRBFS samples was 150–300 m [26]. Prior to combination, the MRBFS and PSPAP-FC data sets were filtered so that they contained only river segments sampled by both projects; MR sites above Fort Peck Dam and Yellowstone River sites were eliminated from the MRBFS data, and, for the combined data set only, KR sites were removed from the PSPAP-FC data. The combined MRBFS+PSPAP-FC data set had 14.7% MRBFS sites (*n* = 580) and 85.3% PSPAP-FC sites (*n* = 3346) after all filtering.

As with the PSPAP data, the support for using MRBFS BT data for our study is provided by the fact that the great majority of sicklefin and sturgeon chub collections came from BT deployments. Across all pre-filtered MRBFS data, the BT accounted for >96% and >87% of samples containing sicklefin and sturgeon chub, respectively. Of all MRBFS sites sampled with BT where sicklefin chub were detected in any gear, 97.8% of these sites had sicklefin chub detected in BT; this was 95.5% for sturgeon chub.

After initial filtering, the potential list of macrohabitats was consolidated to the following five common to all data sets: inside bend, outside bend, channel crossover, tributary mouth, and secondary connected channel. These were the only macrohabitats randomly sampled using BT in the MRBFS data. To match macrohabitats in the MRBFS data and consolidate similar macrohabitats with limited sample size in the PSPAP data, we lumped small and large secondary connected channel habitats as "secondary connected channel", and we lumped large and small tributary mouth habitats as "tributary mouth." After such lumping, we excluded macrohabitats represented by fewer than 10 sites—i.e., a unique segment, bend, macrohabitat, and year site combination—in the PSPAP-FC data set. This condition excluded four macrohabitats: dam tailwater, dendritic channel, deranged channel, i.e., channels with no discernable branching pattern, and secondary non-connected channel. Additionally, we included only macrohabitats where at least 10 sites had detections of at least one of the focal chub species (i.e., two additional macrohabitats were excluded: braided channel and confluence). All five selected macrohabitats met these sample size and capture rate requirements in the MRBFS data.

Only sites located in the mainstem MR or in tributaries that were consistently sampled within a data set were included, i.e., MR tributaries or HUC 8 subbasins that were part of standard sampling protocols [26,27]. Among the data used, the only consistently sampled tributary was the KR for the PSPAP data; other tributaries were excluded. Reservoirs were also excluded, as they were not consistently sampled and did not represent standard river habitat. All sites between Garrison Dam (near Riverdale, North Dakota, USA) and Lake Oahe (North and South Dakota, USA) were excluded as well, as neither focal chub species was observed in this stretch of river by either the MRBFS or the PSPAP. Additionally, this stretch was only sampled by the PSPAP for two sampling years, 2012 and 2013. Each site was assigned to a subbasin designated by a U.S. Geological Survey-standardized 8-digit Hydrologic Unit Code ([30]; HUC 8, Figure 1). The HUC 8 subbasins were chosen as a universal spatial designation similar in size to the different segment systems used by the MRBFS [13,26] and PSPAP [3,27].

We used additional processing steps to make the environmental variables meet the independence assumption. First, within each data set at each site (i.e., macrohabitat), we calculated the mean value across visits for environmental variables of water temperature, depth, velocity, and turbidity. Second, prior to calculating these means, the data were graphically inspected for impossible values. Sites with impossibly high recorded water temperatures ($>60\,°C$) were excluded, resulting in the exclusion of one to two sites from each data set.

References

1. Albers, J.L. Sicklefin chub: *Macrhybopsis meeki* (Jordan and Evermann 1896). In *Kansas Fishes*; University Press of Kansas: Lawrence, KS, USA, 2014; pp. 184–186. ISBN 978-0-7006-1961-0.
2. Albers, J.L. Sturgeon chub: *Macrhybopsis gelida* (Girard 1856). In *Kansas Fishes*; University Press of Kansas: Lawerence, KS, USA, 2014; pp. 178–180. ISBN 978-0-7006-1961-0.
3. Wildhaber, M.L.; Yang, W.H.; Arab, A. Population trends, bend use relative to available habitat and within-river-bend habitat use of eight indicator species of Missouri and Lower Kansas River benthic fishes: 15 years after baseline assessment. *River Res. Appl.* **2016**, *32*, 36–65. [CrossRef]
4. William, B. *Endangered and Threatened Wildlife and Plants*; 12-Month Finding for a Petition to List the Sicklefin Chub (Macrhybopsis meeki) and the Sturgeon Chub (Macrhybopsis gelida) as Endangered; 01-9443; U.S. Fish and Wildlife Service: Washington, DC, USA, 2001; pp. 19910–19914. Available online: https://www.govinfo.gov/content/pkg/FR-2001-04-18/html/01-9443.htm (accessed on 2 November 2023).
5. U.S. Fish and Wildlife Service. *Species Status Assessment for the Sturgeon Chub (Macrhybopsis gelida) and Sicklefin Chub (Macrhybopsis meeki)*; U.S. Fish and Wildlife Service, Region 6: Lakewood, CO, USA, 2023. Available online: https://ecos.fws.gov/ServCat/DownloadFile/238509 (accessed on 29 September 2023).
6. Gerrity, P.C.; Guy, C.S.; Gardner, W.M. Juvenile pallid sturgeon are piscivorous: A call for conserving native cyprinids. *Trans. Am. Fish. Soc.* **2006**, *135*, 604–609. [CrossRef]
7. Huenemann, T. *Central Lowlands and Interior Highlands Pallid Sturgeon Spawning and Stocking Summary 1992–2017*; Middle Basin Pallid Sturgeon Workgroup: Lincoln, NE, USA, 2010; Available online: http://meridian.allenpress.com/jfwm/article-supplement/433030/pdf/10_3996022018-jfwm-013_s4 (accessed on 2 November 2023).
8. Quist, M.C.; Boelter, A.M.; Lovato, J.M.; Korfanta, N.M.; Bergman, H.L.; Korschgen, C.E.; Galat, D.L.; Latka, D.C.; Krentz, S.; Oetker, M.; et al. *Research and Assessment Needs for Pallid Sturgeon Recovery in the Missouri River: Final Report to the U.S. Geological Survey, U.S. Army Corps of Engineers, U.S. Fish and Wildlife Service, and U.S. Environmental Protection Agency*; William D. Ruckelshaus Institute of Environment and Natural Resources, University of Wyoming: Laramie, WY, USA, 2004; p. 82. Available online: https://pallidsturgeon.org/wp-content/uploads/2018/01/28-Quist-et-al-2004.pdf (accessed on 19 December 2023).
9. Rahel, F.J.; Thel, L.A. *Sturgeon Chub (Macrhybopsis gelida): A Technical Conservation Assessment*; USDA Forest Service, Rocky Mountain Region: Laramie, WY, USA, 2004; pp. 1–49. Available online: https://www.fs.usda.gov/Internet/FSE_DOCUMENTS/stelprdb5206786.pdf (accessed on 16 November 2023).

10. Welker, T.L.; Drobish, M.R.; Williams, G.A. *Pallid Sturgeon Population Assessment Project, Guiding Document, Volume 2.0*; U.S. Army Corps of Engineers, Omaha District: Yankton, SD, USA, 2020.

11. Ridenour, C.J.; Starostka, A.B.; Doyle, W.J.; Hill, T.D. Habitat used by chubs associated with channel modifying structures in a large regulated river: Implications for river modification. *River Res. Appl.* **2009**, *25*, 472–485. [CrossRef]

12. Ridenour, C.J.; Wrasse, C.J.; Meyer, H.A.; Doyle, W.J.; Hill, T.D. *Variation in Fish Assemblages among Four Habitat Complexes during a Summer Flood: Implications for Flow Refuge*; U.S. Fish and Wildlife Service, Columbia Fish and Wildlife Conservation Office: Columbia, MO, USA, 2012.

13. Wildhaber, M.L.; Gladish, D.W.; Arab, A. Distribution and habitat use of the Missouri River and Lower Yellowstone River benthic fishes from 1996 to 1998: A baseline for fish community recovery. *River Res. Appl.* **2012**, *28*, 1780–1803. [CrossRef]

14. Steffensen, K.D.; Shuman, D.A.; Stukel, S. The status of fishes in the Missouri River, Nebraska: Shoal chub (*Macrhybopsis hyostoma*), sturgeon chub (*M. gelida*), sicklefin chub (*M. meeki*), silver chub (*M. storeriana*), Flathead chub (*Platygobio gracilis*), plains minnow (*Hybognathus placitus*), western silvery minnow (*H. argyritis*), and brassy minnow (*H. hankinsoni*). *Trans. Neb. Acad. Sci.* **2014**, *34*, 49–67. Available online: https://digitalcommons.unl.edu/tnas/470/ (accessed on 16 January 2024).

15. Braaten, P.J.; Fuller, D.B.; Haddix, T.M.; Hunziker, J.R.; Colvin, M.E.; Holmquist, L.M.; Wilson, R.H. Catch rates for sturgeon chubs and sicklefin chubs in the Upper Missouri River 2004–2016 and correlations with biotic and abiotic variables. *J. Fish Wildl. Manag.* **2021**, *12*, 322–337. [CrossRef]

16. MacKenzie, D.I.; Nichols, J.D.; Hines, J.E.; Knutson, M.G.; Franklin, A.B. Estimating site occupancy, colonization, and local extinction when a species is detected imperfectly. *Ecology* **2003**, *84*, 2200–2207. [CrossRef]

17. Gu, W.; Swihart, R.K. Absent or undetected? Effects of non-detection of species occurrence on wildlife–habitat models. *Biol. Conserv.* **2004**, *116*, 195–203. [CrossRef]

18. Royle, J.A.; Nichols, J.D.; Kéry, M. Modelling occurrence and abundance of species when detection is imperfect. *Oikos* **2005**, *110*, 353–359. [CrossRef]

19. Specht, H.M.; Reich, H.T.; Iannarilli, F.; Edwards, M.R.; Stapleton, S.P.; Weegman, M.D.; Johnson, M.K.; Yohannes, B.J.; Arnold, T.W. Occupancy surveys with conditional replicates: An alternative sampling design for rare species. *Methods Ecol. Evol.* **2017**, *8*, 1725–1734. [CrossRef]

20. Schloesser, J.T.; Paukert, C.P.; Doyle, W.J.; Hill, T.D.; Steffensen, K.D.; Travnichek, V.H. Heterogeneous detection probabilities for imperiled Missouri River Fishes: Implications for large-river monitoring programs. *Endanger. Species Res.* **2012**, *16*, 211–224. [CrossRef]

21. Budy, P.; Conner, M.M.; Salant, N.L.; Macfarlane, W.W. An occupancy-based quantification of the highly imperiled status of desert fishes of the Southwestern United States. *Conserv. Biol.* **2015**, *29*, 1142–1152. [CrossRef] [PubMed]

22. MacKenzie, D.I.; Bailey, L.L.; Nichols, J.D. Investigating species co-occurrence patterns when species are detected imperfectly. *J. Anim. Ecol.* **2004**, *73*, 546–555. [CrossRef]

23. Rota, C.T.; Ferreira, M.A.R.; Kays, R.W.; Forrester, T.D.; Kalies, E.L.; McShea, W.J.; Parsons, A.W.; Millspaugh, J.J. A multispecies occupancy model for two or more interacting species. *Methods Ecol. Evol.* **2016**, *7*, 1164–1173. [CrossRef]

24. Lamothe, K.A.; Dextrase, A.J.; Drake, D.A.R. Characterizing species co-occurrence patterns of imperfectly detected stream fishes to inform species reintroduction efforts. *Conserv. Biol.* **2019**, *33*, 1392–1403. [CrossRef]

25. Holt, A.R.; Gaston, K.J.; He, F.L. Occupancy-abundance relationships and spatial distribution: A review. *Basic Appl. Ecol.* **2002**, *3*, 1–13. [CrossRef]

26. Berry, C.R.; Young, B.A. Introduction to the benthic fishes studies: Population structure and habitat. In *Use of Benthic Fishes along the Missouri and Lower Yellowstone Rivers*; United States Army Corps of Engineers, CENWO-PM-AE: Omaha, NE, USA, 2001; pp. 1–52.

27. Welker, T.L.; Drobish, M.R.; Williams, G.A. *Pallid Sturgeon Population Assessment Project, Guiding Document, Volume 1.8*; U.S. Army Corps of Engineers, Omaha District: Yankton, SD, USA, 2017; Available online: https://meridian.allenpress.com/jfwm/article-supplement/433030/pdf/10_3996022018-jfwm-013_s14/ (accessed on 8 November 2023).

28. Welker, T.L.; Drobish, M.R.; Williams, G.A. *Pallid Sturgeon Population Assessment Project, Guiding Document, Volume 1.1*; U.S. Army Corps of Engineers, Omaha District: Yankton, SD, USA, 2006.

29. Wildhaber, M.L.; West, B.M.; Bennett, K.B.; Beaman, Z.D. *Captures and Habitat Classification of Benthic Fishes along the Missouri and Lower Yellowstone Rivers, 1996–1998*; U.S. Geological Survey Data Release: Columbia, MO, USA, 2023. [CrossRef]

30. Jones, K.A.; Niknami, L.S.; Buto, S.G.; Decker, D. Federal standards and procedures for the national watershed boundary dataset (WBD). In *Section A, Federal Standards, Book 11, Collection and Delineation of Spatial Data*; U.S. Geological Survey: Reston, VA, USA, 2022; ISBN 2328-7055.

31. Michaletz, P.H.; Dillard, J.G. Fisheries management—A survey of catfish management in the United States and Canada. *Fisheries* **1999**, *24*, 6–11. [CrossRef]

32. Braun, A.P.; Phelps, Q.E. Channel catfish habitat use and diet in the Middle Mississippi River. *Am. Midl. Nat.* **2016**, *175*, 47–54. [CrossRef]

33. Wildhaber, M.L.; Albers, J.L. Laboratory studies of potential competition for food and substrate among early juvenile Missouri River sturgeon and sympatric chub species. *N. Am. J. Fish. Manag.* **2023**. [CrossRef]

34. Grohs, K.L.; Klumb, R.A.; Chipps, S.R.; Wanner, G.A. Ontogenetic patterns in prey use by pallid sturgeon in the Missouri River, South Dakota and Nebraska. *J. Appl. Ichthyol.* **2009**, *25*, 48–53. [CrossRef]

35. Wildhaber, M.L. Pallid Sturgeon: *Scaphirhynchus albus* (Forbes and Richardson 1905). In *Kansas Fishes*; University Press of Kansas: Lawrence, KS, USA, 2014; pp. 99–101. ISBN 978-0-7006-1961-0.

36. Bellgraph, B.J. Competition Potential between Sauger and Walleye in Non-Native Sympatry: Historical Trends and Resource Overlap in the Middle Missouri River, Montana. Master's Thesis, Montana State University, Bozeman, MT, USA, 2006. Available online: https://scholarworks.montana.edu/xmlui/handle/1/901 (accessed on 20 December 2023).

37. Kelly, B.; Siepker, M.J.; Weber, M.J. Factors associated with detection and distribution of native Brook Trout and introduced Brown Trout in the driftless area of Iowa. *Trans. Am. Fish. Soc.* **2021**, *150*, 388–406. [CrossRef]

38. MacKenzie, D.I.; Nichols, J.D.; Royle, J.A.; Pollock, K.H.; Bailey, L.L.; Hines, J.E. *Occupancy Estimation and Modeling: Inferring Patterns and Dynamics of Species Occurrence—Softcover*, 2nd ed.; Academic Press: Cambridge, MA, USA, 2018; ISBN 9780128146910.

39. R Core Team. Available online: https://www.R-project.org/ (accessed on 17 January 2023).

40. Fiske, I.J.; Chandler, R.B. Unmarked: An R Package for fitting hierarchical models of wildlife occurrence and abundance. *J. Stat. Softw.* **2011**, *43*, 1–23. [CrossRef]

41. Neebling, T.E.; Quist, M.C. Comparison of boat electrofishing, trawling, and seining for sampling fish assemblages in Iowa's Nonwadeable Rivers. *N. Am. J. Fish. Manag.* **2011**, *31*, 390–402. [CrossRef]

42. Steffensen, K.D.; Wilhelm, J.J.; Haas, J.D.; Adams, J.D. Conditional capture probability of pallid sturgeon in benthic trawls. *N. Am. J. Fish. Manag.* **2015**, *35*, 626–631. [CrossRef]

43. Smith, C.D.; Quist, M.C.; Hardy, R.S. Detection probabilities of electrofishing, hoop nets, and benthic trawls for fishes in two Western North American Rivers. *J. Fish Wildl. Manag.* **2015**, *6*, 371–391. [CrossRef]

44. Everett, S.R.; Scarnecchia, D.L.; Ryckman, L.F. Distribution and habitat use of sturgeon chubs (*Macrhybopsis gelida*) and sicklefin chubs (*M. meeki*) in the Missouri and Yellowstone Rivers, North Dakota. *Hydrobiologia* **2004**, *527*, 183–193. [CrossRef]

45. Livernois, M.C.; Powers, S.P.; Albins, M.A.; Mareska, J.F. Habitat associations and co-occurrence patterns of two estuarine-dependent predatory fishes. *Mar. Coast. Fish.* **2020**, *12*, 64–77. [CrossRef]

46. Fawcett, T. An introduction to roc analysis. *Pattern Recognit. Lett.* **2006**, *27*, 861–874. [CrossRef]

47. Robin, X.; Turck, N.; Hainard, A.; Tiberti, N.; Lisacek, F.; Sanchez, J.C.; Müller, M. pROC: An open-source package for r and s plus to analyze and compare ROC curves. *BMC Bioinform.* **2011**, *12*, 77. [CrossRef]

48. Hosmer, D.W.; Lemeshow, S.; Sturdivant, R.X. *Applied Logistic Regression*, 3rd ed.; John Wiley & Sons: Hoboken, NJ, USA, 2013; ISBN 9780470582473.

49. U.S. Geological Survey. *USGS Water Data for the Nation: U.S. Geological Survey National Water Information System Database*; U.S. Geological Survey: Reston, VA, USA, 2023. [CrossRef]

50. Wildhaber, M.L.; Tabor, V.M.; Whitaker, J.E.; Allert, A.L.; Mulhern, D.W.; Lamberson, P.J.; Powell, K.L. Ictalurid populations in relation to the presence of a main-stem reservoir in a Midwestern warmwater stream with emphasis on the threatened Neosho madtom. *Trans. Am. Fish. Soc.* **2000**, *129*, 1264–1280. [CrossRef]

51. Herman, P.; Plauck, A.; Utrup, N.; Hill, T. *Three Year Summary Age and Growth Report for Sicklefin Chub (Macrhybopsis meeki)*; U.S. Fish and Wildlife Service: Columbia, MO, USA, 2008; pp. 1–73. Available online: https://meridian.allenpress.com/jfwm/article-supplement/465707/pdf/jfwm-20-086.s5/ (accessed on 16 November 2023).

52. Wildhaber, M.L.; DeLonay, A.J.; Papoulias, D.M.; Galat, D.L.; Jacobson, R.B.; Simpkins, D.G.; Braaten, P.J.; Korschgen, C.E.; Mac, M.J. *A Conceptual Life-History Model for Pallid and Shovelnose Sturgeon*; U.S. Geological Survey: Reston, VA, USA, 2007; p. iv. 18p, ISBN 9781411319059.

53. Wildhaber, M.L.; DeLonay, A.J.; Papoulias, D.M.; Galat, D.L.; Jacobson, R.B.; Simpkins, D.G.; Braaten, P.J.; Korschgen, C.E.; Mac, M.J. Identifying structural elements needed for development of a predictive life-history model for pallid and shovelnose sturgeons. *J. Appl. Ichthyol.* **2011**, *27*, 462–469. [CrossRef]

54. Wildhaber, M.L.; Dey, R.; Wilke, C.K.; Moran, E.H.; Anderson, C.J.; Franz, K.J. A stochastic bioenergetics model-based approach to translating large river flow and temperature into fish population responses: The pallid sturgeon example. *Geol. Soc. Publ. Lond.* **2017**, *408*, 101–118. [CrossRef]

55. Bellgraph, B.J.; Guy, C.S.; Gardner, W.M.; Leathe, S.A. Competition potential between saugers and walleyes in nonnative sympatry. *Trans. Am. Fish. Soc.* **2008**, *137*, 790–800. [CrossRef]

56. Wildhaber, M.L.; Albers, J.L.; Green, N.S.; Moran, E.H. A fully-stochasticized, age-structured population model for population viability analysis of fish: Lower Missouri River endangered pallid sturgeon example. *Ecol. Model.* **2017**, *359*, 434–448. [CrossRef]

57. Muff, S.; Nilsen, E.B.; O'Hara, R.B.; Nater, C.R. Rewriting results sections in the language of evidence. *Trends Ecol. Evol.* **2022**, *37*, 203–210. [CrossRef] [PubMed]

58. Wildhaber, M.L.; West, B.M.; Green, N.S.; Albers, J.L.; Bennett, K.R.; May, J.H. *Occupancy Model Coefficients and Observed Co-Occurrence Simulations for Sicklefin Chub, Sturgeon Chub, and Associated Fishes in the Missouri River*; U.S. Geological Survey Data Release: Columbia, MO, USA, 2023. [CrossRef]

59. Galat, D.L.; Berry, C.R.; Peter, E.J.W.; White, R.G. Missouri River Basin. In *Rivers of North America*, 2nd ed.; Delong, M.D.B., Benke, A.C., Jardine, T.D., Cushing, C.E., Eds.; Elsevier: Oxford, UK, 2005; pp. 410–461. ISBN 978-0-12-818847-7.

60. Cross, F.B. Occurrence of the sturgeon chub *Hybopsis gelida* (Girard) in Kansas. *Trans. Kans. Acad. Sci.* **1953**, *56*, 90–91. [CrossRef]

61. Dieterman, D.J.; Galat, D.L. Large-scale factors associated with sicklefin chub distribution in the Missouri and Lower Yellowstone Rivers. *Trans. Am. Fish. Soc.* **2004**, *133*, 577–587. [CrossRef]

62. Pflieger, W.L.; Grace, T.B. Changes in the fish fauna of the Lower Missouri River, 1940–1983. In *Community and Evolutionary Ecology of North Ameircan Stream Fishes*; Matthews, W.J., Heins, D.C., Eds.; University of Oklahoma Press: Norman, OK, USA, 1987; pp. 166–177. ISBN 978-0806122359.

63. Gelwicks, G.; Graham, K.; Galat, D. *Status Survey for Sicklefin Chub, Sturgeon Chub, and Flathead Chub in the Missouri River, Missouri: Final Report*; Missouri Department of Conservation, Fish and Wildlife Research Center: Columbia, MO, USA, 1996.

64. Grady, J.M.; Milligan, J. *Status of Selected Cyprinid Species and Gear Selectivity at Historic Lower Missouri River Sampling Sites*; U.S. Fish and Wildlife Service, Columbia Fisheries Resources Office: Columbia, MO, USA, 1998; p. 52.

65. Pflieger, W.L.; Smith, P.A. *The Fishes of Missouri*; Missouri Department of Conservation: Jefferson City, MO, USA, 1997; p. 343. ISBN 1887247114.

66. Welker, T.L.; Scarnecchia, D.L. Habitat use and population structure of four native minnows (family Cyprinidae) in the Upper Missouri and Lower Yellowstone Rivers, North Dakota (USA). *Ecol. Freshw. Fish* **2004**, *13*, 8–22. [CrossRef]

67. Wildhaber, M.L.; West, B.M.; Ditter, K.K.; Adrian, A.M.; Peterson, A.S. A review of grass carp and related species literature on diet, behavior, toxicology, and physiology focused on informing development of controls for invasive grass carp populations in North America. *Fishes* **2023**, *8*, 547. [CrossRef]

68. Tornabene, B.J.; Smith, T.W.; Tews, A.E.; Beattie, R.P.; Gardner, W.M.; Eby, L.A. Trends in river discharge and water temperature cue spawning movements of blue sucker, *Cycleptus elongatus*, in an impounded Great Plains river. *Copeia* **2020**, *108*, 151–162. [CrossRef]

69. Dieterman, D.J. Spatial Patterns in Phenotypes and Habitat Use of Sicklefin Chub, *Macrhybopsis meeki*, in the Missouri and Lower Yellowstone Rivers. Ph.D. Dissertation, University of Missouri, Columbia, MO, USA, 2000.

 *fishes*

Article

Integrative Taxonomy Clarifies the Historical Flaws in the Systematics and Distributions of Two *Osteobrama* Fishes (Cypriniformes: Cyprinidae) in India

Boni Amin Laskar [1], Dhriti Banerjee [1,2], Sangdeok Chung [3], Hyun-Woo Kim [4,5], Ah Ran Kim [5,*] and Shantanu Kundu [6,*]

[1] High Altitude Regional Centre, Zoological Survey of India, Solan 173211, Himachal Pradesh, India; boniamin.laskar@gmail.com (B.A.L.); dhritibanerjee@gmail.com (D.B.)

[2] Zoological Survey of India, Prani Vigyan Bhawan, Kolkata 700053, West Bengal, India

[3] Distant Water Fisheries Resources Research Division, National Institute of Fisheries Science, Busan 46083, Republic of Korea; sdchung@korea.kr

[4] Department of Marine Biology, Pukyong National University, Busan 48513, Republic of Korea; kimhw@pknu.ac.kr

[5] Research Center for Marine Integrated Bionics Technology, Pukyong National University, Busan 48513, Republic of Korea

[6] Institute of Fisheries Science, Pukyong National University, Busan 48513, Republic of Korea

* Correspondence: ahrankim@pukyong.ac.kr (A.R.K.); shantanu1984@pknu.ac.kr or shantanu1984@gmail.com (S.K.)

Citation: Laskar, B.A.; Banerjee, D.; Chung, S.; Kim, H.-W.; Kim, A.R.; Kundu, S. Integrative Taxonomy Clarifies the Historical Flaws in the Systematics and Distributions of Two *Osteobrama* Fishes (Cypriniformes: Cyprinidae) in India. *Fishes* **2024**, *9*, 87. https://doi.org/10.3390/fishes9030087

Academic Editors: Robert L. Vadas, Jr. and Robert M. Hughes

Received: 6 February 2024
Revised: 21 February 2024
Accepted: 26 February 2024
Published: 27 February 2024

Abstract: The taxonomy and geographical distributions of *Osteobrama* species have historically posed challenges to ichthyologists, leading to uncertainties regarding their native ranges. While traditional taxonomy has proven valuable in classification, the utility of an integrated approach is restricted for this particular group due to limitations in combining information from biogeography, morphology, and genetic data. This study addresses the taxonomic puzzle arising from the recent identification of *Osteobrama tikarpadaensis* in the Mahanadi and Godavari Rivers, casting doubt on the actual distribution and systematics of both *O. tikarpadaensis* and *Osteobrama vigorsii*. The research reveals distinctions among specimens resembling *O. vigorsii* from the Krishna and Godavari riverine systems. Notably, specimens identified as *O. vigorsii* from the Indian Museum exhibit two pairs of barbels, while those from the Godavari River in this study are identified as *O. tikarpadaensis*. Inter-species genetic divergence and maximum likelihood phylogeny provide clear delineation between *O. vigorsii* and *O. tikarpadaensis*. The study suggests that *O. vigorsii* may be limited to the Krishna River system in southern India, while *O. tikarpadaensis* could potentially extend from the Mahanadi River in central India to the Godavari River in southern India. Proposed revision to morphological features for both species, accompanied by revised taxonomic keys, aim to facilitate accurate differentiation among *Osteobrama* congeners. The data generated by this research provide a resource for future systematic investigations into cyprinids in India and surrounding regions. Further, the genetic diversity information obtained from various riverine systems for *Osteobrama* species will be instrumental in guiding aquaculture practices and formulating effective conservation action plans.

Keywords: cyprinids; distribution; genetic divergence; key characters; phylogeny; systematics

Key Contribution: The current investigation resolves a longstanding taxonomic quandary concerning two Indian cyprinids through a comprehensive morphological reassessment; fortified by corroborative molecular data. Furthermore, the research contributes updated morphological keys for the identification of species within the genus *Osteobrama*, offering valuable tools for subsequent systematic studies in the future.

1. Introduction

Freshwater fishes represent a crucial zoogeographical group due to their confinement to drainage systems, conceptualized as dendritic water islands surrounded by land and bordered by saltwater barriers [1]. According to the Animal Discoveries of India 2022 [2], the country hosts 3439 fish species, encompassing both freshwater and marine varieties, with approximately 206 being endemic and ~18 introduced species. A global fish database search (https://www.fishbase.org, accessed on 15 January 2024) has identified approximately 1064 freshwater fish species reported from India and its islands [3]. Accurate identification of organisms at lower taxonomic levels is crucial for ecosystem understanding and conservation applications, but existing identification systems need advancement to address gaps and enhance precision [4,5]. The DNA barcoding technique, standardized for lower-level taxonomic identification, employs a partial mitochondrial cytochrome oxidase c subunit-I gene (mtCOI), consisting of approximately 648 base pairs near the 5′ end of the gene [6,7]. The molecular technique, proven efficient in biodiversity assessment globally, has resolved taxonomic issues in Indian riverine systems [8–15]. It complements traditional taxonomy by swiftly comparing specimens with reference sequences in global databases [16]. Furthermore, the technique has advanced with various species-level delineation methods.

Fishes belonging to the genus *Osteobrama* Heckel 1843 (Cypriniformes: Cyprinidae), with the type species being *Cyprinus cotio* Hamilton 1822, exhibit a laterally compressed body, an elevated dorsum, the absence of a procumbent pre-dorsal spine, a rounded abdomen in front of the pelvic fin, and a keeled abdominal edge from the pelvic-fin origin to the vent. Additionally, they possess a long anal fin with more than 10 branched rays [17]. The genus currently consists of 11 described species [18]. Notably, *O. cotio* (Hamilton, 1822) is widely distributed in eastern India and Bangladesh, extending to northern and central India up to Pakistan. In southern India, five species—*O. peninsularis* Silas, 1952, *O. vigorsii* (Sykes, 1839), *O. dayi* (Hora and Misra, 1940), *O. neilli* (Day, 1873), and *O. bakeri* (Day, 1873)—are found, while three species—*O. cunma* (Day, 1888), *O. belangeri* (Valenciennes, 1844), and *O. feae* Vinciguerra, 1890—are distributed in Southeast Asia, Myanmar, and China [17,19,20]. A recent addition to the genus is *O. tikarpadaensis* (Shangningam, Rath, Tudu and Kosygin, 2020), described from the Mahanadi River in Odisha, central India, and reported in the Erai River, Godavari drainages, Maharashtra [21,22]. Although *O. alfredianus* (Valenciennes, 1844) was originally documented in Mysore, peninsular India [18], subsequent taxonomic assessments have synonymized it with *O. vigorsii* [17]. Later, *O. alfredianus* has been reported in the Salween Basin, Southeast Asia, which is a location distant from its type locality; however, comprehensive taxonomic descriptions are lacking. The absence of compelling literature supporting the validity of *O. alfredianus* and its accurate distribution restricts any definitive statements within the scope of this study.

In the realm of systematics, the presence or absence of barbels stands as a crucial taxonomic trait in *Osteobrama* [21,23]. When present, these barbels may manifest as a single pair of maxillary barbels or include both maxillary and rostral varieties, sometimes being minute or rudimentary in certain species. Rostral barbels may either remain concealed in a groove or be visible only under microscopic examination, while in other species, they can be significantly longer, extending to the base of maxillary barbels [17,19,21]. The type locality of *O. vigorsii* is the Bhima River (a tributary of the Krishna River) at Pairgaon, Maharashtra, but there are also reports in the Krishna and Godavari Rivers [24,25]. However, the recent discovery of *Osteobrama* in the Mahanadi and Godavari Rivers has introduced a taxonomic quandary for ichthyologists regarding the actual distribution and systematics of *O. tikarpadaensis* and *O. vigorsii*. This study aims to resolve this taxonomic challenge through development and presentation of revised taxonomic keys and genetic information. In this study, discrepancies in prevailing diagnostic features for *O. vigorsii* were noted in specimens from the Krishna and Godavari riverine systems, including those examined at the Indian Museum. Genetic divergence analysis on partial mtCOIs among *Osteobrama* species from south India revealed distinctions between specimens resembling *O. vigorsii* from the Krishna and Godavari Rivers. Consequently, based on existing morphological descriptions, distinguishing between the two distinct species from the Krishna and Godavari Rivers as *O. vigorsii* is perplexing. This study identifies specimens resembling *O. vigorsii* from the Godavari River as *O. tikarpadaensis*. Therefore, we hypothesized that *O. vigorsii* is limited to the Krishna River system in southern India, whereas *O. tikarpadaensis* is distributed from the Mahanadi River in central India to the Godavari River in south India.

2. Material and Methods

2.1. Material Examined

The following specimens were taken for morphological investigations—*O. vigorsii*: FBRC/ZSI/F3550, (n = 1), 119.5 mm SL; India: Telangana, Nagarkurnool District, Krishna River: near Somasila, 16°2′46″ N 78°19′34″ E; B. A. Laskar, 28 Jul 2020. FBRC/ZSI/F3551, (n = 2), 116.0–132.0 mm SL and FBRC/ZSI/F3552, (n = 3), 90.5–109.0 mm SL; collection details same as F3550. FBRC/ZSI/F2783, (n = 1), 95.0 mm SL; India: Telangana, Nagarkurnool District, Krishna River: near Somasila, 16°01′12″ N 78°19′37″ E; B. A. Laskar, 18 July 2018 (Figure 1A). The study specimens of *O. vigorsii* were collected from the same river basin as its type locality in the Bhima River, a tributary of the Krishna River. *O. cotio*: FBRC/ZSI/DNA907/F3880, (n = 1), 40.0 mm SL; India: Telangana, Jurala project: Krishna River Basin, Kistampally. FBRC/ZSI/F/2707, (n = 2), 62.0–64.0 mm SL; India: Maharashtra, Darna River: near Bhagur. *O. cotio* iconotype figure from Hamilton plate 207 [26] (Figure S1A). *O. neilli*: FBRC/ZSI/F/3548, (n = 2), 68.0–69.0 mm SL; India: Telangana, Nagarkurnool District, Krishna River: near Somasila (Figure S1B). *O. peninsularis*: FBRC/ZSI/F/3549, (n = 1), 68.0 mm SL; India: Telangana, Wyra Lake, Godavari River drainage, Khammam District (Figure S1C). *O. tikarpadaensis*: FBRC_ZSI_F_2616, (n = 4), 101.0–102.0 mm SL; India: Telangana, Godavari River, 17.7431° N, 80.8798° E. FBRC/ZSI/F/3416, (n = 1), 77.0 mm SL; India: Telangana, East Godavari District, confluence of Sabri River and Godavari River. (Figure 2A,B). Specimens of *O. tikarpadaensis* were collected from the Godavari River and morphologically compared with the type specimen from the Mahanadi River. *O. dayi*: FBRC-ZSI-F 3795, India: Telangana, Godavari River (Figure S1D). In the current study, the urohyal bone structure was examined for two specimens each of *O. vigorsii* and *O. tikarpadaensis*.

2.2. Sampling and Morphological Investigation

Morphometric and meristic data were documented in accordance with the methodology established in prior investigations [19,21,25]. Measurements were obtained using digital calipers, with precision up to 0.1 mm, with the exception of fin rays and scale counts, which were conducted under transmitted light utilizing a stereomicroscope. Enumeration of all pored scales was undertaken to report the number of lateral line scales. The various components of the body are expressed as a percentage of standard length (SL), while subunits of the head are presented as a percentage of head length (HL). Notably, morphometric data and scale counts for two specimens (voucher No. FBRC_ZSI_F2783_DNA301, (n = 1), 95.0 mm SL; FBRC_ZSI_F3551, DNA814, (n = 1), 116.0 mm SL) were omitted due to injuries sustained during the collection process. Nonetheless, their DNA data have been included in the subsequent analysis. Additionally, DNA data for one specimen were not generated, as it was promptly preserved in formalin. The specimens examined have been deposited at the Freshwater Biology Regional Centre, Zoological Survey of India (ZSI), Hyderabad, India.

2.3. Molecular Experiments

Tissue samples were procured from seven recently collected specimens of *O. vigorsii* and one specimen of *O. neilli* from the Krishna River; and two specimens each of *O. tikarpadaensis* and *O. cotio* and one specimen of *O. peninsularis* from the Godavari River. Genomic DNA extraction was performed using the QIAamp DNA Mini Kit (Qiagen, Valencia, CA) following the manufacturer's protocol. The previously published primer pair [27] FishF1-5′–TCAACCAACCACAAAGACATTGGCAC–3′ and FishR1-5′–TAGACTTCTGGGTGGCCAAAGAATCA–3′ was employed to amplify a partial segment of mtCOI. The PCR mixture (30 μL) comprised 10 pmol of each primer, 100 ng of DNA

Table 1. *Cont.*

Species	Museum Registration	Locality	GenBank Accession Number	BOLD-IDs
Osteobrama neilli	FBRC_ZSI_DNA833_F3548	Krishna River at somasila near temple, Telangana, 16.046° N, 78.326° E	MT896378	BOLD:ACR7173
Osteobrama peninsularis	FBRC_ZSI_DNA864_F3549	Wyra lake, Telangana, 17.252° N, 80.384° E	MT896379	BOLD:ACJ3278
Osteobrama vigorsii	FBRC_ZSI_DNA861_F3550	Krishna River somasila near temple, Telangana, 16.046° N, 78.326° E	MT896380	BOLD:ACM5411
Osteobrama vigorsii	FBRC_ZSI_F2783_DNA301	Tungabhadra River, Andhra Pradesh, 16.02° N, 78.327° E	MK336909	BOLD:ACM5411
Osteobrama vigorsii	FBRC_ZSI_DNA814_F3551	Krishna River at somasila, Telangana, 16.048° N, 78.334° E	MT896381	BOLD:ACM5411
Osteobrama vigorsii	FBRC_ZSI_DNA862_F3552	Krishna River somasila near temple, Telangana, 16.046° N, 78.326° E	MT896382	BOLD:ACM5411
Osteobrama vigorsii	FBRC_ZSI_DNA863_F3552	Krishna River somasila near temple, Telangana, 16.046° N, 78.326° E	MT896383	BOLD:ACM5411
Osteobrama vigorsii	FBRC_ZSI_DNA836_F3552	Krishna River somasila near temple, Telangana, 16.046° N, 78.326° E	MT896384	BOLD:ACM5411
Osteobrama vigorsii	FBRC_ZSI_DNA897_F3872	Jurala project, Kistampally, Telangana, 16.370° N, 77.694° E	MW506815	-
Rasbora daniconius	FBRC_ZSI_DNA326_F3464	Andhra Pradesh, 18.0733° N, 82.9505° E	MK681752	-

2.4. Dataset Preparation and Genetic Analyses

The representative COI sequences of three genera within the Smiliogastrinae subfamily, namely *Osteobrama*, *Rohtee* (Sykes, 1839), and *Mystacoleucus* (Günther, 1868), were acquired from the GenBank database. Consistent with prior research [20], uncertain sequences of *O. cotio* from the Narmada River Basin, Karnafuli, and Sangu Rivers were excluded from the dataset. Additionally, a maximum of five representative sequences from three congeners (*O. belangeri*, *O. cunma*, and *O. feae*) were incorporated into the dataset [20]. The COI sequences for *O. dayi*, sourced from GenBank, were included in the study. However, no COI data for *O. bakeri* were found in the database, and no specimens could be collected for the study. GenBank accession numbers are indicated alongside the organism's name in the phylogenetic tree, as well as detailed specimen information, including accession numbers for de novo sequences (Table 1). The dataset was aligned using CLUSTAL X, and genetic distances were estimated using MEGA X [28,29]. The model 'GTR + G + I' was chosen based on the lowest Bayesian information criterion (BIC) scores determined using PartitionFinder 2 [30] on the CIPRES Science Gateway v3.3 [31] and JModelTest v2 [32]. The PhyML 3.0 [33] was employed to construct the maximum likelihood (ML) phylogeny, with 1000 bootstrap support. Furthermore, the Bayesian (BA) tree was created using Mr. Bayes 3.1.2 [34], employing one cold and three hot Metropolis-coupled Markov chain Monte Carlo (MCMC) chains. The analysis extended over 10,000,000 generations, with tree sampling occurring every 100th generation, and 25% of the samples were discarded as burn-in. Visualization of both ML and BA trees was carried out using the iTOL v4 web server (https://itol.embl.de/login.cgi, accessed on 15 January 2024) [35].

3. Results and Discussion

3.1. Morphological Amendment of O. vigorsii

Prior to this study, the taxonomic characters of *O. vigorsii* were followed after Hora and Misra [23]. In dealing with the taxonomy of *O. vigorsii*, Hora and Misra [23] examined specimens from diverse locations, including the Darna River, the Mutha-Mula River, and the Kistna (now Krishna) River (ZSI Cat No. 888), and from Deccan and Odisha (with no precise locality specified). The data presented for *O. vigorsii* were derived from specimens in the Indian Museum collection obtained from various locales within the Krishna River Basin, with the exception of one specimen from Orissa, which was acquired from Dr. F. Day. The dorsal profile of *O. vigorsii* was characterized by a distinct concavity extending from the snout to over the nape, consistent with the original descriptions [23]. Additionally, the taxonomic key to species highlighted the presence of only two rudimentary maxillary barbels in *O. vigorsii*. Subsequently, a new species, *O. dayi* [23], characterized by two rudimentary maxillary barbels, was proposed based on specimens collected from the Godavari River. These two species are distinguishable by differences in anal fin length and the number of lateral line scales. Notably, *O. vigorsii* has since been consistently characterized by the presence of two maxillary barbels, among other morphological characters.

In the absence of any designated types, the original illustration proves highly valuable [36] (Figure 1B), exhibiting a notable similarity to the specimens of *O. vigorsii* in the present study (Figure 1A), as well as to Day's illustration of *O. vigorsii* from 1889 [37]. Jayaram (1995) reproduced Day's illustration [24]. Subsequently, Singh and Yazdani (1992) claimed a striking resemblance between their newly identified species, *Osteobrama bhimensis*, and *O. vigorsii*. However, they primarily differentiated the two based on the absence of barbels, the number of transverse scales, and the shape of the urohyal bone [38]. Notably, Singh and Yazdani [38] did not directly examine specimens of *O. vigorsii* displaying the purported resemblance to their new species, but rather utilized measurements and counts from a previous study [23]. Although they differentiated the two species based on urohyal shape, they failed to specify which specimen of *O. vigorsii* was studied for the urohyal [38]. Jadhav et al. [25] criticized the lack of retrievability in Singh and Yazdani's [38] urohyal study. However, the form of urohyal drawn by Singh and Yazdani [38] for *O. bhimensis* was observed in specimens of *O. vigorsii* from near its type locality area, in both Jadhav et al.'s [25] study and the present investigation (Figure 3A). Jadhav et al. [25] did not observe unequal dorsal spreads of the urohyal in their specimens. Surprisingly, Jadhav et al. [25] did not examine the urohyal in freshly collected specimens from the Godavari River drainage. Although the other form of urohyal was not observed by Jadhav et al. [25], it warrants examination in specimens resembling *O. vigorsii* from the Godavari Basin. Jadhav et al. [25] indicated the presence of one pair of barbels in the type specimens of *O. bhimensis* and in comparative materials of *O. vigorsii* in their study. Remarkably, Jadhav et al. [25] identified a significant resemblance among images of the types of *O. bhimensis*, *O. vigorsii* from various sources, and Sykes' illustration [36]. Consequently, *O. bhimensis* is herein regarded as a junior synonym of *O. vigorsii*.

Sykes [36] provided detailed characteristics of *O. vigorsii*, emphasizing the upturned snout, straight upper line of the head, and the lower line curving upwards from below. The mouth structure observed in all specimens of *O. vigorsii* in the current study markedly differs from its congeners, representing a superior mouth type. The lower jaw exhibits strength with a hook-like structure at its distal tip, fitting into a small concavity at the distal tip of the upper jaw (Figure 3C). Day [37] noted the presence of a very rudimentary pair of maxillary barbels for *O. vigorsii*. The revision by Hora and Misra [23] provided additional insights, characterizing the species by the presence of a distinct concavity from the snout to over the nape and two rudimentary maxillary barbels. Simultaneously, Hora and Misra [23] described another new species, *O. dayi*, possessing two maxillary barbels. The presence or absence of barbels is considered a significant taxonomic trait in *Osteobrama* [21,23]. Based on this, Hora and Misra [23] grouped the species into three categories: (i) with four well-defined barbels, (ii) with two rudimentary maxillary barbels, and (iii) without barbels.

Morphological features for *O. vigorsii* have mostly been derived from Hora and Misra [23]. Despite Jadhav et al.'s [25] extensive examination of *O. vigorsii* specimens from the Krishna Basin, they failed to detect the presence of rostral barbels. In contrast, the present study reveals the presence of both maxillary and rostral barbels in *O. vigorsii* (Figure 3C), placing it in Group-(i) alongside *O. bakeri*, *O. feae*, *O. neilli*, and *O. tikarpadaensis*. Consequently, there is a need to revise the key to species within the genus *Osteobrama*. A revised key, adapted from Shangningam et al. [21], is provided below. Furthermore, the body morphometrics of *O. vigorsii* from the Krishna River (this study) are presented in Table 2.

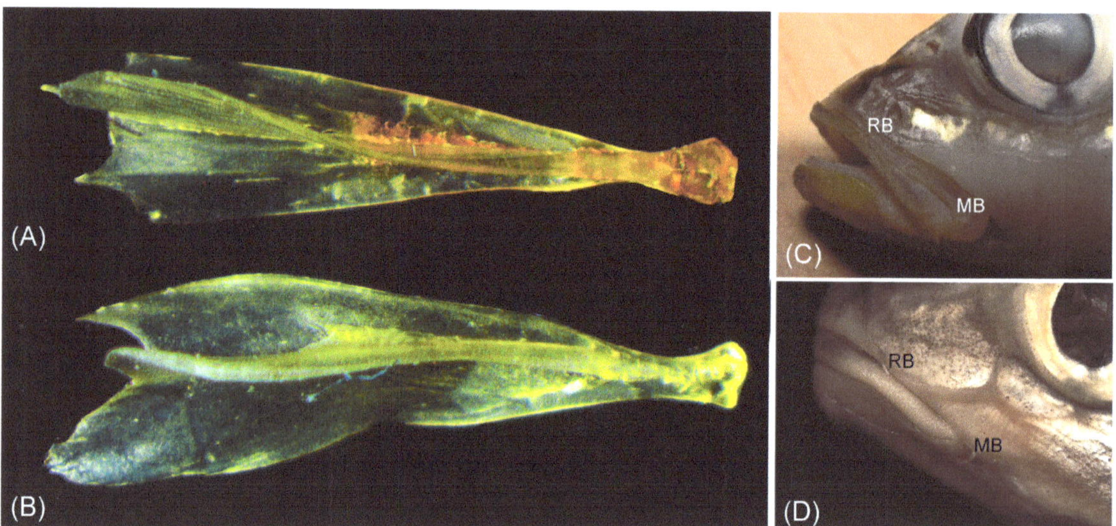

Figure 3. Dorsal view of urohyal bone in (**A**) *O. vigorsii* Krishna River, and (**B**) *O. tikarpadaensis* Godavari River, (**C**) *O. vigorsii*, FBRC/ZSI/ F3550, 119.5 mm SL; India: Telangana, Krishna River, showing the presence of barbels, (**D**) reproduced from Shangningam et al. [21] *O. tikarpadaensis* showing presence of barbels, reproduced with permission from the copyright holder ©Magnolia Press, and authorization for the utilization of the photograph was secured through direct communication with the corresponding author, Shibananda Rath. RB = rostral barbel, MB = maxillary barbel.

Table 2. Morphometric data of *O. vigorsii* from Krishna River (current study). SE, standard error.

Parameters	Range	Mean ± SE
Standard Length	90.5–132.0 mm	
In % SL		
Head length	24.1–28.0	25.8 ± 0.95
Head depth	17.6–22.0	18.9 ± 1.03
Head width	9.9–10.6	7.8 ± 2.59
Mouth width	6.3–7.4	5.2 ± 1.73
Body depth	31.5–35.2	33.2 ± 0.84
Body width	8.6–11.0	9.8 ± 0.49
Pre-dorsal length	51.8–56.9	55.0 ± 1.15
Pre-anal length	59.4–64.3	61.3 ± 1.14
Pre-pelvic length	32.6–42.0	38.6 ± 2.12
Pre-pectoral length	24.8–28.2	26.6 ± 0.75
Pelvic–anal distance	16.5–21.6	19.0 ± 1.21
Dorsal fin base length	11.3–12.4	11.8 ± 0.22
Anal fin base length	22.9–27.5	24.9 ± 0.93

Table 2. *Cont.*

Parameters	Range	Mean $\pm$ SE
Caudal peduncle length	12.6–16.5	13.9 $\pm$ 0.87
Caudal peduncle depth	11.5–13.8	12.1 $\pm$ 0.56
Snout length	6.4–8.3	7.3 $\pm$ 0.40
Eye diameter	7.0–7.4	7.3 $\pm$ 0.07
Inter-orbital distance	6.0–6.4	6.3 $\pm$ 0.11
Inter-narial space	4.3–4.9	4.6 $\pm$ 0.13
Dorsal fin height	28.3–34.3	30.5 $\pm$ 1.40
Pectoral fin length	19.8–20.2	20.0 $\pm$ 0.08
Anal fin height	16.5–19.9	18.1 $\pm$ 0.70
Pelvic fin length	20.5–23.2	21.4 $\pm$ 0.61
In % HL		
Eye diameter	26.3–30.0	28.3 $\pm$ 0.82
Interorbital width	23.0–26.7	24.6 $\pm$ 0.77
Head depth	62.7–82.8	73.8 $\pm$ 4.14
Head width	37.9–40.9	39.1 $\pm$ 0.73
Mouth width	24.8–30.5	26.6 + 1.34

3.2. Note on O. tikarpadaensis, with Urohyal Features

Shangningam et al. [21] delineated *O. tikarpadaensis* as a novel species, highlighting its unique features, particularly the oblique black streak on the anterior body immediately posterior to the opercle, parallel to the upper opercular margin, which distinguished it from all congeners. In the current investigation, it was noted that none of the *Osteobrama* specimens, with the exception of those resembling *O. vigorsii* from the Godavari River Basin, exhibited the distinct oblique black streak precisely described for *O. tikarpadaensis* (Figure 2A). The urohyal morphology in these *O. vigorsii*-like specimens from the Godavari River displayed two unequal ends posteriorly, with the left side being longer and thickened (Figure 3B), akin to one of the urohyal forms illustrated by Singh and Yazdani [38]. Consequently, the *O. vigorsii*-like specimens from the Godavari River Basin differ from their counterparts in the Krishna River Basin due to the presence of the oblique black streak immediately posterior to the opercle, urohyal characteristics with two unequal ends posteriorly, and a combination of other morphological features. These variations, previously overlooked, challenge the previous taxonomic assessments. Singh and Yazdani [38], despite noting some variations, failed to accurately identify true *O. vigorsii*, rendering their proposed new species (*O. bhimensis*) invalid. Following the discovery of *O. tikarpadaensis* by Shangningam et al. [21], it is evident that the taxonomic characteristics of *O. vigorsii* were confounded by the representation of characters from two distinct species. Consequently, the *O. vigorsii*-like specimens from the Godavari River are now identified as *O. tikarpadaensis*, a distinction further supported by genetic divergence analysis highlighting the dissimilarity between *O. tikarpadaensis* and *O. vigorsii*.

Despite the documented pre-dorsal distance for *O. tikarpadaensis* being reported as 37.8–40.4% of standard length (SL) in its descriptions [21], this measurement seems either exceptionally shorter compared to congeners (e.g., 53.5–56.1% in *O. feae*, 53.0–56.5% in *O. neilli*, 55.8–56.1% in *O. belangeri*, 51.2–52.2% in *O. cotio*, 51.8–56.9% in *O. vigorsii* in this study) or may represent inaccurate data. Notably, Shangningam et al. [21] did not examine any specimens from the Godavari River Basin. Rath et al. [22] identified *O. tikarpadaensis* from an old collection from the Erai River, Chandrapur District, Maharashtra. However, discrepancies in body morphometry, including the pre-dorsal distance, that were reported by Rath et al. [22] compared closely to the descriptions of *O. tikarpadaensis*. The body morphometrics of *O. tikarpadaensis* from the Godavari River in this study, along with a reproduction of the data from Shangningam et al. [21], are presented to facilitate a comprehensive understanding (Table 3).

Table 3. Morphometric data of *O. tikarpadaensis* from the Godavari River (current study) and from Shangningam et al. [21].

Parameters	Range	Shangningam et al. [21]
Standard Length	101.0–102.0 mm	
In % SL	Specimens from Godavari River	
Head length	26.6–26.6	24.5–28.8
Head depth	11.7–12.7	16.4–18.6
Head width	11.0–12.0	13.2–14.4
Mouth width	5.2–5.7	5.6–7.1
Body depth	32.3–35.1	34.5–39.5
Body width	9.7–10.8	9.3–11.7
Pre-dorsal length	50.0–53.2	37.8–40.4
Pre-anal length	53.2–59.7	60.0–61.7
Pre-pelvic length	40.5–41.6	39.9–43.1
Pre-pectoral length	26.0–26.6	24.7–26.6
Pelvic–anal distance	13.9–15.6	19.7–21.3
Dorsal fin base length	11.7–12.0	13–14.2
Anal fin base length	27.8–28.6	29.5–32
Caudal peduncle length	12.0–15.6	14.5–15.6
Caudal peduncle depth	10.1–11.0	10.3–12.2
Snout length	7.1–7.6	7.1–8.3
Eye diameter	6.5–7.0	6.7–8.3
Inter-orbital distance	7.8–8.2	8.7–10.0
Inter-narial space	5.2–5.7	5.0–6.0
Dorsal fin height	27.2–27.3	24.6–29.4
Pectoral fin length	18.8–19.0	19.2–21.2
Anal fin height	13.0–13.3	29.7–31.7
Pelvic fin length	17.1–18.8	17.6–18.9

3.3. Genetic Inferences

The estimated genetic divergence (K2P) between the groups (genera) ranged from 17.3% to 18.4%. In both ML and BA phylogenetic trees (Figures 4 and S2), the de novo sequences of *O. vigorsii* from the Krishna River, including four database sequences labelled *O. cotio*, constitute a cohesive cluster with pairwise genetic distances (K2P) ranging from 0.00 to 0.77%. By analyzing COI data, we confirm that the four database sequences (KX946745 to KX946748) collected from Kolhapur, Maharashtra, likely from the Dhamna River, a tributary of the Krishna River, are indeed conspecific with *O. vigorsii*. The de novo sequence of *O. cotio* from Maharashtra State is placed in a distinct cluster, previously identified as *O. cotio* in an earlier study, and comprises conspecific sequences from the Brahmaputra and Meghna Rivers [20]. Rahman et al. [20] demonstrated that COI sequences identified as *O. cotio* from the Narmada River Basin, used in studies by Khedkar et al. [10] and Singh et al. [39], exhibited greater genetic distance from *O. cotio* from the Barak and Brahmaputra River Basins. We identified *O. peninsularis* from the Godavari River Basin, maintaining a 5.4% K2P genetic divergence with *O. cotio* from the Barak and Brahmaputra River Basins. The de novo sequence of *O. peninsularis* formed a distinct cluster with database sequences (KF550101 to KF550103) with pairwise genetic distances ranging from 0.0 to 0.62% and maintaining 5.28 to 5.68% genetic distances within the cluster of *O. cotio*. Sequences (KF550101 to KF550103) previously misidentified as *O. cotio* by Khedkar et al. [10] and Singh et al. [39] aligned with one of the subclades of Clade A, as referred to in Rahman et al. [20]. Similarly, the de novo sequence of *O. neilli* formed a cohesive cluster with one database sequence of *Osteobrama* sp. sampled from Kolhapur, Maharashtra, maintaining only a 0.8% genetic divergence within the group.

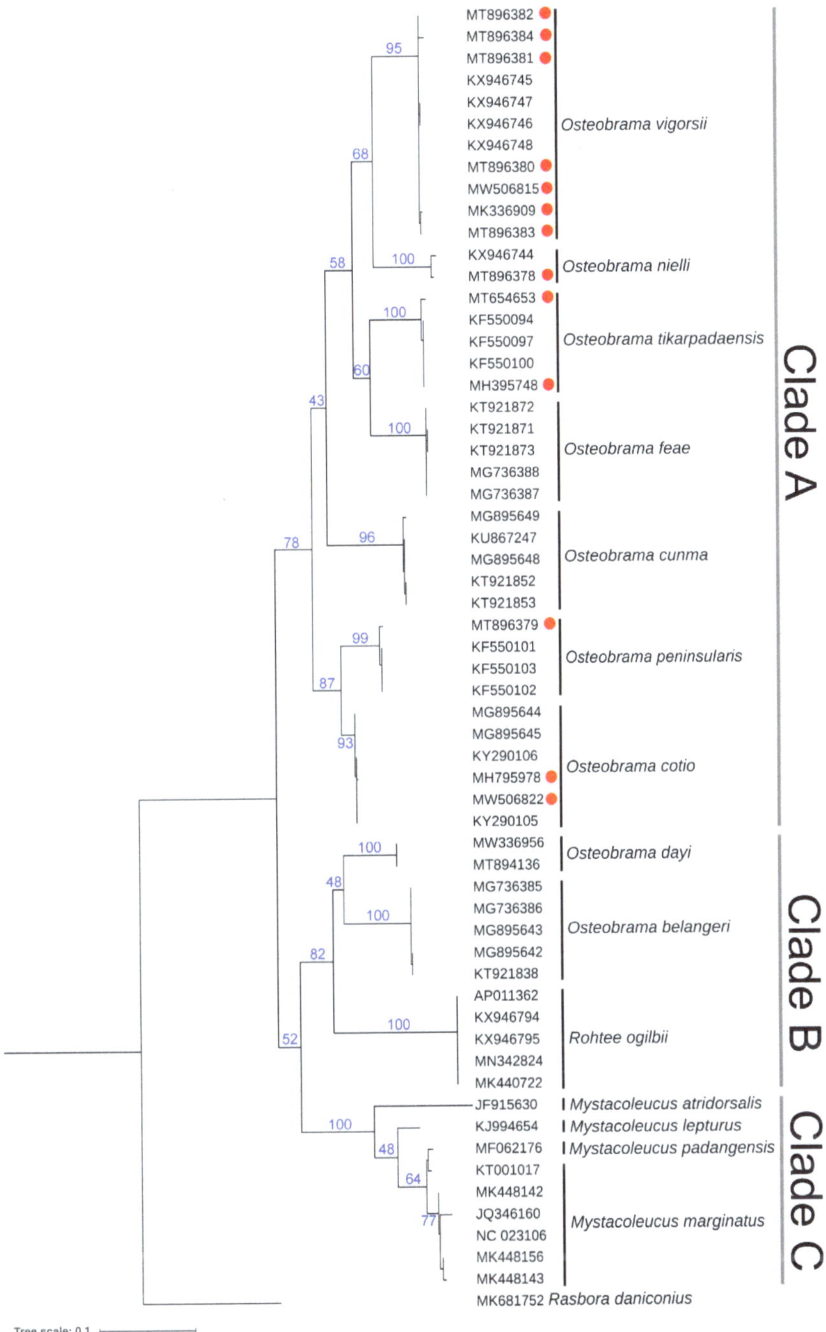

Figure 4. Maximum likelihood phylogeny of *Osteobrama* congeners, utilizing mtCOI data, distinctly separating *O. vigorsii* and *O. tikarpadaensis*. The sequences generated in this study are highlighted with red dots, and the species names with their corresponding GenBank accession numbers are indicated on the tree. Bootstrap support values are indicated at each node in blue font.

O. vigorsii from the Krishna River exhibits characteristics that are distinct from its congeners, as indicated by a K2P genetic divergences ranging from 9.31% to 17.50% in the partial COI gene sequence (Table 4). Its lowest genetic divergence (9.31%) was observed with *O. neilli*, while the highest (17.50%) was with *O. belangeri*. The sequences of *O. vigorsii* form a well-defined cluster, which also incorporated four database sequences with locality information in India, specifically Maharashtra, Kolhapur, the Northern Western Ghats of Kolhapur, and Gavashi, likely from the Dhamana River, a tributary of the Krishna River, situated at coordinates 16.605 N 73.987 E. Among the generated sequences of *Osteobrama*, two specimens from the Godavari River formed a cohesive cluster with certain database sequences identified as *O. vigorsii* but lacking locality information. These specimens maintained a significant interspecies genetic distance (10.64 to 12.35%) from sequences of *O. vigorsii* from the Krishna River Basin. Notably, the specimens from the Godavari River, characterized by having two pairs of barbels, represented a distinct species identified as *O. tikarpadaensis* based on morphological traits (Figure 2A,B).

Table 4. The estimated K2P genetic distance among the respective *Osteobrama* congeners.

Species	Between Species (%)								Within Species (%)
O. cotio									0.2
O. peninsularis	5.28–5.68								0.3
O. cunma	10.64–11.23	10.46–11.25							0.3
O. feae	12.23–13.23	12.58–13.88	13.34–14.94						0.1
O. tikarpadaensis	12.54–13.43	13.90–15.17	13.33–14.33	9.25–9.77					0.3
O. neilli	12.14–12.75	12.50–12.96	11.91–12.58	11.48–12.23	11.53–11.80				0.8
O. vigorsii	10.95–11.95	12.91–14.47	13.88–15.31	10.96–11.97	10.64–12.35	9.31–10.26			0.3
O. belangeri	13.50–14.37	15.22–16.16	15.33–16.24	18.92–20.08	16.48–17.73	16.98–17.60	15.32–17.50		0.1
O. dayi	16.2–16.5	16.7–16.9	15.5–15.8	17.6–17.8	16.5–17.3	16.3–16.5	17.5–18.5	11.1–11.4	0.0

The current phylogenetic analysis robustly distinguished the two targeted species, *O. vigorsii* and *O. tikarpadaensis*, with high bootstrap support. However, among the congeners, the phylogenetic analysis revealed inconsistent clustering in certain instances. Differently named sequences occasionally exhibited conspecific clustering, as discussed in the preceding section. Although the genus *Rohtee* is presently regarded as monotypic [18,23], *Rohtee ogilbii* Sykes 1839 exhibited a cohesive clustering with *O. dayi* and *O. belangeri* in the current phylogeny (Clade B), irrespective of the presence or absence of barbels. These three species in Clade B shared similarities as deep-bodied and large-growing, surpassing the sizes of any species in Clade A. Both Clade A and Clade B species exhibited similarities in possessing a long anal fin with more than 11 branched anal fin rays, distinguishing them from species in *Mystacoleucus* (Clade C) due to the length of the anal fin. *R. ogilbii*, while showing similarity to *Mystacoleucus* species through the presence of a procumbent pre-dorsal spine, formed a separate clade with a mean genetic divergence of 17.3%. The procumbent pre-dorsal spine in *R. ogilbii* is reduced and somewhat concealed by scales compared to other *Mystacoleucus* species. However, additional sampling with multiple gene markers would be necessary to assess the potential merging of the two genera, *Rohtee* and *Osteobrama*.

Taxonomic investigations involving an ample number of specimens from diverse taxonomic lineages, along with their DNA data, have proven effective in illuminating the diversity and phylogeographic structure associated with biogeography [40,41]. The results from this comprehensive research indicate that *O. vigorsii* may have a limited distribution primarily within the Krishna River system in southern India, whereas *O. tikarpadaensis* might potentially have a larger range from the Mahanadi River in Central India to the Godavari River in southern India (Figure 5). Further, considering the present phylogeny and genetic distances among *Osteobrama* congeners, this study underscores the need for additional genetic data from various riverine systems in the Indian subcontinent to unravel the true species diversity of these cyprinids, thereby informing future conservation implications.

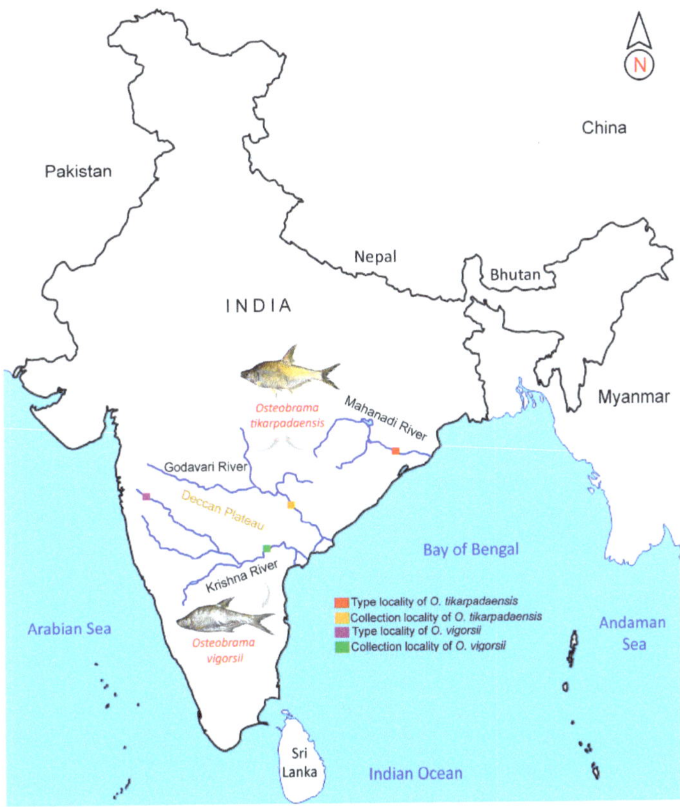

Figure 5. The true distributions of *O. vigorsii* and *O. tikarpadaensis* across diverse riverine systems in India.

3.4. Revised Key to Species of the Genus Osteobrama

1. Barbels absent	2.
- Barbels present	5.
2. Lateral line scales 42–63, pre-dorsal scales 21–30	3.
- Lateral line scales 71–76, pre-dorsal scales 30–32	*O. belangeri.*
3. Branched pectoral fin rays 14–15, lateral line scales 55–63	4.
- Branched pectoral fin rays 12, lateral line scales 42–53	*O. cunma.*
4. Lateral-line scales 55–60	*O. peninsularis.*
- Lateral-line scales 62–63	*O. cotio.*
5. Both rostral and maxillary barbels present	6.
- Only maxillary barbels present, branched anal fin rays 16–18, lateral line scales 68–70	*O. dayi.*
6. Barbels prominent	7.
- Barbels minute	9.
7. Branched anal fin rays 11–18	8.
- Branched anal fin rays 22–27, pre-dorsal scales 34–38, branched pectoral fin rays 14	*O. feae.*
8. Pre-dorsal scales 15, lateral line scales 44, branched anal fin ray 11	*O. bakeri.*
- Pre-dorsal scales 19–22, lateral line scales 52–57, branched anal fin rays 16–18	*O. nielli.*
9. Branched anal fin rays 25–27, branched pectoral fin rays 15–16, presence of oblique black streak on the body immediately posterior to the operculum, lateral line scales 59–71	*O. tikarpadaensis.*
- Branched anal fin rays 21–23, branched pectoral fin rays 13–14, lateral line scales 74–84, no oblique black streak on the body	*O. vigorsii.*

4. Conclusions

Prior to this investigation, the systematic and phylogenetic relationships between two *Osteobrama* species, *O. vigorsii* and *O. tikarpadaensis*, presented challenges to ichthyologists, causing confusion. Proposed amendments to the morphological characters of both *O. vigorsii* and *O. tikarpadaensis,* accompanied by revised taxonomic keys for distinguishing *Osteobrama* congeners, aim to address these challenges. Although the urohyal bone structure offers insights into these two *Osteobrama* species, vertebra and rib counts are expected to provide more informative data for future investigations. The inter-species genetic divergence and maximum likelihood phylogeny distinctly differentiate *O. vigorsii* and *O. tikarpadaensis.* The study's findings indicate that *O. vigorsii* may have a restricted distribution in the Krishna River system in southern India, while *O. tikarpadaensis* could potentially extend from the Mahanadi River in central India to the Godavari River in southern India. The genetic diversity information obtained from various riverine systems for *Osteobrama* species will be pivotal in guiding aquaculture practices and formulating effective conservation action plans. A similar integrated approach with morphological and molecular data provides a resource for future investigations of cyprinids in India and neighboring regions.

Supplementary Materials: The following supporting information can be downloaded at: https://www.mdpi.com/article/10.3390/fishes9030087/s1, Figure S1: (A) *Osteobrama cotio,* reproduced from Hamilton (1822) plate 207; (B) *Osteobrama neilli,* FBRC/ZSI/F/3548, 68.0 mm SL; India: Telangana, Nagarkurnool District, Krishna River: near Somasila; (C) *Osteobrama peninsularis,* FBRC/ZSI/F/3549, India; Telangana, Wyra lake, Khamam District; (D) *Osteobrama dayi,* FBRC-ZSI-F 3795, India: Telangana, Godavari River. Photo credit @Boni Amin Laskar; Figure S2: Bayesian phylogeny of *Osteobrama* congeners distinctly separating *O. vigorsii* and *O. tikarpadaensis.* The sequences generated in this study are highlighted with red dots, and the species names with their corresponding GenBank accession numbers are indicated on the tree. Posterior probability values are indicated at each node.

Author Contributions: Conceptualization: B.A.L. and S.K.; methodology: B.A.L., S.C. and A.R.K.; software: B.A.L., A.R.K. and S.K.; validation: D.B. and H.-W.K.; formal analysis: B.A.L., A.R.K. and S.K.; investigation: S.C., A.R.K. and S.K.; resources: B.A.L., D.B. and H.-W.K.; data curation: B.A.L., S.C. and A.R.K.; writing—original draft: B.A.L. and S.K.; writing—review and editing: B.A.L., A.R.K. and S.K.; visualization: D.B., S.C. and H.-W.K.; supervision: D.B. and H.-W.K.; project administration: B.A.L., A.R.K. and S.K.; funding acquisition: H.-W.K., A.R.K. and S.K. All authors have read and agreed to the published version of the manuscript.

Funding: This research was supported by the Basic Science Research Program through the National Research Foundation of Korea (NRF) funded by the Ministry of Education (2021R1A6A1A03039211) and supported by the National Institute of Fisheries Science, Ministry of Oceans and Fisheries, Korea (R2023003).

Institutional Review Board Statement: Given that the specimens examined in this article are edible freshwater fishes typically captured from unprotected water bodies, the fish specimens are exempt from any provisions of animal ethics. Consequently, the authors declare that the study does not involve any act of animal ethics.

Informed Consent Statement: Not applicable.

Data Availability Statement: The DNA sequence data that support the findings of this study are available in NCBI GenBank (https://www.ncbi.nlm.nih.gov/nuccore) and BOLD systems (https://www.boldsystems.org/), under the accession numbers and BOLD-IDs presented in Table 1.

Acknowledgments: The authors are grateful to the Director of the Zoological Survey of India, Kolkata for her continued support and enthusiasm and for providing necessary facilities throughout the study. The authors further thank the Officer-in-Charge at the Freshwater Biology Regional Centre, Zoological Survey of India, Hyderabad, for providing necessary facilities to carry out this study.

Conflicts of Interest: The authors declare no conflicts of interest.

References

1. Berra, T.M. *Freshwater Fish Distribution;* The University of Chicago Press: Chicago, IL, USA, 2007.
2. Banerjee, D.; Raghunathan, C.; Rizvi, A.N.; Das, D. *Animal Discoveries 2022: New Species and New Records;* Zoological Survey of India: Kolkata, India, 2023; pp. 1–349.

3.	Froese, R.; Pauly, D. (Eds.) FishBase. World Wide Web Electronic Publication. Version (08/2022). Available online: www.fishbase. org (accessed on 15 January 2024).

4.	Hubert, N.; Hanner, R.; Holm, E.; Maandrak, N.E.; Taylor, E.; Burridge, M.; Watkinson, D.; Dumont, P.; Curry, A.; Bentzen, P.; et al. Identifying Canadian freshwater fishes through DNA barcodes. *PLoS ONE* **2008**, *3*, e2490. [CrossRef] [PubMed]

5.	Steinke, D.; Zemlak, T.S.; Hebert, P. Barcoding Nemo: DNA-based identifications for the ornamental fish trade. *PLoS ONE* **2009**, *4*, e6300. [CrossRef] [PubMed]

6.	Hebert, P.D.N.; Cywinska, A.; Ball, S.L.; deWaard, J.R. Biological identifications through DNA barcodes. *Proc. Biol. Sci.* **2003**, *270*, 313–321. [CrossRef] [PubMed]

7.	Collins, R.A.; Armstrong, K.F.; Meier, R.; Yi, Y.; Brown, S.D.J.; Cruickshank, R.H.; Keeling, S.; Johnston, C. Barcoding and border biosecurity: Identifying cyprinid fishes in the aquarium trade. *PLoS ONE* **2012**, *7*, e28381. [CrossRef] [PubMed]

8.	Bhattacharjee, M.J.; Laskar, B.A.; Dhar, B.; Ghosh, S.K. Identification and re-evaluation of freshwater catfishes through DNA barcoding. *PLoS ONE* **2012**, *7*, e49950. [CrossRef] [PubMed]

9.	Laskar, B.A.; Bhattacharjee, M.J.; Dhar, B.; Mahadani, P.; Kundu, S.; Ghosh, S.K. The species dilemma of northeast Indian mahseer (Actinopterygii: Cyprinidae): DNA barcoding in clarifying the riddle. *PLoS ONE* **2013**, *8*, e53704. [CrossRef]

10.	Khedkar, G.D.; Jamdade, R.; Naik, S.; David, L.; Haymer, D. DNA barcodes for the fishes of the Narmada, one of India's longest rivers. *PLoS ONE* **2014**, *9*, e101460. [CrossRef]

11.	Lakra, W.S.; Singh, M.; Goswami, M.; Gopalakrishnan, A.; Lal, K.K.; Mohindra, V.; Sarkar, U.K.; Punia, P.P.; Singh, K.V.; Bhatt, J.P.; et al. DNA barcoding Indian freshwater fishes. *Mitochondrial DNA Part A DNA Mapp. Seq. Anal.* **2016**, *27*, 4510–4517. [CrossRef]

12.	Barman, A.S.; Singh, M.; Singh, S.K.; Saha, H.; Singh, Y.J.; Laishram, M.; Pandey, P.K. DNA barcoding of freshwater fishes of Indo-Myanmar biodiversity hotspot. *Sci. Rep.* **2018**, *8*, 8579. [CrossRef]

13.	Laskar, B.A.; Kumar, V.; Kundu, S.; Tyagi, K.; Chandra, K. Taxonomic quest: Validating two mahseer fishes (Actinopterygii: Cyprinidae) through molecular and morphological data from biodiversity hotspots in India. *Hydrobiologia* **2018**, *815*, 113–124. [CrossRef]

14.	Kundu, S.; Chandra, K.; Tyagi, K.; Pakrashi, A.; Kumar, V. DNA barcoding of freshwater fishes from Brahmaputra River in Eastern Himalaya biodiversity hotspot. *Mitochondrial DNA B Resour.* **2019**, *4*, 2411–2419. [CrossRef]

15.	Laskar, B.A.; Kumar, V.; Kundu, S.; Darshan, A.; Tyagi, K.; Chandra, K. DNA barcoding of fishes from River Diphlu within Kaziranga National Park in northeast India. *Mitochondrial DNA Part A DNA Mapp. Seq. Anal.* **2019**, *30*, 126–134. [CrossRef]

16.	Ward, R.D.; Hanner, R.; Hebert, P.D. The campaign to DNA barcode all fishes, FISH-BOL. *J. Fish Biol.* **2009**, *74*, 329–356. [CrossRef]

17.	Talwar, P.K.; Jhingran, A.G. *Inland Fishes of India and Adjacent Countries*; Oxford & IBH Publishing Co.: New Delhi/Bombay/Calcutta, India, 1991; Volume 1, 541p.

18.	Fricke, R.; Eschmeyer, W.N.; Van der Laan, R. (Eds.) Eschmeyer's Catalog of Fishes: Genera, Species, References. 2023. Available online: http://researcharchive.calacademy.org/research/ichthyology/catalog/fishcatmain.asp (accessed on 15 January 2024).

19.	Jayaram, K.C. *The Freshwater Fishes of the Indian Region*, 2nd ed.; Narendra Publishing House: Delhi, India, 2010; 616p.

20.	Rahman, M.M.; Norén, M.; Mollah, A.R.; Kullander, S. The identity of *Osteobrama cotio*, and the status of "*Osteobrama serrata*" (Teleostei: Cyprinidae: Cyprininae). *Zootaxa* **2018**, *4504*, 105–118. [CrossRef]

21.	Shangningam, B.; Rath, S.; Tudu, A.K.; Kosygin, L. A new species of *Osteobrama* (Teleostei: Cyprinidae) from the Mahanadi River, India with a note on the validity of *O. dayi*. *Zootaxa* **2020**, *4722*, 68–76. [CrossRef]

22.	Rath, S.; Tudu, A.K.; Shangningam, B. First record of *Osteobrama tikarpadaensis* (Teleostei: Cyprinidae) from Maharashtra India. *Int. J. Fish. Aquat. Stud.* **2021**, *9*, 250–252. [CrossRef]

23.	Hora, S.L.; Misra, K.S. Notes on fishes in the Indian Museum. XL. On fishes of the genus *Rohtee* Sykes. *Rec. Indian Mus.* **1940**, *42*, 155–172.

24.	Jayaram, K.C. *The Krishna River System Bioresources Study*; Occasional Paper, No. 160; Zoological Survey of India: Kolkata, India, 1995.

25.	Jadhav, S.; Paingankar, M.; Dahanukar, N. *Osteobrama bhimensis* (Cypriniformes: Cyprinidae): A junior synonym of *O. vigorsii*. *J. Threat. Taxa* **2011**, *3*, 2078–2084. [CrossRef]

26.	Hamilton, F. *An Account of the Fishes Found in the River Ganges and Its Branches*; A. Constable & Co.: Edinburgh/London, UK, 1822; vii + 405p, 39 pls.

27.	Ward, R.D.; Zemlak, T.S.; Innes, B.H.; Last, P.R.; Hebert, P.D.N. DNA barcoding Australia's fish species. *Philos. Trans. R. Soc. B Biol. Sci.* **2005**, *360*, 1847–1857. [CrossRef] [PubMed]

28.	Kumar, S.; Stecher, G.; Li, M.; Knyaz, C.; Tamura, K. MEGA X: Molecular Evolutionary Genetics Analysis across computing platforms. *Mol. Biol. Evol.* **2018**, *35*, 1547–1549. [CrossRef] [PubMed]

29.	Thompson, J.D.; Gibson, T.J.; Plewniak, F.; Jeanmougin, F.; Higgins, D.G. The CLUSTAL_X windows interface: Flexible strategies for multiple sequence alignment aided by quality analysis tools. *Nucleic Acids Res.* **1997**, *25*, 4876–4882. [CrossRef] [PubMed]

30.	Lanfear, R.; Frandsen, P.B.; Wright, A.M.; Senfeld, T.; Calcott, B. PartitionFinder 2: New methods for selecting partitioned models of evolution for molecular and morphological phylogenetic analyses. *Mol. Biol. Evol.* **2016**, *34*, 772–773. [CrossRef] [PubMed]

31.	Miller, M.A.; Schwartz, T.; Pickett, B.E.; He, S.; Klem, E.B.; Scheuermann, R.H.; Passarotti, M.; Kaufman, S.; O'Leary, M.A. A RESTful API for access to phylogenetic tools via the CIPRES science gateway. *Evol. Bioinform.* **2015**, *11*, 43–48. [CrossRef]

32.	Darriba, D.; Taboada, G.L.; Doallo, R.; Posada, D. JModelTest 2: More models, new heuristics and parallel computing. *Nat. Methods* **2012**, *9*, 772. [CrossRef]

33. Guindon, S.; Dufayard, J.F.; Lefort, V.; Anisimova, M.; Hordijk, W.; Gascuel, O. New algorithms and methods to estimate maximum-likelihood phylogenies: Assessing the performance of PhyML 3.0. *Syst. Biol.* **2010**, *59*, 307–321. [CrossRef] [PubMed]
34. Ronquist, F.; Huelsenbeck, J.P. MrBayes 3: Bayesian phylogenetic inference under mixed models. *Bioinformatics* **2003**, *19*, 1572–1574. [CrossRef] [PubMed]
35. Letunic, I.; Bork, P. Interactive Tree of Life (iTOL): An online tool for phylogenetic tree display and annotation. *Bioinformatics* **2007**, *23*, 127–128. [CrossRef] [PubMed]
36. Sykes, W.H. On the fishes of the Dukhun. *Trans. Zool. Soc. Lond.* **1841**, *2*, 349–378. [CrossRef]
37. Day, F. *The Fauna of British India, Including Ceylon and Burma*; Taylor and Francis: London, UK, 1889; Volume 2, 509p.
38. Singh, D.F.; Yazdani, G.M. *Osteobrama bhimensis*, a new cyprinid fish from Bhima River, Pune District, Maharahtra. *J. Bombay Nat. Hist. Soc.* **1992**, *89*, 96–99.
39. Singh, M.; Verma, R.; Yumnam, R.; Vishwanath, W. Molecular phylogenetic analysis of genus *Osteobrama* Heckel, 1843 and discovery of *Osteobrama serrata* sp. nov. from northeast India. *Mitochondrial DNA Part A* **2018**, *29*, 361–366. [CrossRef]
40. Moritz, C.; Cicero, C. DNA barcoding: Promise and pitfalls. *PLoS Biol.* **2004**, *2*, e354. [CrossRef] [PubMed]
41. Bickford, D.; Lohman, D.J.; Sodhi, N.S.; Ng, P.K.; Meier, R.; Winker, K.; Ingram, K.K.; Das, I. Cryptic species as a window on diversity and conservation. *Trends Ecol. Evol.* **2007**, *22*, 148–155. [CrossRef] [PubMed]

Article

A Cross-Decadal Change in the Fish and Crustacean Community of Lower Yaquina Bay, Oregon, USA

Scott A. Heppell *, Selina S. Heppell, N. Scarlett Arbuckle and M. Brett Gallagher

Department of Fisheries, Wildlife, and Conservation Sciences, Oregon State University, Corvallis, OR 97330, USA;
selina.heppell@oregonstate.edu (S.S.H.); scarlett.arbuckle@oregonstate.edu (N.S.A.);
gallagher.brett@gmail.com (M.B.G.)
* Correspondence: scott.heppell@oregonstate.edu; Tel.: +1-541-737-1086

Abstract: Natural environmental change, anthropogenic development, and inter-annual variability can affect the ecology of estuarine fish and invertebrates. Yaquina Bay, Oregon, a well-studied estuary, has undergone intense development, as well as deep-draft dredging during the latter half of the 20th century, resulting in the alteration of ~45% of the lower estuary's natural shoreline. In 1967, the United States Environmental Protection Agency (USEPA) conducted a 21-month survey of Yaquina Bay to characterize the demersal fishes and epibenthic crustaceans that occupy the bay. From 2003 to 2005, we conducted a 25-month survey to replicate that work and provide a comparative snapshot of the demersal fish and epibenthic crustacean community in the bay. A comparison of the trawl survey datasets reveals a 91% decline in total catch per unit effort (CPUE) between surveys, as well as a decline in multiple measures of biodiversity. Furthermore, the fishes and crustaceans of Yaquina Bay have experienced a shift in species dominance from demersal fishes in the late 1960s to epibenthic crustaceans in the 2000s, marked most notably by a nine-fold increase in the Dungeness crab CPUE. While this work does not establish a causal relationship between changes in the demersal communities of this West Coast estuary and human or natural events, it does document substantial changes in both the diversity and total abundance of animals in that community over a three-plus decade period of development and environmental variability. Hence, this forms a second baseline for continued long-term monitoring.

Keywords: long-term monitoring; community change; habitat alteration; diversity; abundance; natural variability; estuaries

Key Contribution: Our work documents a multi-decadal change in faunal abundance, diversity, and dominance in a highly developed northeast Pacific estuary. While causal mechanisms are not identified, estuarine development and a changing climate likely contribute to these changes.

Citation: Heppell, S.A.; Heppell, S.S.; Arbuckle, N.S.; Gallagher, M.B. A Cross-Decadal Change in the Fish and Crustacean Community of Lower Yaquina Bay, Oregon, USA. *Fishes* **2024**, *9*, 125. https://doi.org/10.3390/fishes9040125

Academic Editors: Robert L. Vadas, Jr., Robert M. Hughes and José Lino Costa

Received: 27 February 2024
Revised: 26 March 2024
Accepted: 28 March 2024
Published: 30 March 2024

1. Introduction

Shoreline development and watershed urbanization are well-demonstrated drivers of coastal ecosystem change. Physical development, including shoreline armoring and construction of hardened structures, disrupts benthic communities, thereby reducing prey resources for nearshore fish and wildlife [1] and habitat available for epibenthic crustaceans [2]. It is possible that even small, human-built structures influence habitat structure, fish distribution, migration, feeding behavior, and availability of prey resources [3]. Chemical alterations, in the form of nutrient and pollutant inputs, have effects ranging from altering primary productivity and microbial community composition to changing community structure [4–7]. Other activities associated with the urbanization of estuaries, such as channel dredging for ship operations, alter available habitat and tidal flow, with subsequent biotic impacts [8]. These types of impacts are widespread and long occurring [9], and it is likely that ecosystem change continues to occur in estuaries throughout the United States and elsewhere. Long-term monitoring of these types of changes, or even contemporary point comparisons with historical data, however, are rare [10], but see [7], so it is often difficult to understand the long-term effect of urbanization on estuarine biotic communities.

Yaquina Bay, a 15.8 km^2 drowned river mouth estuary on the central Oregon coast, is the fourth largest estuary in Oregon and is classified as "developed, deep draft" by the Oregon Department of Land Conservation and Development [11]. The bay experiences mixed-semidiurnal tides that influence the Yaquina River as far as 42 km upstream [12]. Local river flow is highest during the rainy winter months when discharge reaches nearly 70 m^{3}s^{-1} [13]. Yaquina Bay is also a well-recognized juvenile nursery ground for commercially important species on the central coast, including the Dungeness crab, flatfish and rockfish [14–17].

The port and town of Newport and adjacent South Beach, located in the lower part of the bay, have undergone intense shoreline development since the 1940s. In addition to supporting logging ships and a commercial fishing fleet, the last half of the past century has brought an extended armored inlet, a deeper channel, a public marina, an expanded marine laboratory, waterfront shopping, shoreline condominiums, a liquid natural gas storage plant, and, most recently, a fully revamped, international terminal and facility to support a fleet of scientific research vessels at Oregon State University's Hatfield Marine Science Center [18–20]. These installations in the lower estuary of Yaquina Bay have resulted in the alteration of at least 45% of the natural shoreline (as calculated directly from the Oregon Coastal Management Plan's Coastal Atlas [21] and may have changed river and tidal hydrology, salinity, dissolved oxygen, and temperatures as well [22]. Each of these installations required shoreline development, environmental mitigation, and continued maintenance that may have sustained impacts on the quality of the habitat in Yaquina Bay for both resident and seasonal inhabitants of the estuary. Newport, as a deep-draft estuary, has a main channel regularly dredged to a depth of 13.2 m, deepened from 8.5 m in 1969 [20], with dredge spoil deposited in USACE/USEPA designated locations in the nearshore area adjacent to the mouth of the bay.

In 1967–1968, the United States Environmental Protection Agency (USEPA) conducted a 21-month trawl survey of Yaquina Bay to investigate the spatio-temporal fluctuations in the distribution and abundance of demersal fishes and epibenthic crustaceans [23]. This survey included 42 bi-weekly otter trawls across 10 stations in Yaquina Bay, spanning five previously identified salinity and temperature ranges [24,25]. Yaquina Bay has since been the subject of consistent monitoring, experimentation, and evaluation, with over 1400 theses, dissertations, white papers, and peer-reviewed documents referencing Yaquina Bay being published since 1968 (bibliographic database maintained by the Oregon State University Hatfield Marine Science Center Guin Library [26]). Throughout the 1970s, fish research in the bay focused on fish ecology, with an emphasis on the effect of local upwelling conditions and the bay's role in recruitment and capacity as a nursery ground [14,15,17,27–32]. However, no long-term repeated assessments of the demersal community at the scale of De Ben et al. [23] have been conducted since that original study. In 2003, we began a 25-month trawl survey intended to replicate that work for the first three of the five salinity and temperature ranges previously identified in Yaquina Bay. Our goal was to provide an updated snapshot following multiple decades of shoreline development and environmental change and to determine what, if any, impacts those changes may have had on community structure and diversity.

2. Materials and Methods

We conducted bi-weekly otter trawls at five sampling locations in Yaquina Bay from January 2003 to October 2005 (Figure 1), resulting in a total of 139 net sets, compared with the 126 net sets during the 1967–1968 survey. Estuary sections sampled based on salinity and temperature regimes followed the original survey [23]; however, the 2003–2005 trawl sampling was confined to sections 1–3 of Yaquina Bay. All sampling of demersal fish and epibenthic crustaceans occurred during daylight hours, following slack high tide in water depths ranging from 2 to 10 m, with a mean of 3.7 m. Trawl samples were collected using a two-seam, 3.5 m treated nylon shrimp trawl with 3.4 cm body netting and 1.3 cm cod end netting, the same style net used by De Ben and co-workers in the original study [23]. To retain smaller organisms, a 9.5-mm cod end liner was sewn into the net. All trawling was conducted against the prevailing current, and trawl duration was standardized to ten minutes, which facilitated catch per unit effort (CPUE) comparisons with the 1967–1968 survey data published by [23].

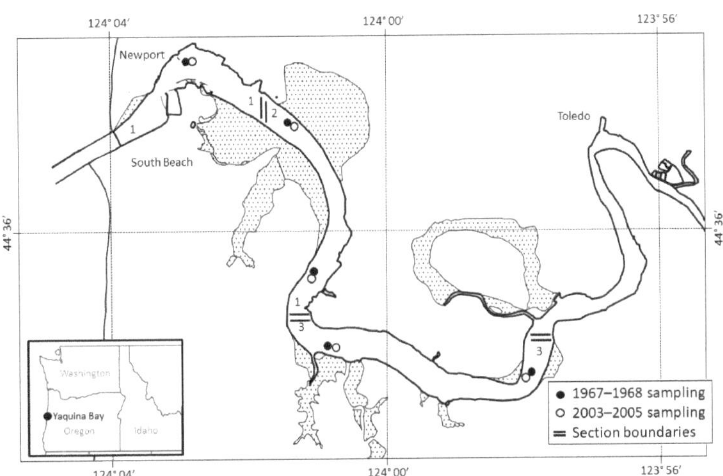

Figure 1. Map of Yaquina Bay, Oregon, with sampling locations for both the 1967–1968 (closed circles) and 2003–2005 (open circles) surveys. Sampling sections 1, 2, and 3, as designated by [23] and replicated for this study, are indicated by solid lines separating sections of the estuary. The present study collected comparative data for the three lower bay sections, chosen to facilitate direct comparison with historical data.

All captured organisms were identified to the lowest possible taxonomic level, measured, and released at the point of capture. Fish and shrimp were measured to the nearest millimeter total length, and crabs were measured across the widest point of the carapace in millimeters. Bottom water temperature and salinity were measured before each trawl using a YSI model 85 handheld multi-meter. Environmental and CPUE data for sections 1–3 of Yaquina Bay were compared between the 2003–2005 survey and the 1967–1968 survey [23].

Catch per unit effort (CPUE) was calculated by dividing the catch of species X by the number of trawls Y from each survey. Additionally, measures of biodiversity were compared between the two time periods using species richness, Simpson's index [33], the Shannon–Wiener index of diversity [34], Margalef's species richness index [35], and Shannon's evenness index [36]. As the original data analyzed by De Ben et al. [23] were unavailable, the comparisons presented herein were made using the data from their manuscript alone. All work was conducted under an approved Animal Care and Use Protocol issued by the Oregon State University Institutional Animal Care and Use Committee.

3. Results

Most environmental data were similar across the two time periods, with the exception that 1968 was a wetter year than 2003–2005 for Yaquina Bay, and summer (June–September) bottom water temperatures were higher in the 1967–1968 survey (Table 1).

Thirty species of fish and invertebrates from 19 families were captured in sections 1–3 of Yaquina Bay during the 2003–2005 trawl survey. This is a sharp decline from the catch reported in the 1967–1968 survey, where 60 species from 30 families were captured in the same sections (Table 2). The CPUE of all individuals in the top 95% of the catch declined by 91% between surveys, with demersal fish CPUE declining by 96% and epibenthic crustacean CPUE (excluding Dungeness crab; see below) declining by ~67% (Table 3). Section 2 of Yaquina Bay contained the largest percentage of the total catch in the 2003–2005 survey, followed closely by sections 3 and 1. This pattern was identical to the 1967–1968 survey. These sections also showed the smallest change in diversity indices between the two sampling periods.

Table 1. Environmental data during the 1967–1968 and 2003–2005 survey periods. Summer is defined here as 1 June to 30 September. Precipitation data for 1967–1968 from U.S. National Weather Service Applied Climate Information System. Precipitation data for 2003–2005 from the OSU Hatfield Marine Science Center weather station archives.

Environmental Data	1967, 1968	2003, 2004, 2005
Mean Monthly Salinity Range	8–28	14–36
Maximum Salinity	34	36.4
Summer Monthly Salinity Range	16–34	24–36.4
Annual Rainfall (cm)	158, 282	176, 150, 158
Summer Bottom Temperature Range (°C)	21–23	12–17
Coldest Bottom Temperature (°C)	6	9

Table 2. Summary table of total number of fish and epibenthic crustaceans caught by species in each of three sections of lower Yaquina Bay in 1967–1968 and 2003–2005. Species are listed in the 1967–1968 rank order of total catch. DeBen et al. (1990) conducted 126 net sets in sections 1–3. A total of 139 net sets were conducted for this study.

Common Name	Scientific Name	1967–1968 Section 1	2	3	2003–2005 Section 1	2	3
English sole	*Parophrys vetulus*	379	8349	2267	76	996	714
Snake prickleback	*Lumpenus sagitta*	37	2547	445	0	10	13
Shiner surfperch	*Cymatogaster aggregata*	105	1116	1119	30	66	300
Dungeness crab	*Cancer magister*	323	867	298	130	1780	2189
Blacktail bay shrimp	*Crangon nigricauda*	155	1326	3	27	297	265
Pile surfperch	*Rhacochilus vacca*	58	829	394	4	218	139
Buffalo sculpin	*Enophrys bison*	1170	85	1	0	8	6
Starry flounder	*Platichthys stellatus*	211	692	317	2	72	42
White seaperch	*Phanerodon furcatus*	53	587	121	11	1	0
California bay shrimp	*Crangon franciscorum*	11	87	603	0	0	0
Opossum shrimp	*Neomysis mercedis*	4	42	590	0	0	0
Pacific staghorn sculpin	*Leptocottus armatus*	30	395	96	0	57	156
Surf smelt	*Hypomesus pretiosus*	4	28	333	16	161	36
Sand sole	*Psettichthys melanosticus*	189	114	36	0	0	0
Speckled sanddab	*Citharichthys stigmaeus*	239	96	1	1	252	30
Striped seaperch	*Embiotoca lateralis*	202	120	1	0	0	0
Walleye surfperch	*Hyperprosopon argenteum*	1	53	226	0	0	0
Kelp greenling	*Hexagrammos decagrammus*	154	20	1	0	6	1
Saddleback gunnel	*Pholis ornata*	2	85	9	1	6	4
Pacific tomcod	*Microgadus proximus*	43	40	4	0	57	9
Northern anchovy	*Engraulis mordax*	0	2	80	0	0	0
Longfin smelt	*Spirinchus thaleichthys*	2	1	76	0	0	0
Red rock crab	*Cancer productus*	41	3	0	1	28	26
Pacific herring	*Clupea pallasi*	0	8	35	0	19	102
Black rockfish	*Sebastes melanops*	22	14	5	2	27	1
American shad	*Alosa sapidissima*	1	0	36	0	0	27
Bay pipefish	*Syngnathus leptorhynchus*	3	10	21	0	10	3
Lingcod	*Ophiodon elongatus*	13	9	2	0	5	3
California coastal shrimp	*Heptacarpus paludicola*	8	9	0	0	0	0
Tube snout	*Aulorhynchus flavidus*	1	14	0	10	1	0
Cabezon	*Scorpaenichthys marmoratus*	11	4	0	0	1	0
Rock greenling	*Hexagrammos lagocephalus*	7	1	1	0	0	0
Redtail surfperch	*Amphistichus rhodoterus*	2	6	1	0	0	0
C-O sole	*Pleuronichthys coenosus*	6	0	0	0	0	0
Three-spined stickleback	*Gasterosteus aculeatus*	0	0	6	0	0	0
Shortspine shrimp	*Heptacarpus brevirostris*	1	5	0	0	0	0
Whitebait smelt	*Allosmerus elongatus*	3	1	0	0	0	0
Tubenose poacher	*Pallasina barbata*	3	1	0	0	0	0
Big skate	*Raja binoculata*	1	2	0	0	0	0
Penpoint gunnel	*Apodichthys flavidus*	3	0	0	0	0	0
California spot prawn	*Pandalus platyceros*	1	0	1	0	0	0

Table 2. *Cont.*

| Common Name | Scientific Name | 1967–1968 | | | 2003–2005 | | |
| | | Section | | | Section | | |
		1	2	3	1	2	3
Bay shrimp	*Lissocrangon stylirostris*	2	0	0	0	0	0
Bay goby	*Lepidogobius lepidus*	0	2	0	0	0	0
Pacific sand lance	*Ammodytes hexapterus*	1	0	1	0	0	0
Longnose skate	*Raja rhina*	1	1	0	0	0	0
Copper rockfish	*Sebastes caurinus*	1	0	0	0	1	1
Arrow goby	*Clevlandia ios*	0	1	0	0	0	0
Scalyhead sculpin	*Artedius harringtoni*	1	0	0	0	0	0
Whitespotted greenling	*Hexagrammos stelleri*	1	0	0	0	0	0
Tidepool snailfish	*Liparis florae*	1	0	0	0	0	0
Green sturgeon	*Acipenser medirostris*	0	0	1	0	0	0
Pacific hake	*Merluccius productus*	0	1	0	0	0	0
Wolf eel	*Anarrhichthys ocellatus*	1	0	0	0	0	0
Brown Irish lord	*Hemilepidotus spinosus*	1	0	0	0	0	0
Silver surfperch	*Hyperprosopon elliptiicum*	1	0	0	0	0	0
Vermilion rockfish	*Sebastes miniatus*	1	0	0	0	0	0
Chinook salmon	*Oncorhynchus tshawytscha*	0	0	1	0	0	0
Coho salmon	*Oncorhynchus kisutch*	0	0	1	0	0	0
Prickly sculpin	*Cottus asper*	0	0	1	0	0	0
Dock shrimp	*Pandalus danae*	1	0	0	0	0	0
Porcelain crab	*Petrolisthes cinctipes*	0	0	0	0	2	12
Hermit crab	*Pagurus sp.*	0	0	0	0	0	2
Kelp crab	*Pugettia producta*	0	0	0	0	6	1
Red Irish lord	*Hemilepidotus hemilepidotus*	0	0	0	0	2	0
Pacific sanddab	*Citharichthys sordidus*	0	0	0	0	1	0
Eulachon	*Thaleichthys pacificus*	0	0	0	0	1	0
Totals		3512	17,573	7134	311	4091	4082

Table 3. Catch per unit effort (CPUE) of species comprising the top 95% of the total catch from 1967–1968 and 2003–2005.

CPUE of Species in Top 95% of Catch	1967–1968	2003–2005
All Individuals	644.64	57.99
Epibenthic Crustaceans (all)	134.6	146.5
Epibenthic Crustaceans (minus Dungeness crab)	102.6	33.73
Demersal Fishes	542.05	24.27

The families Embiotocidae and Cottidae were the most represented families in the 2003–2005 trawl survey; Embiotocidae was the most represented family in 1967–1968. English sole (*Parophrys vetulus*; #1 in abundance in 1967–1968) and Dungeness crab (*Metacarcinus magister*; #4 in abundance in 1967–1968) were the dominant species in the 2003–2005 survey; however, the relative abundance of these two species to each other changed from 7.39:1 (*P. vetulus*:*C. magister*) during 1967–1968 to 1:2.25 in 2003–2005, a proportional change of approximately 1600%. Dungeness crab was the only one of the top 20 species to see an increase in total catch and CPUE between the two time periods. When the Dungeness crab is included in the epibenthic crustacean CPUE calculations, epibenthic crustacean CPUE between the sampling periods increases by ~9%. Males dominated the Dungeness crab catch in 2003–2005, and 93% of the Dungeness crabs captured were sublegal in size [37] with a mean carapace width of 74 mm. Male Dungeness crabs were also dominant during the 1967–1968 survey, where 100% of the Dungeness crabs caught at that time were reported as sublegal in size [23], although actual sizes were not reported for that survey. The minimum legal size for Dungeness crab in Oregon, established in 1964 (Oregon Department of Fish and Wildlife), is 146 mm (5.25″) and 159 mm (6.25″) for recreational and commercial harvest, respectively.

Measures of biodiversity were calculated for both the 2003–2005 trawl survey and the 1967–1968 survey (Table 4). With the exception of Simpson's index and Shannon's evenness index, estimates of biodiversity decreased in all bay sections between surveys. This decline in biodiversity, when paired with a ~10% increase in total epibenthic crustacean catch in the 2003–2005 survey, provides evidence of a shift in community dominance. This shift was driven principally by the Dungeness crab in the 2003–2005 survey, which comprised 86% of the epibenthic crustacean catch and 48% of the total catch.

Table 4. Biodiversity indices for all fish and epibenthic crustaceans in three sections of lower Yaquina Bay from 1967–1968 and 2003–2005.

	1967–1968			2003–2005		
Bay Section	1	2	3	1	2	3
Species Richness	50	40	36	13	28	24
Simpson's Index	0.15	0.26	0.15	0.26	0.26	0.33
Shannon's Index	2.43	1.90	2.29	1.70	1.80	1.62
Margalef's Index	6.00	3.99	3.94	2.09	3.25	2.77
Shannon's Evenness	0.62	0.52	0.64	0.66	0.54	0.51

4. Discussion

Our re-visitation of De Ben and co-worker's [23] comprehensive trawl survey provides an updated assessment of the benthic community in a northeast Pacific coastal estuary during the latter half of the past century. While we cannot establish direct causal mechanisms for any differences observed in presence/absence, diversity, or abundance, several compelling trends of change are evident between the two time periods covered by this work.

Most significantly, total CPUE dropped by 91% between the two survey periods. This highlights a tremendous decline in estuarine species abundance and richness in the section of Yaquina Bay that has been most altered by development in the last four decades (Figure 2). Similar declines in abundance have been observed following waterfront development in the southeast United States, where demersal fish and epibenthic crustaceans were less abundant in stretches of shoreline altered by rubble or bulkheads [38], while a broader synthesis aimed at evaluating the ecological impacts of overwater structures (OWSs) in the U.S. Pacific Northwest [3] found that OWSs reduce natural habitat (seagrassess) and prey (abundance and diversity of invertebrate species), and affect the movement, migration, and feeding behaviors of salmonids. Changes to Yaquina Bay have also affected community composition, as most measures of biodiversity declined between surveys. Bilkovic and Roggero [39] found analogous results in Virginia's James River, a tidal tributary of the Chesapeake Bay, where estuarine communities populating natural shoreline habitat tended to be more diverse than highly developed sites where a few generalist species dominated. Kimball and co-workers [40] worked over a similar decadal time span but evaluated different metrics. They saw a decrease in ichthyofaunal abundance and a shift from pelagic to benthic finfish dominance. Recent work in Yaquina Bay indicates, too, that natural seagrass habitat supports a richer and more abundant fish community than anthropogenic structures [41].

English sole and Dungeness crab were the most abundant fish and invertebrate species in both surveys, but the relative dominance of finfish to crustaceans changed dramatically between the two time periods. Although the total catch proportion by section represented by these two species was similar in both surveys, the ratio of finfish to crustaceans was drastically lower in 2003–2005 (0.8:1) than in 1967–1968 (5.5:1). Dungeness crab comprised over 48% of the total catch in 2003–2005 but only 5% of the catch in 1967–1968. Despite the potential for minor variations between studies in exact sampling technique (distance towed, tow speed, sampling gear type) to inflate the 2003–2005 crab catch, we used the same gear type and attempted to replicate the sampling effort as closely as possible to that of De Ben and co-workers [23], so any sampling approach differences likely cannot explain the magnitude of change in finfish and Dungeness crab catch.

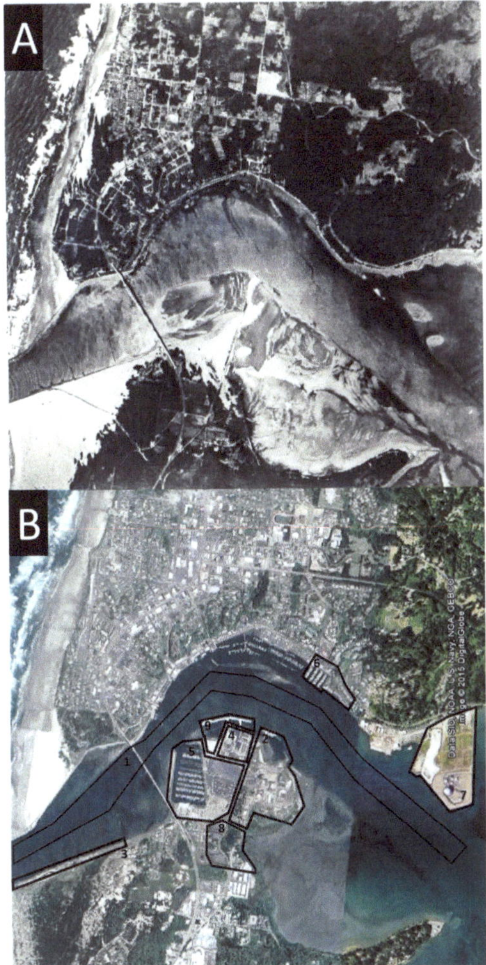

Figure 2. Aerial images of lower Yaquina Bay from (**A**) 1939 and (**B**) 2015. Numbered blocks indicate major changes in bay structure and dredging since 1967–1968 as follows: (1) 1969: Channel depth change from 26′ to 40′ deep, (2) 1970s–1990s: Expansion of Oregon State University's Hatfield Marine Science Center to contemporary footprint, (3) 1972: South Jetty 1800′ extension, (4) 1978: Oregon Aqua Foods facility, (5) 1978–1979: South Beach Marina construction, (6) 1976: Liquid natural gas storage facility, (7) 1974–1977: Embarcadero condominiums, (8) 1992: Oregon Coast Aquarium, and (9) 2011–2012: NOAA Marine Operations Center Pacific (MOC-P), which was built at the site of Oregon Aqua Foods and included the dredging, filling, and construction of a ship pier. Note that the NOAA MOC-P site was not in place at the time of our survey. Image A courtesy of Hatfield Marine Science Center. Image B from [42].

Commercial Dungeness crab catch has increased over the last several decades [43], which indicates a potential overall abundance increase in the ocean region surrounding Yaquina Bay (Figure 3). Oregon's annual Dungeness crab landings during 1967–1968 averaged 4756 ± 555 metric tons (mean $\pm$ SD), whereas in 2003–2005, catch rose significantly to $10{,}423 \pm 2196$ metric tons (two-tailed t-test $p = 0.0507$), with 2004 having the highest catch of any year in the intervening period. Unfortunately, there are no estimates of biomass or CPUE for the Oregon Dungeness crab population, nor are there any effort controls (or measures of effort) on the fishery; therefore, increased harvest could also be explained by increased exploitation rate. In contrast, this does not satisfactorily explain the decline in finfish catch, for which we have no explicit explanation.

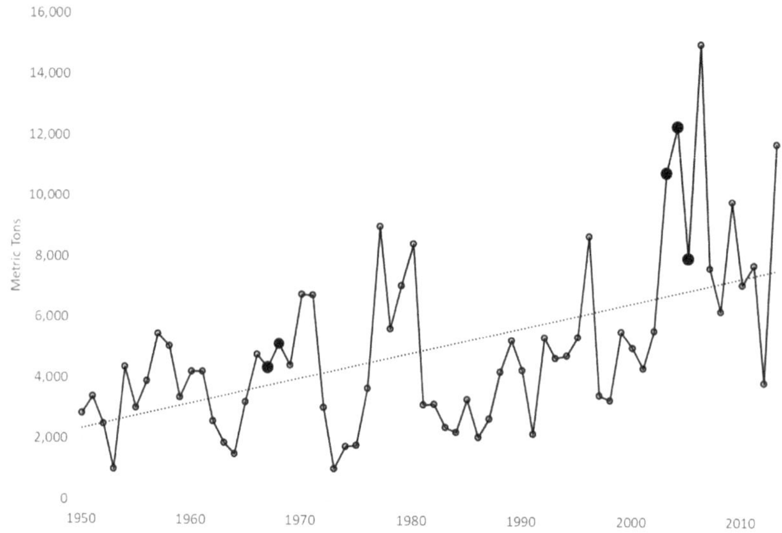

Figure 3. Annual commercial landings (metric tons) of Dungeness crab in Oregon from 1950 to 2013. Emphasized (solid) symbols indicate catch during 1967–1968 and 2003–2005. Data from the NOAA Fisheries Commercial Fisheries Statistics online database are accessible online at http://www.st.nmfs. noaa.gov/commercial-fisheries/commercial-landings/ (accessed on 17 August 2023).

Whereas other studies have identified shoreline development as a driver of community change in estuarine systems [38,39], we acknowledge that our study is only indirectly indicative of these changes. Further evaluation of whether development is the driver of community change in Oregon estuaries like Yaquina Bay would require a different approach, either through tracking changes in community composition during experimental restoration efforts or a multi-estuary study across estuaries experiencing different levels of development. Scientific research in the bay has proliferated since our 2003–2005 survey. Questions about the patterns of dissolved oxygen, nutrient transport, and coastal water mass coupling have addressed the physical ecology of the bay [44,45]. Invertebrate research has centered on the cultivation of the Pacific oyster (*Crassostrea gigas*) and the ecology of mud shrimp [46–48], whereas submerged aquatic vegetation research has monitored the interaction of eelgrass and macroalgae relative to shoreline erosion and tracked the production of the invasive eelgrass, *Zostera japonica* [49–51]. Marine fish research has investigated the habitat preferences of juvenile lingcod and rockfishes, as well as concerns about the effects of upland contaminants on out-migrating salmon smolts [17,32,41,52–54]. Spencer et al. [54] conducted a camera sled survey of Yaquina Bay to estimate the abundance of juvenile flatfishes, but that is the only other research (besides ours) to catalog the benthic marine fish community of the bay on a broad scale since the 1967–1968 survey [23].

5. Conclusions

This work is a meaningful contribution to our understanding of how the animal community in estuaries changes. Long-term monitoring of any system is difficult, but this work can be viewed as a more recent baseline for the benthic marine community structure of Yaquina Bay. It is also valuable to the field of coastal and estuarine science as an endpoint study of community change in a developing port. As shoreline development continues around the world, this work is one example of the potential changes that may result over several decades' time.

Our work measures fish and epibenthic crustacean community structure in Yaquina Bay, Oregon, USA, as a snapshot survey of the change that is possible following an approximately four-decade period. Our work is a partial replicate of De Ben et al.'s [23] survey design, and we use our results to show ecological changes in the benthic marine community of Yaquina Bay following prolonged and intense development of adjacent natural shoreline, dredging of the main channel, and across a time course of environmental change. Overall, our results suggest that Yaquina Bay has experienced both a substantial decline in the total abundance of demersal species and a shift in benthic community dominance in the last half of the 20th century. The largely similar abiotic conditions during both surveys (excluding 1968 rainfall), the similar patterns of catch rank, and the shift to a dominant crustacean allude to a persistent and local driver of community change. Shoreline development and channelization of Yaquina Bay accelerated during the time that elapsed between the two surveys, and habitat degradation and anthropogenic nutrient input resulting from shoreline development seem likely catalysts of community reorganization. The development of well-designed, multi-metric indicators for estuaries like Yaquina Bay may help track the ecosystem-level impacts of these changes, allowing us to better understand the broader impacts of human development and environmental change on these systems.

Author Contributions: Conceptualization, S.A.H. and S.S.H.; methodology, S.A.H., S.S.H. and M.B.G.; fieldwork, S.A.H., S.S.H., M.B.G. and N.S.A.; formal analysis, S.A.H., S.S.H. and N.S.A.; resources, S.A.H. and S.S.H.; data curation, N.S.A. and M.B.G.; writing—original draft preparation, S.A.H. and N.S.A.; writing, S.A.H., S.S.H. and N.S.A.; visualization, S.A.H. and N.S.A.; supervision, S.A.H., S.S.H., N.S.A. and M.B.G.; project administration, S.A.H. and S.S.H.; funding acquisition, S.A.H. and S.S.H. All authors have read and agreed to the published version of the manuscript.

Funding: This work was funded in part by recurring funding from the Oregon State University Cooperative Institute for Marine Resources Studies (CIMRS) and by Oregon Sea Grant project #NA16RG1039 from the National Oceanic and Atmospheric Administration's National Sea Grant College Program.

Institutional Review Board Statement: The animal study protocol (Title: Monitoring fish and invertebrates in the Yaquina River and Yaquina Bay from Toledo, Oregon to the Pacific Ocean) was approved by the Institutional Animal Care and Use Committee of Oregon State University Project, ACUP Number: 4402 (Original approval 5 January 2003).

Informed Consent Statement: Not applicable.

Data Availability Statement: The raw data supporting the conclusions of this article will be made available by the authors upon reasonable request.

Acknowledgments: The authors would like to thank Oregon State University's *Marine Team* student research group of undergraduate and graduate student volunteers for the collection of much of the trawl data and measurement of physical parameters. Marissa Litz and Andrew Claiborne assisted with edits on an earlier version of this manuscript. Ted Dewitt of the USEPA provided excellent discussions early on in the project's formation. We would also like to acknowledge Dewitt's attempts to track down the original De Ben data, only to find a pile of water-damaged, worm-eaten computer punch cards. Finally, thanks to two anonymous reviewers whose efforts have helped improve this manuscript.

Conflicts of Interest: The authors declare no conflicts of interest.

References

1. Sobocinski, K.L.; Cordell, J.R.; Simenstad, C.A. Effects of shoreline modifications on supratidal macroinvertebrate fauna on Puget Sound, Washington beaches. *Estuaries Coasts* **2010**, *33*, 699–711. [CrossRef]
2. Long, C.W.; Grow, J.N.; Majoris, J.E.; Hines, A.H. Effects of anthropogenic shoreline hardening and invasion by *Phragmites australis* on habitat quality for juvenile blue crabs (*Callinectes sapidus*). *J. Exp. Mar. Biol. Ecol.* **2011**, *409*, 215–220. [CrossRef]
3. Lambert, M.; Ojala-Barbour, R.; Vadas, R., Jr.; McIntyre, A.A.; Quinn, T. Do small overwater structures impact marine habitats and biota? *Pac. Con. Biol.* **2024**, *30*, PC22037. Available online: https://www.publish.csiro.au/PC/PC22037. [CrossRef]
4. Engle, V.D.; Summers, J.K.; Gaston, G.R. A benthic index of environmental condition of Gulf of Mexico estuaries. *Estuaries* **1994**, *17*, 372–384. [CrossRef]

5. Cao, Y.; Cherr, G.N.; Córdova-Kreylos, A.L.; Fan, T.W.-M.; Green, P.G.; Higashi, R.M.; LaMontagne, M.G.; Scow, K.M.; Vines, C.A.; Yuan, J.; et al. Relationships between sediment microbial communities and pollutants in two California salt marshes. *Microb. Ecol.* **2006**, *52*, 619–633. [CrossRef]

6. Carpenter, S.R.; Caraco, N.F.; Correll, D.L.; Howarth, R.W.; Sharpley, A.N.; Smith, V.H. Non-point pollution of surface waters with phosphorus and nitrogen. *Ecol. Appl.* **1998**, *8*, 559–568. [CrossRef]

7. Rothenberger, M.B.; Swaffield, T.; Calomeni, A.J.; Cabrey, C.D. Multivariate analysis of water quality and plankton assemblages in an urban estuary. *Estuaries Coasts* **2014**, *37*, 695–711. [CrossRef]

8. Van Dolah, R.F.; Calder, D.R.; Knott, D.M. Effects of dredging and open-water disposal on benthic macroinvertebrates in a South Carolina estuary. *Estuaries* **1984**, *7*, 28–37. [CrossRef]

9. Lotze, H.K.; Lenihan, H.S.; Bourque, B.J.; Bradbury, R.H.; Cooke, R.G.; Kay, M.C.; Kidwell, S.M.; Kirby, M.X.; Peterson, C.H.; Jackson, J.B.C. Depletion, degradation, and recovery potential of estuaries and coastal seas. *Science* **2006**, *312*, 1806–1809. [CrossRef]

10. Wolfe, D.A.; Champ, M.A.; Flemer, D.A.; Mearns, A.J. Long-term biological data sets: Their role in research, monitoring, and management of estuarine and coastal marine systems. *Estuaries* **1987**, *10*, 181–193. [CrossRef]

11. DLCD (Oregon Department of Land Conservation and Development). The Oregon Estuary Plan Book. 1987. Available online: www.inforain.org/oregonestuary/ (accessed on 30 June 2015).

12. Pearcy, W.G.; Myers, S.S. Larval fishes of Yaquina Bay, Oregon: A nursery ground for marine fishes? *Fish Bull.* **1974**, *72*, 201–213.

13. Emmett, R.L.; Llanso, R.; Newton, J.; Thom, R.; Hornberger, M.; Morgan, C.; Levings, C.; Copping, A.; Fishman, P. Geographic signatures of North American West Coast estuaries. *Estuaries* **2000**, *23*, 765–792. [CrossRef]

14. Krygier, E.E.; Pearcy, W.G. The role of estuarine and offshore nursery areas for young English sole, *Paraphrys vetulus* Girard, of Oregon. *Fish Bull.* **1986**, *84*, 119–132.

15. Boehlert, G.W.; Mundy, B.C. Recruitment dynamics of metamorphosing English sole, *Parophorys vetulus*, to Yaquina Bay, Oregon. *Estuar. Est. Coast. Shelf Sci.* **1987**, *25*, 261–281. [CrossRef]

16. Armstrong, D.A.; Rooper, C.; Gunderson, D.R. Estuarine production of juvenile Dungeness crab (*Cancer magister*) and contribution to the Oregon-Washington coastal fishery. *Estuaries* **2003**, *26*, 1174–1189. [CrossRef]

17. Dauble, A.D.; Heppell, S.A.; Johansson, M.L. Settlement patterns of young-of-the-year rockfish among six Oregon estuaries experiencing different levels of human development. *Mar. Ecol. Prog. Ser.* **2012**, *448*, 143–154. [CrossRef]

18. Bottin, R.R.; Briggs, M.J. *Newport North Marina, Yaquina Bay, Oregon, Design for Wave Protection: Coastal Model Investigation*; United States Army Corps of Engineers Technical Report CERC-96-2; 1996.

19. Good, J.W. Estuary development with implications for management: A case study of Northwest Natural Gas Company's liquified natural gas (LNG) project at Yaquina Bay, Oregon. Department of Oceanography, Oregon State University; 1975 Corvallis, OR. Port of Newport. History of the Port of Newport. 2012. Available online: www.portofnewport.com (accessed on 3 July 2012).

20. USACE. U.S. Army Corps of Engineers: Yaquina Bay. U.S. Army Corps of Engineers. 2012. Available online: www.nwp.usace. army.mil/locations/yaquinabay.asp (accessed on 3 July 2012).

21. Oregon Coast Management Program (OCMP) (2016) Estuary Planning. Available online: https://www.oregon.gov/lcd/OCMP/ Pages/Estuary-Planning.aspx (accessed on 28 March 2024).

22. Brown, C.A.; Power, J.H. Historic and recent patterns of dissolved oxygen in the Yaquina Estuary (Oregon, USA): Importance of anthropogenic activities and oceanic conditions. *Estuar. Coast Shelf Sci.* **2011**, *92*, 446–455. [CrossRef]

23. De Ben, W.A.; Clothier, W.D.; Ditsworth, G.R.; Baumgartner, D.J. Spatio-temporal fluctuations in the distribution and abundance of demersal fish and epibenthic crustaceans in Yaquina Bay, Oregon. *Estuaries* **1990**, *13*, 469–478. [CrossRef]

24. Frolander, H.F. Biological and chemical features of tidal estuaries. *J. Water Pollut. Control Fed.* **1964**, *36*, 1037–1048.

25. Wares, P.G. *Biology of the Pile Perch, Rhacochilus vacca, in Yaquina Bay, Oregon*; United States Bureau of Sports Fisheries and Wildlife Technical Paper 57; U.S. Fish and Wildlife Service: Washington, DC, USA, 1971.

26. Oregon State University Hatfield Marine Science Center Guin Library. Available online: https://yaquina.library.oregonstate.edu/ (accessed on 28 March 2024).

27. Frolander, H.F.; Miller, C.B.; Flynn, M.J. Seasonal cycles of abundance in zooplankton populations of Yaquina Bay Oregon. *Mar. Biol.* **1973**, *21*, 277–288. [CrossRef]

28. Krygier, E.E.; Johnson, W.C.; Bond, C.E. Records of California tonguefish, threadfin shad and smooth alligatorfish from Yaquina Bay, Oregon. *Cal. Fish Game.* **1973**, *59*, 140–142.

29. Richardson, S.L.; Pearcy, W.G. Coastal and oceanic fish larvae in an area of upwelling off Yaquina Bay, Oregon. *Fish Bull.* **1977**, *75*, 125–145.

30. Barton, M. Comparative distribution and habitat preferences of 2 species of stichaeoid fishes in Yaquina Bay, Oregon. *J. Exp. Mar. Biol. Ecol.* **1982**, *59*, 77–87. [CrossRef]

31. Bond, C.E.; Kan, T.T.; Myers, K.W. Notes on the marine life of the river lamprey, *Lampetra ayresi* in Yaquina Bay, Oregon, and the Columbia River estuary. *Fish Bull.* **1983**, *81*, 165–167.

32. Gallagher, M.B.; Heppell, S.S. Essential habitat identification for age-0 rockfish along the central Oregon coast. *Mar. Coastal. Fish* **2010**, *2*, 60–72. [CrossRef]

33. Simpson, E.H. Measurement of diversity. *Nature* **1949**, *163*, 688. [CrossRef]

34. Shannon, C.E.; Weaver, W. *The Mathematical Theory of Communication*; University of Illinois Press: Champaign, IL, USA, 1949.

35. Margalef, R. *Perspectives in Ecological Theory*; University of Chicago Press: Chicago, IL, USA, 1968.
36. Pielou, E.C. *An Introduction to Mathematical Ecology*; Wiley: New York, NY, USA, 1969.
37. Pauley, G.B.; Armstrong, D.A.; Van Citter, R.; Thomas, G.L. Species profiles: Life histories and environmental requirements of coastal fishes and invertebrates (Pacific Southwest)—Dungeness Crab. *USFWS Biol. Rep.* **1989**, *82*, TR EL-82-4.
38. Peterson, M.S.; Comyns, B.H.; Hendon, J.R.; Bond, P.J.; Duff, G.A. Habitat use by early life-history stages of fishes and crustaceans along a changing estuarine landscape: Differences between natural and altered shoreline sites. *Wetl. Ecol. Manag.* **2000**, *8*, 209–219. [CrossRef]
39. Bilkovic, D.M.; Roggero, M.M. Effects of coastal development on nearshore estuarine nekton communities. *Mar. Ecol. Prog. Ser.* **2008**, *358*, 27–39. [CrossRef]
40. Kimball, M.E.; Pfirrmann, B.W.; Allen, D.M.; Ogburn-Matthews, V.; Kenny, P.D. Intertidal creek pool nekton assemblages: Long-term patterns in diversity and abundance in a warm-temperate estuary. *Estuaries Coasts* **2023**, *46*, 860–877. [CrossRef]
41. Schwartzkopf, B.D.; Dauble, A.D.; Lindsley, A.J.; Heppell, S.A. Temporal and habitat differences in the juvenile demersal fish community at a marine-dominated northeast Pacific estuary. *Fish Res.* **2020**, *227*, 105557. [CrossRef]
42. NOAA. Fisheries Statistics: Commercial landings, Dungeness crab in Oregon from 1967–2010, NOAA Fisheries OST. 2012. Available online: www.st.nmfs.noaa.gov/st1/commercial/landings/annual_landings.html (accessed on 3 July 2012).
43. Sigleo, A.C.; Frick, W.E. Seasonal variations in river discharge and nutrient export to a Northeastern Pacific estuary. *Coast. Shelf Sci.* **2007**, *73*, 368–378. [CrossRef]
44. Brown, C.A.; Ozretich, R.J. Coupling between the coastal ocean and Yaquina Bay, Oregon: Importance of oceanic inputs relative to other nitrogen sources. *Estuaries Coasts* **2009**, *32*, 219–237. [CrossRef]
45. Camara, M.D.; Evans, S.; Langdon, C.J. Parental relatedness and survival of Pacific oysters from a naturalized population. *J. Shellfish Res.* **2008**, *27*, 323–336. [CrossRef]
46. D'Andrea, A.F.; DeWitt, T.H. Geochemical ecosystem engineering by the mud shrimp *Upogebia pugettensis* (Crustacea: Thalassinidae) in Yaquina Bay, Oregon: Density-dependent effects on organic matter remineralization and nutrient cycling. *Limnol. Oceanogr.* **2009**, *54*, 1911–1932. [CrossRef]
47. Repetto, M.; Griffen, B.D. Physiological consequences of parasite infection in the burrowing mud shrimp, *Upogebia pugettensis*, a widespread ecosystem engineer. *Mar. Fresh Res.* **2012**, *63*, 60–67. [CrossRef]
48. Kaldy, J.E. Production ecology of the non-indigenous seagrass, dwarf eelgrass (*Zostera japonica* Ascher & Graeb), in a Pacific Northwest estuary, USA. *Hydrobiologia* **2006**, *553*, 201–217.
49. Boese, B.L.; Robbins, B.D. Effects of erosion and macroalgae on intertidal eelgrass (*Zostera marina*) in a northeastern Pacific estuary (USA). *Bot. Mar.* **2008**, *51*, 247–257. [CrossRef]
50. Hessing-Lewis, M.L.; Hacker, S.D.; Menge, B.A. Context-dependent eelgrass-macroalgae interactions along an estuarine gradient in the Pacific Northwest, USA. *Estuaries Coasts* **2011**, *34*, 1169–1181. [CrossRef]
51. Petrie, M.E.; Ryer, C.H. Laboratory and field evidence for structural habitat affinity of young-of-the-year lingcod. *Trans. Am. Fish Soc.* **2006**, *135*, 1622–1630. [CrossRef]
52. Johnson, L.L.; Ylitalo, G.M.; Arkoosh, M.R. Contaminant exposure in outmigrant juvenile salmon from Pacific Northwest estuaries of the United States. *Environ. Monit. Assess* **2007**, *124*, 167–194. [CrossRef] [PubMed]
53. Schwartzkopf, B.D.; Heppell, S.A. A feeding ecology-based approach to evaluating nursery potential of estuaries for Black Rockfish. *Mar. Coast. Fish* **2020**, *12*, 124–141. [CrossRef]
54. Spencer, M.L.; Stoner, A.W.; Ryer, C.H.; Munk, J.E. A towed camera sled for estimating abundance of juvenile flatfishes and habitat characteristics: Comparison with beam trawls and divers. *Est. Coast. Shelf. Sci.* **2005**, *64*, 497–503. [CrossRef]

 *fishes*

 MDPI

Article

A Fish-Based Tool for the Quality Assessment of Portuguese Large Rivers

António Tovar Faro [1],*, Maria Teresa Ferreira [1] and João Manuel Oliveira [2,3,4]

1   Forest Research Centre, Associate Laboratory TERRA, School of Agriculture, University of Lisbon, 1349-017 Lisbon, Portugal; terferreira@isa.ulisboa.pt
2   CERNAS–Research Centre for Natural Resources, Environment and Society, Santarém Polytechnic University, Quinta do Galinheiro–S. Pedro, 2001-904 Santarém, Portugal; joao.oliveira@esa.ipsantarem.pt
3   Agrarian School of Santarém, Santarém Polytechnic University, Quinta do Galinheiro–S. Pedro, 2001-904 Santarém, Portugal
4   cE3c–Centre for Ecology, Evolution and Environmental Changes & CHANGE–Global Change and Sustainability Institute, Faculdade de Ciências, Universidade de Lisboa, Campo Grande, 1749-016 Lisboa, Portugal
*   Correspondence: antoniotfaro@isa.ulisboa.pt

Abstract: Multimetric indices play a pivotal role in assessing river ecological quality, aligning with the European Water Framework Directive (EU WFD) requirements. However, indices developed specifically for large rivers are uncommon. Our objective was to develop a fish-based tool specifically tailored to assess the ecological quality in Portuguese large rivers. Data were collected from seven sites in each of three Portuguese large rivers (Minho, Guadiana, and Tagus). Each site was classified using an environmental disturbance score, combining different pressure types, such as water chemistry, land use, and hydromorphological alterations. The Fish-based Multimetric Index for Portuguese Large Rivers (F-MMIP-LR) comprises four metrics: % native lithophilic individuals; % alien individuals; % migrant individuals; and % freshwater native individuals, representing compositional, reproductive, and migratory guilds. The index showed good performance in separating least- and most-disturbed sites. Least-disturbed sites were rated 'high' or 'good' by F-MMIP-LR, contrasting with no such classification for most-disturbed sites, highlighting index robustness. The three rivers presented a wide range of F-MMIP-LR values across the gradient of 'bad' to 'high', indicating that, on a large spatial extent, the biological condition was substantially altered. The F-MMIP-LR provides vital information for managers and decision-makers, guiding restoration efforts and strengthening conservation initiatives in line with the WFD.

Keywords: ecological quality; large rivers; water framework directive; MMI; fishes; freshwater ecosystems

Key Contribution: Our study is significant in developing a new fish-based tool specifically tailored for assessing the ecological quality of Portuguese large rivers. This tool offers valuable insights to enhance river management and conservation efforts, in alignment with the EU WFD.

Citation: Faro, A.T.; Ferreira, M.T.; Oliveira, J.M. A Fish-Based Tool for the Quality Assessment of Portuguese Large Rivers. *Fishes* **2024**, *9*, 149. https://doi.org/10.3390/fishes9050149

Academic Editors: Robert L. Vadas, Jr., Robert M. Hughes and Bror Jonsson

Received: 6 March 2024
Revised: 15 April 2024
Accepted: 20 April 2024
Published: 23 April 2024

1. Introduction

Large rivers and their riparian zones are vital features of the Earth's hydrological systems, providing many ecosystem services and being globally recognized as hot spots of biodiversity [1]. However, most European large rivers have been severely degraded by human interventions that include channelization, dam construction, wastewater discharges, and introduction of non-native species, among others [2,3]. Likewise, Portuguese large rivers (Minho, Tagus, and Guadiana; catchment areas $\geq 10,000$ km^2) have been altered by humans for centuries [4], causing the degradation of the riverbed and the riparian areas, river connectivity, flow regimes, and water quality. The number of non-native species in these systems has also increased exponentially as a result of introductions seeking to improve fisheries [5]. The ecological condition of Portuguese large rivers has been markedly affected by these historical intensive uses, thus jeopardizing the structure of the aquatic biotic communities. Because they are connected to the sea, these rivers also support several endangered freshwater and diadromous fishes, making them important and valuable resources for conservation and fisheries [6].

The Water Framework Directive (WFD) was implemented in 2000 and set the goal of "good ecological status" for all European inland waters [7]. With this aim, EU member states must assess the ecological status of rivers, lakes, and transitional and coastal water bodies in their territory, and establish programmes of measures to reduce substantial anthropogenic pressures. Ecological status is assessed based on biological quality elements, such as fish assemblages, and their supporting physico-chemical and hydromorphological quality elements, which indicate the condition of an aquatic ecosystem in response to a variety of human-caused stressors. Given the increasing seriousness of the environmental degradation of European waters in general, and large rivers in particular [8], the need for effective ecological and biodiversity monitoring programs has never been higher [9,10].

Multimetric indices (MMIs) are common methods for assessing the biological quality of rivers and evaluating the rehabilitation of aquatic communities [11,12]. These tools are based on the premise that biological communities respond to human-caused pressures in expectable and measurable ways, facilitating the estimation of the relationship between the biological community and the amount of environmental degradation [13]. MMIs are composed of a set of metrics related to the species composition and functional attributes of biological assemblages, such as taxa richness, trophic and habitat niche, and abundance. This method has been adapted to a wide range of lotic aquatic ecosystems in European waters to assess ecological status in accordance with the EU WFD [14–20]. However, most methods were not specifically developed for large rivers, which demonstrates the need for the development of new studies and tools focused on bioassessment of these systems [21,22]. In fact, large rivers are complex and very diverse ecosystems [23], presenting unique challenges to their biological assessment, such as the selection of efficient sampling techniques, seasonal changes of fish assemblage composition, and the low number of minimally disturbed sites needed to set reference conditions [2,21].

The WFD requires EU member states to develop typologies for surface waters based on a set of environmental variables that represent the fixed abiotic conditions, e.g., altitude, size, and basin geology, to explain the natural variability of the ecosystems [8]. These typologies categorize water bodies into distinct groups (river types) characterized by similar geomorphological, hydrological, physico-chemical, and biological attributes. This paper aimed to assess the spatial variations of the biological quality of one of those Portuguese river types—large rivers—based on an MMI specifically developed for them. Thus, we expected to detect changes in fish assemblages according to the environmental conditions of the river segments. Such an MMI can serve as a valuable resource for managers and decision-makers to assess the biological quality of Portuguese large rivers, helping to direct rehabilitation efforts towards the most severely disturbed sites, and to strengthen the conservation of the least-disturbed ones. It can aid in identifying areas with the greatest impairments, potentially establishing the underlying causes of these impairments, and recommending mitigation strategies. Furthermore, an MMI can enable the tracking of improvements in fish assemblages over time, thereby evaluating the success or failure of rehabilitation projects and facilitating the implementation of adaptive management strategies, where interventions can be adjusted based on observed outcomes. In fact, the ability of these tools to track fish assemblages over time represents a significant contribution for the effective management and conservation planning of river ecosystems.

2. Materials and Methods

2.1. Study Area and Sampling Sites

According to Borgwardt et al. [24], Portuguese large rivers are classified into a unique large river type (LRT), the Mediterranean rivers. These rivers are characterized by having catchment areas $\geq$ 10,000 km^2 and where most of the river's course has not been impounded by large dams to any significant extent. By this definition, Portugal has three large rivers: Minho, Tagus, and Guadiana, corresponding to deep and wide fluvial channels with gentle slopes, and generally with wide floodplains, although they may also cross areas of narrow, rocky valleys. Although the Douro River is one of the Portuguese "big rivers", its sequence of dams excluded it from this work. The Minho River is in northwestern Iberia and extends ~300 km through Spain to Portugal with the last 75 km of river defining the border between both countries. This international section is classified as an LRT and begins immediately downstream of the Spanish Frieira Dam (Figure 1). The Tagus River is in middle Iberia, between the Douro and Guadiana basins, and it extends ~1100 km through Spain and Portugal, sustaining a series of dams during its course. In Portugal, only the lower 170 km are free flowing waters, a fluvial segment classified as an LRT that extends from the river mouth to the first hydroelectric structure, Belver Dam. Lastly, the Guadiana River is also an Iberian watercourse that flows 820 km into the Atlantic Ocean at the southern border between the two countries. In Portugal, the Guadiana River is classified as an LRT upstream and downstream of the Alqueva/Pedrogão system, which is an important multiple-use water supply system that is located ~150 km from the estuary.

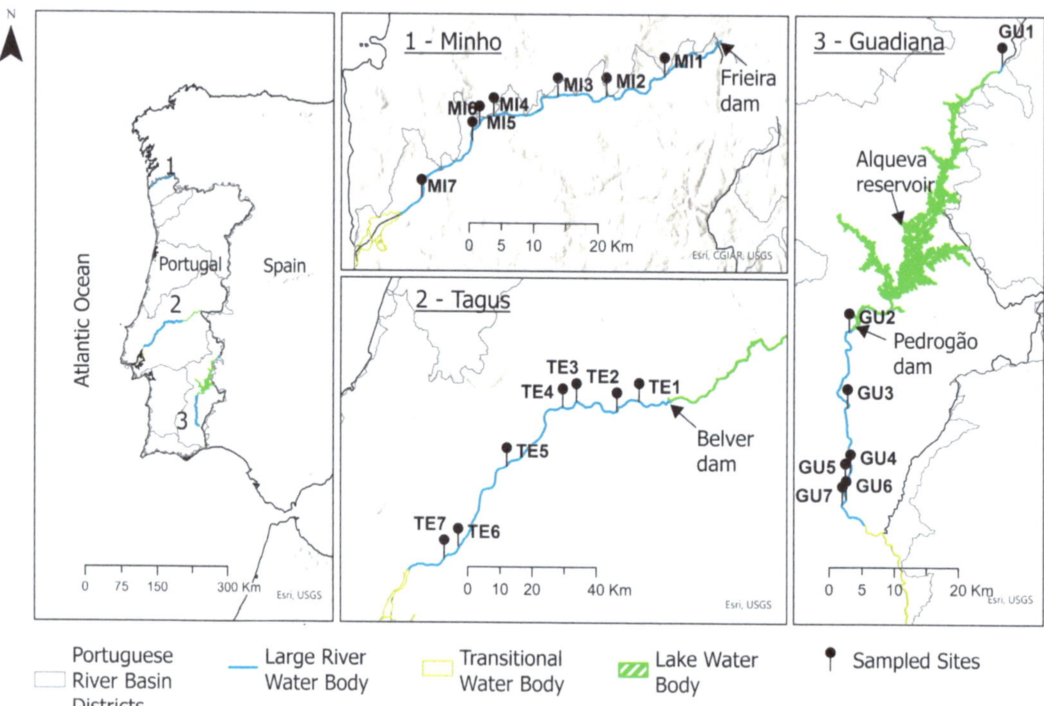

Figure 1. Study area and location of the sampled sites.

The Convention on Cooperation for the Protection and Sustainable Use of Waters in Portuguese-Spanish River Basins (Albufeira Convention) is the instrument of cooperation between Portugal and Spain, for the protection and sustainable use of water in these basins [25]. This convention is designed to provide a framework for bilateral cooperation in the context of the WFD, for the protection of water bodies, aquatic and terrestrial ecosystems, and for the sustainable use of water resources.

Iberian large rivers have endured a long-history of human interventions and natural disturbances in the fluvial corridors and on the surrounding valleys [4], including highly modified river flows [26]. The fish assemblages in these rivers are dominated by a mixture of larger potamodromous species, several non-native species, and diadromous fish populations with life-cycles spanning marine and freshwater ecosystems [6].

We selected 7 sites in each of the international Minho River, the Tagus River downstream from the Belver Dam, and the Guadiana River upstream of the Alqueva Dam and downstream of the Pedrogão Dam (Figure 1). Although the sites were not chosen randomly, we tried to ensure that they encompassed the range of natural conditions and human stressors occurring in the study areas. Except for two sites in the Guadiana River, all samples were taken between 2019 and 2022. Because of the lack of recent samples from least-disturbed reaches in the Guadiana River, we included two sites that were sampled there in 1996 and 1998, prior to the construction of the Alqueva Dam.

2.2. Anthropogenic Disturbance and Site Classification

There are no "near natural" reaches in our large rivers, but only lotic segments that present least-disturbed conditions (i.e., the presently best available condition). This is a common situation in many large rivers of the world, leading to the use of least-disturbed sites, with considerable levels of human influence, as "reference conditions", which is key for the development of most biotic indices [12,27,28].

To classify each site, we developed an environmental disturbance score (EDS) based on a wide range of pressure types, namely, nutrient enrichment, non-natural land uses, channel morphology modifications, riparian disturbance, and flow regulation (Table 1). Disturbance scores for each variable were based on professional judgement and on adaptations of the classifications proposed by Oliveira et al. and Weigel and Dimick [29,30], who developed biological indices for large rivers. Variables were scored to the degree from which they deviated from the least-disturbed conditions (from 1 for no deviation, to 4 for highly deviated; Table 1).

Table 1. Criteria to score variables related to human disturbance. Variables were scored to the degree they deviated from least-disturbed conditions (from 1 for no deviation, to 4 for highly degraded); TP—Total Phosphorus; TN—Total Nitrogen.

Class	Agricultural Land Use	Artificial Land Use	TP (mg/L)	TN (mg/L)	Channel Morphology/Riparian Disturbance	Flow Regulation
1	<10% agriculture, and <3% intensive farming	<5%	<0.13	<1.0	No or minor impacts	Infrequent or no hydropeaking
2	10–30% agriculture, and <10% intensive farming	5–15%	0.14–0.26	1.0–2.0	Most of natural channel form maintained, and >70% of the streambank vegetation in natural state	Regular hydropeaking and distance > 30 km from a large hydroelectric power plant (LHPP)
3	31–70% agriculture and <15% intensive farming	16–25%	0.27–0.39	2.1–3.0	Channelized (some natural habitat types missing), and/or 50–70% of the streambank vegetation in natural state	Regular hydropeaking and distance < 30 km from an LHPP
4	>70% agriculture and/or >15% intensive farming	>25%	>0.39	>3.0	Strongly channelized (most natural habitat types missing), and/or <50% of the streambank vegetation in natural state	Regular hydropeaking and marked seasonal dewatering of the river

For land use data, we used three CORINE Land Cover (CLC) inventories: 2000, 2010, and 2018 (depending on the sampling date) produced within the framework of the Copernicus Land Monitoring Service [31]. For each inventory, we grouped the categories already defined in CLC in three land use classes: artificial (mostly urban), intensive agriculture, and non-irrigated agriculture. ArcGis Pro data were extracted using a buffer with a 12-km radius, with the buffer centroids being placed exactly 10 km upstream from each sampling site. Agricultural land use was estimated as less than both 10% of agriculture and 3% of intensive farming (1) to more than 70% agriculture or more than 15% intensive farming (4). Artificial land use was estimated as <5% (1) to >25% (4). Chemical data were obtained from SNIRH (National Water Resources Information System) [32], and total P (TP) and total N (TN) were calculated as the mean of the available values in the last five years, i.e., considering the sampling year/month and the previous four years. Analyzing a set of data over time offers a more thorough understanding of the chemical composition of water in a site compared to relying only on a single sample [33]. Based on Weigel and Dimick [30], TP and TN ranged from, <0.13 mg/L and 1.0 mg/L (1) to >0.39 mg/L and 3.0 mg/L (4), respectively. Morphological modifications and riparian disturbance were evaluated in the field and from direct observation in Google Earth, on a river reach extending 1 km upstream from each sampling site. Channel morphology and riparian condition were evaluated from no or minor impacts (1) to strongly channelized river (most natural habitat types missing) and/or <50% of the streambank vegetation in natural state (4). Flow regulation was evaluated as a function of the influence of large hydroelectric power plants (LHPPs) upstream from the site (the operation of these structures is similar, imposing an "on–off" pattern of flow that depends on electricity demands). Thus, flow regulation was evaluated as infrequent or no hydropeaking (1) to regular hydropeaking and marked seasonal dewatering of the river (4).

A composite score of the six disturbance measures (i.e., the sum of scores (1–4) across the 6 measures) was calculated for each site (EDS), and the two lowest scoring sites from each of the three rivers were selected as the least-disturbed sites (i.e., a total of six LD sites); an additional condition for a site to be classified as LD was to have a classification of 1 or 2 on at least five pressure variables. The remaining 15 sites were classified as most-disturbed (MD) sites.

The sites spanned a considerable gradient of environmental disturbance as indicated by TN concentrations (0.82–3.67 mg/L), agricultural areas (19–67%), irrigated agriculture (0–44%), channel morphology and riparian condition (1–4), and flow regulation (1–4) (Table S1). These results indicate a clear anthropogenic pressure gradient and environment conditions that are determined independently from the aquatic biota [30].

2.3. Fish Sampling

Except for the two sites sampled before 2000 in the Guadiana River (GR sites < 2000), all other fish assemblages were sampled according to the WFD protocol for Portuguese rivers [34]. Each site was boat-electrofished once during spring–summer base flow. Electrofishing distances were at least 10 times the mean wetted width of the channel and both banks were surveyed. This method was complemented by gill netting in the pelagic zone of the channel, which included the placement of one surface and one deep pelagic multi-mesh net; both nets were 30-m long by 1.5-m deep and were composed of 2.5-m long segments with 12 different mesh sizes (ranging from 5 mm to 55 mm). The nets were fished for 3 h in all segments. The GR sites < 2000 were electrofished in a similar way but no gill nets were used. Fish were identified and measured in the field; native specimens were returned alive to the water, and non-natives were killed, in accordance with Portuguese legislation. For analytical purposes, the total captures resulting from both electrofishing and gill netting were aggregated.

2.4. Index Development

The F-MMIP-LR was developed following Whittier et al., Krause et al., and Gonino et al. [35–37]. First, we selected fish metrics from the literature based on species composition or related to the percentage of fish guilds grouped into ecological functions. Although we used a standardized sampling in most of the sites, fish species abundance in large rivers is particularly reliant on sample size or effort, and to account for this, species abundance in each site was quantified in terms of relative abundance (%) rather than absolute numbers. On the other hand, most diadromous species are widely distributed throughout Portuguese larger rivers, but some freshwater species are restricted to one or a few basins (Table 2). For example, two *Luciobarbus* species occur only in the Guadiana River. Because of this heterogeneity in the number of species between the studied large rivers, our metrics were only based on the relative abundance of individuals. Thus, we considered fourteen metrics grouped into six groups, following Noble et al. and Oliveira et al. [38,39] (Table 2): (1) compositional metrics (freshwater natives—FNAT, aliens—ALIE, and threatened fishes—THRE (taxa classified as at least vulnerable on the Portuguese Red Book of Freshwater and Diadromous Fishes [40])); (2) overall tolerance guilds, based on species ability to endure a wide range of environmental conditions (non-tolerant—NOTO and tolerant—TOLE); (3) trophic guilds, based on the diet of adult individuals (native invertivore—INVE and omnivorous—OMNI); (4) habitat guilds, based on the preferred feeding and living habitats (benthic—BENT and native water column—PELA); (5) reproduction guilds, based on spawning substrate (native lithophilic—LITH and generalist—GENE); (6) migratory behavior guilds (diadromous (species that migrate between marine and freshwater habitats)—DIAD, potamodromous (species that migrate amongst multiple freshwater environments)—POTA, and migrant (the sum of DIAD and POTA)). Biological characteristics of fish species were based on the European EFI+ project [41] with a few modifications supported by additional published data [39], and best professional judgment (Table 2).

Table 2. Species distribution by basin (M—Minho; T—Tagus; G—Guadiana), frequency of occurrence (FO) (%), and compositional and functional guilds (FNAT—freshwater native; ALIE—alien; THRE—threatened; NOTO—non-tolerant; TOLE—tolerant; INVE—native invertivore; OMNI—omnivorous; BENT—benthic; PELA—native water column; LITH—native lithophilic; GENE—generalist; DIAD—diadromous; POTA—potamodromous).

Family	Species	Basin	FO	Guilds
Anguillidae	*Anguilla anguilla*	M; T; G	61.9%	THRE, TOLE, OMNI, BENT, DIAD
Atherinidae	*Atherina boyeri*	M; T; G	38.1%	NOTO, INVE, PELA, GENE, DIAD
Centrarchidae	*Lepomis gibbosus*	M; T; G	76.2%	ALIE, TOLE, GENE
	Micropterus salmoides	M; T; G	52.4%	ALIE, TOLE, GENE
Cichlidae	*Australoheros facetus*	G	28.6%	ALIE, TOLE, GENE
Clupeidae	*Alosa alosa*	M; T; G	9.5%	THRE, NOTO, PELA, LITH, DIAD
	Alosa fallax	M; T; G	4.8%	THRE, NOTO, PELA, LITH, DIAD
Cobitidae	*Cobitis paludica*	T; G	23.8%	FNAT, TOLE, INVE, BENT, GENE
Cyprinidae	*Carassius auratus*	M; T; G	47.6%	ALIE, TOLE, BENT
	Carassius gibelio	T; G	4.8%	ALIE, TOLE, BENT
	Cyprinus carpio	M; T; G	42.9%	ALIE, TOLE, BENT
	Luciobarbus bocagei	M; T	47.6%	FNAT, TOLE, OMNI, BENT, LITH, POTA
	Luciobarbus comizo	T; G	19.0%	FNAT, TOLE, OMNI, BENT, LITH, POTA
	Luciobarbus microcephalus	G	9.5%	FNAT, THRE, TOLE, OMNI, BENT, LITH, POTA
	Luciobarbus sclateri	G	28.6%	FNAT, TOLE, OMNI, BENT, LITH, POTA
	Luciobarbus steindachneri	T; G	9.5%	FNAT, TOLE, OMNI, BENT, LITH, POTA
Gasterosteidae	*Gasterosteus aculeatus*	M; T	9.5%	FNAT, THRE, NOTO, OMNI, PELA
Gobiidae	*Pomatoschistus microps*	M; T; G	4.8%	NOTO, OMNI, BENT, GENE
Gobionidae	*Gobio lozanoi*	M; T	38.1%	ALIE, TOLE, BENT
Ictaluridae	*Ameirus melas*	T; G	9.5%	ALIE, TOLE, BENT
	Ictalurus punctatus	G	4.8%	ALIE, TOLE, BENT

Table 2. *Cont.*

Family	Species	Basin	FO	Guilds
Leuciscidae	*Achondrostoma oligolepis*	M; T	14.3%	FNAT, TOLE, OMNI, PELA, GENE
	Alburnus alburnus	T; G	42.9%	ALIE, TOLE, OMNI
	Pseudochondrostoma duriense	M	28.6%	FNAT, NOTO, OMNI, BENT, LITH, POTA
	Pseudochondrostoma polylepis	T	14.3%	FNAT, NOTO, OMNI, BENT, LITH, POTA
	Pseudochondrostoma willkommii	G	4.8%	FNAT, THRE, NOTO, OMNI, BENT, LITH, POTA
	Squalius carolitertii	M	14.3%	FNAT, NOTO, INVE, PELA, LITH
	Squalius pyrenaicus	T; G	4.8%	FNAT, THRE, NOTO, INVE, PELA, LITH
Moronidae	*Dicentrarchus labrax*	M; T; G	4.8%	NOTO, INVE, PELA
Mugilidae	*Liza ramada*	M; T; G	23.8%	TOLE, OMNI, PELA, DIAD
	Mugil cephalus	M; T; G	9.5%	TOLE, OMNI, PELA, DIAD
Percidae	*Sander lucioperca*	T; G	23.8%	ALIE, TOLE, GENE
Petromyzontidae	*Petromyzon marinus*	M; T; G	4.8%	THRE, NOTO, BENT, LITH, DIAD
Pleuronectidae	*Platichthys flesus*	M; T; G	9.5%	NOTO, INVE, BENT, DIAD
Poecilidae	*Gambusia holbrooki*	M; T; G	52.4%	ALIE, TOLE
Salmonidae	*Salmo trutta*	M; T	9.5%	FNAT, NOTO, INVE, PELA, LITH, POTA
Siluridae	*Silurus glanis*	T	28.6%	ALIE, TOLE, BENT, GENE

Candidate metrics were screened in a four-step process. First, we checked the distribution of metric values across all sites to eliminate those metrics with very small ranges (range test). Second, we performed a Kruskal–Wallis test ($p < 0.1$) to examine the responsiveness of the metrics that passed the first step in distinguishing the least and most disturbed sites. Third, we used the Spearman correlation coefficient to choose metrics lacking redundant information with other metrics ($r_s > 0.70$). In the last step, we conducted a range test for metric values, based on the examination of box plots representing the metric scores (medians) for the LD and MD sites, to determine if most of the values from the two groups did not overlap. Metrics were scored on a continuous scale from 0 to 1. For metric scoring and calculation of the F-MMIP-LR, floor and ceiling values were defined as the 5th and 95th percentiles of metric values across all sites [35]. Metric scores between this range of values were interpolated linearly. For negative metrics, we reversed the floor and ceiling values. The scored metrics were then summed, and the summed score was divided by the number of metrics. Thus, the final value of the index was scaled to a range of 0 to 1, where 0 corresponds to the worst and 1 to the best quality of each site.

Following Hering et al. [42], we defined five quality classes (high, good, moderate, poor, and bad) with equal ranges to provide five ordinal rating categories for assessment of disturbance in accordance with the demands of the WFD. We performed a Kruskal–Wallis test ($p < 0.05$) to verify the ability of our index to discriminate least- from most-disturbed sites; and we used a Spearman's test to check the correlation between the F-MMIP-LR scores and the EDS scores.

3. Results

A total of 9501 individuals comprised of 37 fish species and 20 families were collected (Table 2). Of these, 24 (65%) species were native and 13 were alien (35%). As expected, the most-collected species are widespread throughout the Portuguese large rivers, exploring a great variety of environmental conditions. The alien *Lepomis gibbosus* was the most frequently occurring species, occurring in 16 sites, followed by the native *Anguilla anguilla* (13 sites), and the aliens *Gambusia holbrooki* and *Micropterus salmoides* (both occurring in 11 sites).

Of all candidate metrics, only four metrics were approved in all tests to compose the final F-MMIP-LR: % lithophilics, % migrants, % aliens, and % freshwater natives (Figure 2; Table 3). The F-MMIP-LR clearly discriminated least- from most-disturbed sites (Kruskal–Wallis test: $H = 10.188$; $p < 0.001$; $n = 21$) (Figure 3, Table S1), and we found a significant negative Spearman's correlation between F-MMIP-LR and EDS for all sites ($r_s = -0.639$, $p < 0.0021$) (Figure 4). All but one of the least-impacted sites were classified as 'high' or 'good' by the F-MMIP-LR, and none of the most-impacted sites were classified as 'high' or 'good', (Figure 5; Table S1).

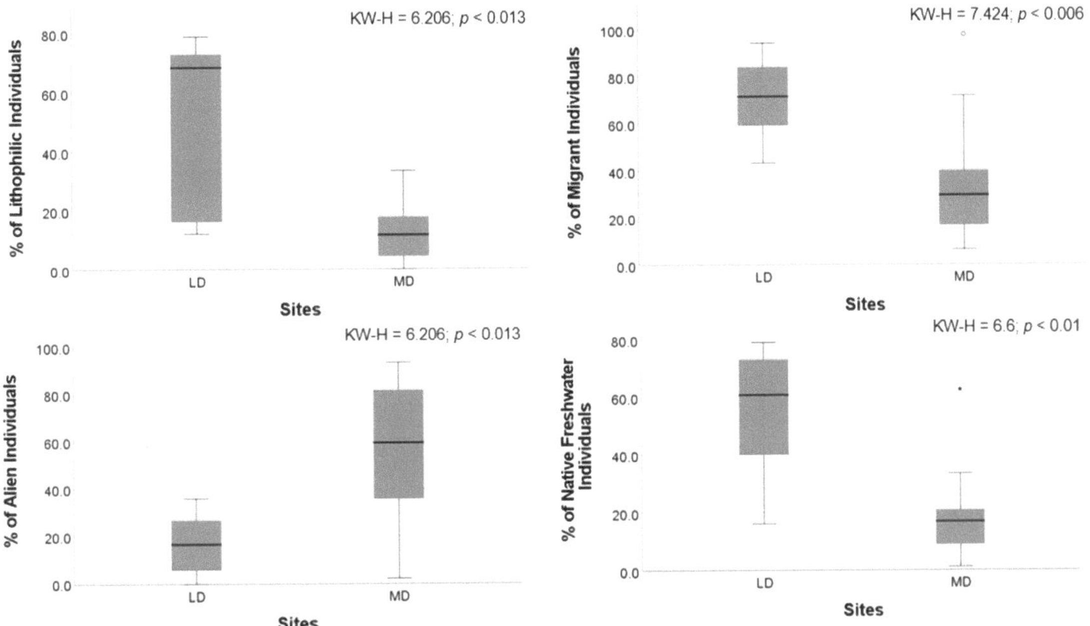

Figure 2. Distribution of metric values from the final range test for the least disturbed (LD) and most disturbed (MD) sites, and results from the Kruskal–Wallis test ($n = 21$).

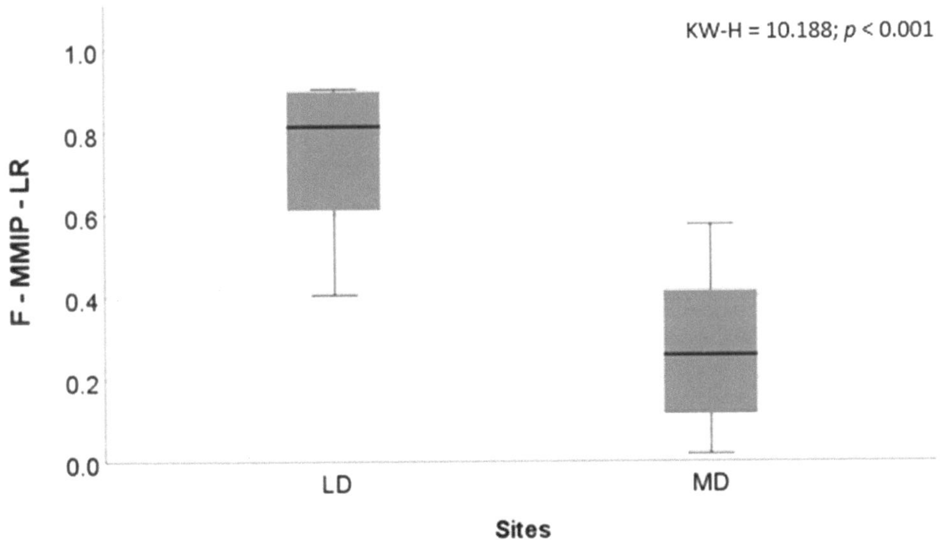

Figure 3. Distribution of F-MMIP-LR scores across least disturbed (LD) and most disturbed (MD) sites, and results from the Kruskal–Wallis test ($n = 21$).

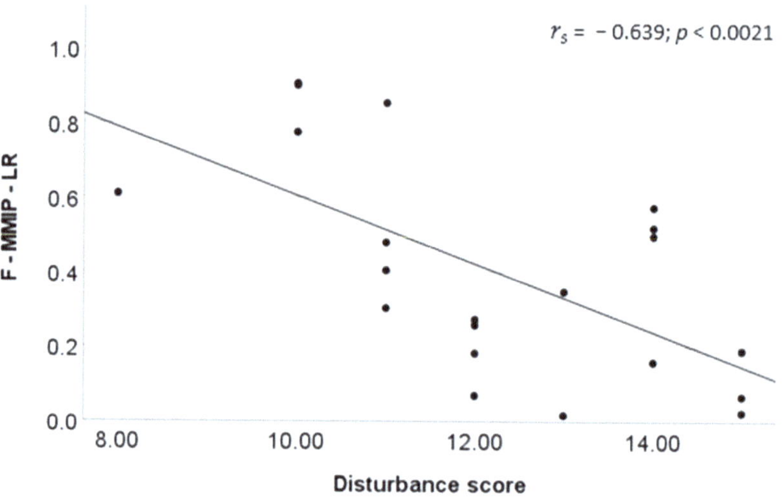

Figure 4. Relationship between the F-MMIP-LR scores and the environmental disturbance scores ($r_s = -0.639$; $n = 21$).

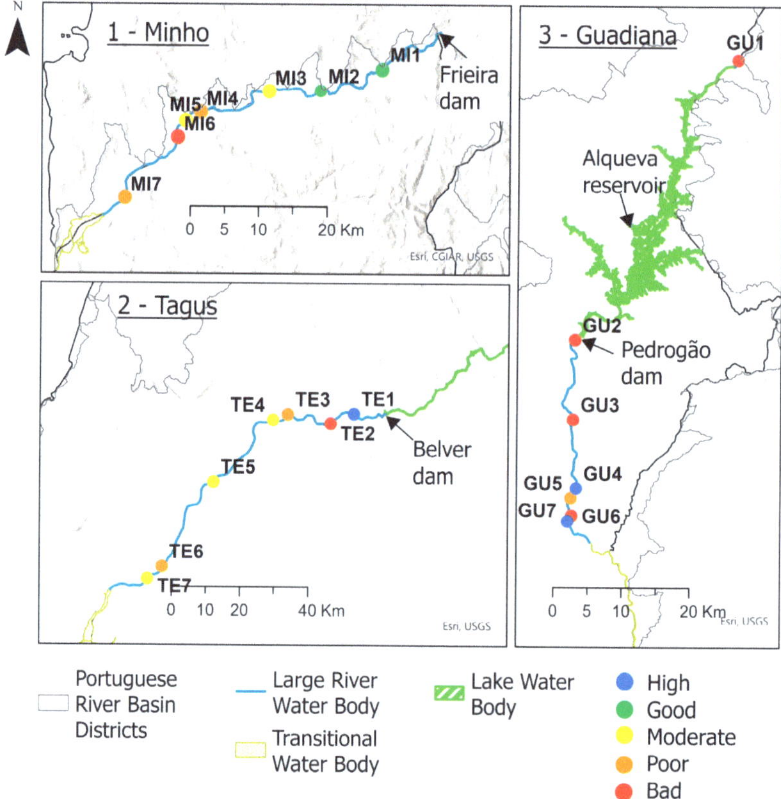

Figure 5. F-MMIP-LR quality classes (High; Good; Moderate; Poor; Bad) across sampled sites.

Table 3. The 5th percentile (P5) and 95th percentile (P95) values for the selected metrics.

Metrics	P5	P95
% of Lithophilic individuals	1	73
% of Migrant individuals	11	94
% of Alien individuals	2	88
% of Native freshwater individuals	4	73

The F-MMIP-LR scores ranged from 0.018 to 0.905 (Table S1). Based on the five quality classes with equal ranges, our index classified three sites (2 in Guadiana and 1 in Tagus) (14%) as 'high', two sites (both in Minho) (10%) as 'good', five sites (2 in Minho and 3 in Tagus) (24%) as 'moderate', four sites (2 in Minho, 1 in Tagus, and 1 in Guadiana) (19%) as 'poor', and seven sites (1 in Minho, 2 in Tagus, and 4 in Guadiana) (33%) as 'bad' (Figure 5). Thus, the three rivers presented a wide range of F-MMIP-LR values across the gradient of 'bad' to 'high', indicating that 76% of sites were in not-good condition, but still showing some sites with less substantial human impacts.

4. Discussion

The development of fish-based indices in assessing the quality of large rivers is a challenging task [2,21,43]. Large rivers require expensive and time-demanding fishing efforts to adequately characterize fish assemblages [21,44]. Moreso than in wadeable streams, sampling fish in large rivers requires striking a balance between accuracy, precision, and cost, as all three factors are critical for effective and practical monitoring programs [45]. The distribution and catchability of fish in large rivers are highly variable because of extensive local movements and seasonal distribution of fish, presence of very deep habitats, variation in water levels, and relatively small sampling units [2]. Thus, all assessment metrics that are applied in large rivers can be based only on proxies of abundances and taxa richness because of gear and habitat selectivity and insufficient sampling effort [2,46–48]. We are aware of these limitations, which obviously extend to the tool we developed to assess the quality of large Portuguese rivers. For example, our sampling period most likely underestimated anadromous species that spawn in winter–early spring, i.e., *P. marinus* and *Alosa* spp. However, we believe that the use of a standardized protocol that also included two sampling techniques (electrofishing and gillnetting) enhanced the robustness of our fish assemblage assessments [48,49] and provided a more accurate picture of the biotic condition of our rivers.

The F-MMIP-LR was composed of four metrics (all metrics as percent relative abundance of individuals): native lithophilics, aliens, native migrants, and freshwater natives. We agree with Karr and Chu [50] that the selection of appropriate metrics is the key step in robustness of these biological indices. To this concept, we also add the need to produce a versatile set of metrics, that can be quickly and easily calculated to provide user-friendly tools for managers and decision makers. The fish data for the calculation of our metrics are easy to collect (e.g., do not require fish measurements or the identification of DELT—deformities, erosion, lesions, and tumors—specimens), and the metrics themselves are easy to apply, interpret, and communicate to broad audiences.

A decrease in lithophilic fish was associated with degradation of the index score, and typically reflects a degradation of the riverbed because they require clean, coarse substrates for reproductive success [51,52]. In our study, the most-impacted sites were generally present in larger or more intensive agricultural areas, with more channel and riparian degradation. Agricultural land use is commonly seen as a key variable in assessing the effects of human activity on stream and river ecosystems and a good predictor of both physical habitat quality and in-stream biotic condition [53–55]. In fact, agricultural activities are the most widespread cause of stream degradation, increasing nonpoint inputs of sediments, and often being a principal factor affecting riparian areas [56–59]. Riparian areas serve crucial ecological functions for river systems, such as bank stability, nutrient and sediment trapping, and habitat availability for fish in the form of woody debris, overhanging vegetation, and rootwads [60–62].

The number of non-native fish species and individuals has been growing exponentially in Portugal (and Iberia) in the last few decades mainly as a result of the growing use of these species for sport fisheries and in the aquarium trade [5]. This is particularly evident in large rivers, because of the natural spread of individuals from Spanish populations [63]. Research has largely revealed that non-native fishes can flourish in degraded conditions, thereby causing substantial negative impacts on natural fish assemblages [64–66], and thus representing one of the main causes of decline in biotic condition. Our index successfully included freshwater fish natives as a positive metric and non-native individuals as a sign of degradation. We also excluded the latter group from the other metrics. In fact, several authors have emphasized the problem of considering non-natives in MMIs, especially in the Iberian Peninsula, as the use of metrics with both native and non-native fish could restrict the ability of the index to detect the effect of non-native intrusions [66–68]. The metrics we developed also suggested a large influence of the proportion of non-natives in the degradation of the biological indices.

Least-disturbed river reaches are likely to support and maintain a wide range of ecological processes and functions, so it is not surprising that they include higher abundances of migrants. In fact, it is reasonable to assume that these river reaches generally present higher water quality, riparian cover, and shelter, together with lower levels of pollutants and sedimentation, creating suitable spawning areas for potamodromous and anadromous species, as well as feeding grounds for catadromous species. Additionally, least-disturbed sites might have better connectivity with other stream reaches, including the tributaries of the main rivers, that are used by different life stages of migratory fish. As emphasized by Jungwirth et al. [69], the ecological condition of large rivers is largely associated with the spatial/temporal connectivity of habitat subsystems, which are viewed as a crucial basis for a wide range of exchange processes and migration opportunities. However, McDowall and Taylor [70] pointed out the problems in establishing relationships between migratory species and environmental quality, as species become rarer with increasing distance inland/elevation. In that case, differences in abundance may not reflect differences in proximal habitat quality. However, we believe that this is not a relevant factor in our study, because historically these species abundantly occupied the habitat network along these rivers, including segments located many kilometres upstream of our study areas [71,72].

We found no clear relationship between the flow-regulated sites and fish biotic condition because sites with higher F-MMIP-LR scores in the Minho and Tagus Rivers were farther upstream, closer to large hydroelectric dams (Figure 5). In contrast, in the Guadiana River, the sites closest to the Pedrogão Dam presented the lowest MMI scores, and these results are aligned with Lyons et al. [73], who observed similar trends in Wisconsin rivers, where hydroelectric-peaking caused fish-habitat degradation and were associated with low fish MMI scores. The differences observed in the Guadiana River, particularly at site "GU4", can be attributed to its proximity to the Pedrogão Dam (<1 km), showing the direct influence of hydroelectric flow regimes, as opposed to the more upstream points of the Tagus and Minho Rivers, which are located ~10 km from the dam immediately upstream. The lack of a clear relationship in our study could be influenced by several factors and be context-dependent [74]. One of these factors may be the better adaptability of native species, mostly cypriniforms, to lotic habitats with frequent high-flow events, compared to some non-native fish which are more successful in stable limnological conditions [75,76]. Native species possess natural adaptive responses to high flows that are detrimental to some non-native species by disrupting their critical life stages.

We believe that our tool is very useful for interpreting, comparing, and conveying the biotic condition of Portuguese large rivers. The F-MMIP-LR showed a significant ecological consistency in relation to the degree of perturbation of a site, and both the metrics and the overall index were able to discriminate between least- and most-disturbed sites. However, limitations should be considered in interpreting our results. First, we used few sites to construct the index—e.g., Yoder and Rankin [77] suggested >30 sites to develop a more robust tool—and second, we did not validate the index with an independent data set to assess its performance [78]. Finally, these river systems are degraded by altered temperature, salinity gradients, flow rates, and toxic chemicals [79], which may not have been fully addressed in our study. Future research should consider the inclusion of these and other factors, which certainly provide additional insights for understanding human impacts on fish assemblages and implications for species management and conservation. Ultimately, we are confident in the usefulness of the F-MMIP-LR for informing managers and decision-makers in evaluating the biological quality of Portuguese large rivers within the framework of the EU WFD.

5. Conclusions

Several EU member states rely on locally developed fish indices customized to their specific regions to assess the biological quality of their rivers in the context of the WFD. In line with this approach, the objective of this study was to pioneer the development of the first fish-based index to assess the biological quality of Portuguese large rivers. The Fish-based Multimetric Index for Portuguese Large Rivers (F-MMIP-LR) incorporates four metrics: native lithophilics, aliens, migrants, and freshwater natives. The fish data for the calculation of the metrics are easy to collect, and the metrics themselves are easy to apply, interpret, and communicate to broad audiences. Our findings demonstrate the effectiveness of the index in discerning between least- and most-disturbed sites and its significant ecological consistency in relation to the degree of perturbation of a site. The research underscores the importance of evaluating both native and non-native fish species when assessing river quality, while also acknowledging the impact of human activities, such as agriculture, on aquatic biodiversity. Furthermore, the study emphasizes the critical role of preserving ecological processes and functions within rivers, as these foster healthier fish assemblages. We conclude that our index could be an effective monitoring tool in the context of the EU WFD and can be used to communicate river health to the public and policy makers.

Supplementary Materials: The following supporting information can be downloaded at: https://www.mdpi.com/article/10.3390/fishes9050149/s1, Table S1: Detailed Environmental Disturbance Score—EDS and F-MMIP-LR classification per site. Coordinate system: World Geodetic System 1984. LAT—Latitude, LNG—Longitude; Total P—Total Phosphorus; Total N—Total Nitrogen.

Author Contributions: A.T.F.: methodology, software, formal analysis, investigation, writing—original draft, writing—review and editing. M.T.F.: writing—review and editing. J.M.O.: conceptualization, methodology, validation, formal analysis, investigation, writing—review and editing, supervision. All authors have read and agreed to the published version of the manuscript.

Funding: This research received funding from the project ALBUFEIRA (0489_ALBUFEIRA_6_E) INTERREG V POCTEP) and by the project "Aquisição de serviços para melhorar e complementar os critérios de classificação do estado das massas de água superficiais interiores" for APA—Agência Portuguesa do Ambiente I.P., supported by the POSEUR—"Programa Operacional Sustentabilidade e Eficiência no Uso de Recursos" through the application POSEUR-03-2013-FC-000001—"Melhoria da Avaliação do Estado das Massas de Água". Forest Research Centre (CEF) is a research unit funded by Fundação para a Ciência e a Tecnologia I.P. (FCT), Portugal (UIDB/00239/2020). The publication in Open Access was funded by the Project UIDB/00239/2020 from CEF (Forest Research Centre). The Associate Laboratory TERRA (LA/P/0092/2020) is also funded by FCT. António Faro was supported by a Ph.D. grant from the FLUVIO–River Restoration and Management program funded by FCT, Portugal (2021.06859.BD). J.M.O. was funded by the Fundação para a Ciência e a Tecnologia I.P. (FCT) under Project UIDP/00681/2020 (https://doi.org/10.54499/UIDP/00681/2020). M.T.F. was funded by the Fundação para a Ciência e a Tecnologia I.P. (FCT) under Project UIDB/00239/2020 (https://doi.org/10.54499/UIDB/00239/2020).

Institutional Review Board Statement: The study was conducted in line with national and international guidelines of animal welfare.

Informed Consent Statement: Not applicable.

Data Availability Statement: Data are contained within the article. The data of fish assemblages are unavailable to the public due to privacy restrictions associated with the involved projects.

Conflicts of Interest: The authors declare no conflicts of interest.

References

1. Ward, J.V.; Tockner, K.; Schiemer, F. Biodiversity of floodplain river ecosystems: Ecotones and connectivity. *Regul. Rivers Res. Manag.* **1999**, *15*, 125–139. [CrossRef]
2. De Leeuw, J.J.; Buijse, A.D.; Haidvogl, G.; Lapinska, M.; Noble, R.; Repecka, R.; Virbickas, T.; Wiśniewolski, W.; Wolter, C. Challenges in developing fish-based ecological assessment methods for large floodplain rivers. *Fish. Manag. Ecol.* **2007**, *14*, 483–494. [CrossRef]
3. Tockner, K.; Uehlinger, U.; Robinson, C.T. *Rivers of Europe*; Elsevier: Amsterdam, The Netherlands, 2009.
4. Fernandes, M.R.; Aguiar, F.C.; Martins, M.J.; Rivaes, R.; Ferreira, M.T. Long-term human-generated alterations of Tagus river: Effects of hydrological regulation and land-use changes in distinct river zones. *CATENA* **2020**, *188*, 104466. [CrossRef]
5. Anastácio, P.M.; Ribeiro, F.; Capinha, C.; Banha, F.; Gama, M.; Filipe, A.F.; Rebelo, R.; Sousa, R. Non-native freshwater fauna in Portugal: A review. *Sci. Total Environ.* **2019**, *650*, 1923–1934. [CrossRef]
6. Collares-Pereira, M.J.; Alves, M.J.; Ribeiro, F.; Domingos, I.; Almeida, P.R.; Costa, L.; Gante, H.; Filipe, A.F.; Aboim, M.A.; Rodrigues, P.M.; et al. *Guia dos Peixes de água doce e Migradores de Portugal Continental*; Edições Afrontamento: Porto, Portugal, 2021.
7. *European Commission: Directive 2000/60/EC of the European Parliament and of the Council of 23 October 2000 Establishing a Framework for the Community Action in the Field of Water Policy*; Official Journal of the European Union: Strasbourg, France, 2000.
8. Solheim, A.-L.; Globevnik, L.; Austnes, K.; Kristensen, P.; Moe, S.J.; Persson, J.; Phillips, G.; Poikane, S.; Van De Bund, W.; Birk, S. A new broad typology for rivers and lakes in Europe: Development and application for large-scale environmental assessments. *Sci. Total Environ.* **2019**, *697*, 134043. [CrossRef]
9. Radinger, J.; Britton, J.R.; Carlson, S.M.; Magurran, A.E.; Alcaraz-Hernández, J.D.; Almodóvar, A.; Benejam, L.; Fernández-Delgado, C.; Nicola, G.G.; Oliva-Paterna, F.J.; et al. Effective monitoring of freshwater fish. *Fish Fish.* **2019**, *20*, 729–747. [CrossRef]
10. Reyjol, Y.; Argillier, C.; Bonne, W.; Borja, A.; Buijse, A.D.; Cardoso, A.C.; Daufresne, M.; Kernan, M.; Ferreira, M.T.; Poikane, S.; et al. Assessing the ecological status in the context of the European Water Framework Directive: Where do we go now? *Sci. Total Environ.* **2014**, *497–498*, 332–344. [CrossRef] [PubMed]
11. Ruaro, R.; Gubiani, É.A.; Hughes, R.M.; Mormul, R.P. Global trends and challenges in multimetric indices of biological condition. *Ecol. Indic.* **2020**, *110*, 105862. [CrossRef]
12. Vadas, R.L.; Hughes, R.M.; Bae, Y.J.; Baek, M.J.; Gonzáles, O.C.B.; Callisto, M.; Carvalho, D.R.D.; Chen, K.; Ferreira, M.T.; Fierro, P.; et al. Assemblage-based biomonitoring of freshwater ecosystem health via multimetric indices: A critical review and suggestions for improving their applicability. *Water Biol. Secur.* **2022**, *1*, 100054. [CrossRef]
13. Karr, J.R.; Fausch, K.; Angermeier, P.L.; Yant, P.R.; Schlosser, I.J. *Assessing Biological Integrity in Running Waters. A Method and Its Rationale*; Illinois Natural History Survey Special Publication: Champaign, IL, USA, 1986.
14. Kesminas, V.; Virbickas, T. Application of an adapted index of biotic integrity to rivers of Lithuania. *Assess. Ecol. Integr. Run. Water* **2000**, 257–270. [CrossRef]
15. Breine, J.; Simoens, I.; Goethals, P.; Quataert, P.; Ercken, D.; Van Liefferinghe, C.; Belpaire, C. A fish-based index of biotic integrity for upstream brooks in Flanders (Belgium). *Hydrobiologia* **2004**, *522*, 133–148. [CrossRef]
16. Pont, D.; Hugueny, B.; Rogers, C. Development of a fish-based index for the assessment of river health in Europe: The European Fish Index. *Fish. Manag. Ecol.* **2007**, *14*, 427–439. [CrossRef]
17. Magalhães, M.F.; Ramalho, C.E.; Collares-Pereira, M.J. Assessing biotic integrity in a Mediterranean watershed: Development and evaluation of a fish-based index. *Fish. Manag. Ecol.* **2008**, *15*, 273–289. [CrossRef]
18. Aparicio, E.; Carmona-Catot, G.; Moyle, P.B.; García-Berthou, E. Development and evaluation of a fish-based index to assess biological integrity of Mediterranean streams. *Aquat. Conserv. Mar. Freshw. Ecosyst.* **2011**, *21*, 324–337. [CrossRef]
19. Adamczyk, M.; Prus, P.; Buras, P.; Wiśniewolski, W.; Ligięza, J.; Szlakowski, J.; Borzęcka, I.; Parasiewicz, P. Development of a new tool for fish-based river ecological status assessment in Poland (EFI+IBI_PL). *Acta Ichthyol. Piscat.* **2017**, *47*, 173–184. [CrossRef]
20. Ramos-Merchante, A.; Prenda, J. Macroinvertebrate taxa richness uncertainty and kick sampling in the establishment of Mediterranean rivers ecological status. *Ecol. Indic.* **2017**, *72*, 1–12. [CrossRef]
21. Seegert, G. Considerations regarding development of index of biotic integrity metrics for large rivers. *Environ. Sci. Pol.* **2000**, *3*, 99–106. [CrossRef]
22. Yoder, C.O.; Kulik, B.H. The development and application of multimetric indices for the assessment of impacts to fish assemblages in large rivers: A review of current science and applications. *Can. Water Resour. J.* **2003**, *28*, 301–328. [CrossRef]
23. Petts, G.E.; Nestler, J.; Kennedy, R. Advancing science for water resources management. *Hydrobiologia* **2006**, *565*, 277–288. [CrossRef]

24. Borgwardt, F.; Leitner, P.; Graf, W.; Birk, S. Ex uno plures–Defining different types of very large rivers in Europe to foster solid aquatic bio-assessment. *Ecol. Indic.* **2019**, *107*, 105599. [CrossRef]
25. Agência Portuguesa do Ambiente: Convenção de Albufeira (Cooperação Luso-Espanhola). Available online: https://apambiente.pt/agua/convencao-de-albufeira-cooperacao-luso-espanhola (accessed on 27 March 2024).
26. Feio, M.J.; Ferreira, V. *Rios de Portugal: Comunidades, Processos e Alterações*; Imprensa da Universidade de Coimbra: Coimbra, Portugal, 2019.
27. Simon, T.P.; Evans, N.T. Environmental quality assessment using stream fishes. In *Methods in Stream Ecology*; Elsevier: Amsterdam, The Netherlands, 2017; Volume 39, pp. 319–334.
28. Stoddard, J.L.; Larsen, D.P.; Hawkins, C.P.; Johnson, R.K.; Norris, R.H. Setting expectations for the ecological condition of streams: The concept of reference condition. *Ecol. Appl.* **2006**, *16*, 1267–1276. [CrossRef] [PubMed]
29. Oliveira, J.M.; Ferreira, M.T.; Morgado, P.; Hughes, R.M.; Teixeira, A.; Cortes, R.M.; Bochechas, J.H. A preliminary fishery quality index for Portuguese streams. *N. Am. J. Fish. Manag.* **2009**, *29*, 1466–1478. [CrossRef]
30. Weigel, B.M.; Dimick, J.J. Development, validation, and application of a macroinvertebrate-based Index of biotic integrity for nonwadeable rivers of Wisconsin. *J. N. Am. Benthol. Soc.* **2011**, *30*, 665–679. [CrossRef]
31. European Environmental Agency: Copernicus Land Monitoring Service. CORINE Land Cover. Available online: https://www.eea.europa.eu/en/datahub/datahubitem-view/a5144888-ee2a-4e5d-a7b0-2bbf21656348 (accessed on 15 October 2022).
32. SNIRH–Sistema Nacional de Informação de Recursos Hídricos: Rede Hidrométrica do Sistema Nacional de Informação de Recursos Hídricos da APA (SNIRH). Localização Geográfica, Classificação e Caracterização. Available online: https://snirh.apambiente.pt/ (accessed on 15 October 2022).
33. Smith, R.A.; Alexander, R.B.; Wolman, M.G. Water-quality trends in the nation's rivers. *Science* **1987**, *235*, 1607–1615. [CrossRef]
34. Instituto da Água I. P. *Manual para a Avaliação Biológica da Qualidade da Água em Sistemas Fluviais Segundo a Directiva Quadro da Água: Protocolo de Amostragem e Análise para a Fauna Piscícola*; Ministério do Ambiente, do Ordenamento do Território e do Desenvolvimento Regional; Instituto da Água: Alfragide, Portugal, 2008.
35. Whittier, T.R.; Hughes, R.M.; Stoddard, J.L.; Lomnicky, G.A.; Peck, D.V.; Herlihy, A.T. A structured approach for developing indices of biotic integrity: Three examples from streams and rivers in the western USA. *Trans. Am. Fish. Soc.* **2007**, *136*, 718–735. [CrossRef]
36. Krause, J.R.; Bertrand, K.N.; Kafle, A.; Troelstrup, N.H. A fish index of biotic integrity for South Dakota's Northern glaciated plains ecoregion. *Ecol. Indic.* **2013**, *34*, 313–322. [CrossRef]
37. Gonino, G.; Benedito, E.; Cionek, V.D.M.; Ferreira, M.T.; Oliveira, J.M. A fish-based index of biotic integrity for neotropical rainforest sandy soil streams—Southern Brazil. *Water* **2020**, *12*, 1215. [CrossRef]
38. Noble, R.A.A.; Cowx, I.G.; Goffaux, D.; Kestemont, P. Assessing the health of European rivers using functional ecological guilds of fish communities: Standardising species classification and approaches to metric selection. *Fish. Manag. Ecol.* **2007**, *14*, 381–392. [CrossRef]
39. Oliveira, J.M.; Segurado, P.; Santos, J.M.; Teixeira, A.; Ferreira, M.T.; Cortes, R.V. Modelling stream-fish functional traits in reference conditions: Regional and local environmental correlates. *PLoS ONE* **2012**, *7*, e45787. [CrossRef]
40. Magalhães, M.F.; Amaral, S.D.; Sousa, M.; Alexandre, C.M.; Almeida, P.R.; Alves, M.J.; Cortes, R.; Farrobo, A.; Filipe, A.F.; Franco, A.; et al. *Livro Vermelho dos Peixes Dulciaquícolas e Diádromos de Portugal Continental*; FCiências.ID & ICNF, I.P.: Lisboa, Portugal, 2023.
41. Pont, D.; Bady, P.; Logez, M.; Veslot, J. EFI+ Project. Improvement and Spatial Extension of the European Fish Index Deliverable 4.1: Report on the Modelling of Reference Conditions and on the Sensitivity of Candidate Metrics to Anthropogenic Pressures. Deliverable 4.2: Report on the Final Development and Validation of the New European Fish Index and Method, Including a Complete Technical Description of the New Method. 6th Framework Programme Priority FP6-2005-SSP-5-A. N° 0044096. 2009. Available online: https://hal.inrae.fr/hal-02592964/document (accessed on 15 October 2022).
42. Hering, D.; Feld, C.K.; Moog, O.; Ofenböck, T. Cook book for the development of a multimetric index for biological condition of aquatic ecosystems: Experiences from the European AQEM and STAR projects and related initiatives. *Hydrobiologia* **2006**, *566*, 311–324. [CrossRef]
43. Emery, E.B.; Simon, T.P.; McCormick, F.H.; Angermeier, P.L.; Deshon, J.E.; Yoder, C.O.; Sanders, R.E.; Pearson, W.D.; Hickman, G.D.; Reash, R.J.; et al. Development of a multimetric index for assessing the biological condition of the Ohio river. *Trans. Am. Fish. Soc.* **2003**, *132*, 791–808. [CrossRef]
44. Lapointe, N.W.R.; Corkum, L.D.; Mandrak, N.E. A comparison of methods for sampling fish diversity in shallow offshore waters of large rivers. *N. Am. J. Fish. Manag.* **2006**, *26*, 503–513. [CrossRef]
45. Hughes, R.M.; Peck, D.V. Acquiring data for large aquatic resource surveys: The art of compromise among science, logistics, and reality. *J. N. Am. Benthol. Soc.* **2008**, *27*, 837–859. [CrossRef]
46. Hughes, R.M.; Kaufmann, P.R.; Herlihy, A.T.; Intelmann, S.S.; Corbett, S.C.; Arbogast, M.C.; Hjort, R.C. Electrofishing distance needed to estimate fish species richness in raftable Oregon rivers. *N. Am. J. Fish. Manag.* **2002**, *22*, 1229–1240. [CrossRef]
47. Hughes, R.M.; Herlihy, A.T.; Peck, D.V. Sampling efforts for estimating fish species richness in western USA river sites. *Limnologica* **2021**, *87*, 125859. [CrossRef] [PubMed]
48. Dunn, C.G.; Paukert, C.P. A flexible survey design for monitoring spatiotemporal fish richness in nonwadeable rivers: Optimizing efficiency by integrating gears. *Can. J. Fish. Aquat. Sci.* **2020**, *77*, 978–990. [CrossRef]

49. Goffaux, D.G. Electrofishing versus gillnet sampling for the assessment of fish assemblages in large rivers. *Arch. Hydrobiol.* **2005**, *162*, 73–90. [CrossRef]

50. Karr, J.R.; Chu, E.W. Sustaining living rivers. *Hydrobiologia* **2000**, *422–423*, 1–14. [CrossRef]

51. Berkman, H.E.; Rabeni, C.F. Effect of siltation on stream fish communities. *Environ. Biol. Fishes* **1987**, *18*, 285–294. [CrossRef]

52. Kemp, P.; Sear, D.; Collins, A.; Naden, P.; Jones, I. The impacts of fine sediment on riverine fish. *Hydrol. Process.* **2011**, *25*, 1800–1821. [CrossRef]

53. Hughes, R.M.; Vadas, R.L. Agricultural effects on streams and rivers: A western USA focus. *Water* **2021**, *13*, 1901. [CrossRef]

54. Kaufmann, P.R.; Hughes, R.M.; Paulsen, S.G.; Peck, D.V.; Seeliger, C.W.; Weber, M.H.; Mitchell, R.M. Physical habitat in conterminous US streams and rivers, Part 1: Geoclimatic controls and anthropogenic alteration. *Ecol. Indic.* **2022**, *141*, 109046. [CrossRef] [PubMed]

55. Herlihy, A.T.; Sifneos, J.C.; Hughes, R.M.; Peck, D.V.; Mitchell, R.M. The relation of lotic fish and benthic macroinvertebrate condition indices to environmental factors across the conterminous USA. *Ecol. Indic.* **2020**, *112*, 105958. [CrossRef] [PubMed]

56. Wang, L.; Lyons, J.; Kanehl, P.; Gatti, R. Influences of watershed land use on habitat quality and biotic integrity in Wisconsin streams. *Fisheries* **1997**, *22*, 6–12. [CrossRef]

57. Allan, J.D. Landscapes and riverscapes: The influence of land use on stream ecosystems. *Annu. Rev. Ecol. Evol. Syst.* **2004**, *35*, 257–284. [CrossRef]

58. Hermoso, V.; Clavero, M.; Blanco-Garrido, F.; Prenda, J. Invasive species and habitat degradation in Iberian streams: An analysis of their role in freshwater fish diversity loss. *Ecol. Appl.* **2011**, *21*, 175–188. [CrossRef] [PubMed]

59. Fierro, P.; Bertrán, C.; Tapia, J.; Hauenstein, E.; Peña-Cortés, F.; Vergara, C.; Cerna, C.; Vargas-Chacoff, L. Effects of local land-use on riparian vegetation, water quality, and the functional organization of macroinvertebrate assemblages. *Sci. Total Environ.* **2017**, *609*, 724–734. [CrossRef] [PubMed]

60. Gregory, S.V.; Swanson, F.J.; McKee, W.A.; Cummins, K.W. An ecosystem perspective of riparian zones. *BioScience* **1991**, *41*, 540–551. [CrossRef]

61. Naiman, R.J.; Decamps, H.; Pollock, M. The role of riparian corridors in maintaining regional biodiversity. *Ecol. Appl.* **1993**, *3*, 209–212. [CrossRef]

62. Pusey, B.J.; Arthington, A.H. Importance of the riparian zone to the conservation and management of freshwater fish: A review. *Mar. Freshw. Res.* **2003**, *54*, 1. [CrossRef]

63. Martelo, J.; Da Costa, L.; Ribeiro, D.; Gago, J.; Magalhães, M.; Gante, H.; Alves, M.; Cheoo, G.; Gkenas, C.; Banha, F.; et al. Evaluating the range expansion of recreational non-native fishes in Portuguese freshwaters using scientific and citizen science data. *BioInvasions Rec.* **2021**, *10*, 378–389. [CrossRef]

64. Kennard, M.J.; Arthington, A.H.; Pusey, B.J.; Harch, B.D. Are alien fish a reliable indicator of river health? *Freshw. Biol.* **2005**, *50*, 174–193. [CrossRef]

65. Ferreira, T.; Caiola, N.; Casals, F.; Oliveira, J.M.; De Sostoa, A. Assessing perturbation of river fish communities in the Iberian ecoregion. *Fish. Manag. Ecol.* **2007**, *14*, 519–530. [CrossRef]

66. Ramos-Merchante, A.; Prenda, J. The ecological and conservation status of the Guadalquivir river basin (s Spain) through the application of a fish-based multimetric index. *Ecol. Indic.* **2018**, *84*, 45–59. [CrossRef]

67. Hermoso, V.; Clavero, M. Revisiting ecological integrity 30 years later: Non-native species and the misdiagnosis of freshwater ecosystem health. *Fish Fish.* **2013**, *14*, 416–423. [CrossRef]

68. Aparicio, E.; Alcaraz, C.; Rocaspana, R.; Pou-Rovira, Q.; García-Berthou, E. Adaptation of the European Fish Index (EFI+) to include the alien fish pressure. *Fishes* **2023**, *9*, 13. [CrossRef]

69. Jungwirth, M.; Muhar, S.; Schmutz, S. Fundamentals of fish ecological integrity and their relation to the extended serial discontinuity concept. *Assess. Ecol. Integr. Run. Waters* **2000**, 85–97. [CrossRef]

70. McDowall, R.M.; Taylor, M.J. Environmental indicators of habitat quality in a migratory freshwater fish fauna. *Environ. Manag.* **2000**, *25*, 357–374. [CrossRef] [PubMed]

71. Duarte, G.; Moreira, M.; Branco, P.; Da Costa, L.; Ferreira, M.T.; Segurado, P. One millennium of historical freshwater fish occurrence data for Portuguese rivers and streams. *Sci. Data* **2018**, *5*, 180163. [CrossRef]

72. Duarte, G.; Branco, P.; Haidvogl, G.; Ferreira, M.T.; Pont, D.; Segurado, P. iPODfish–A new method to infer the historical occurrence of diadromous fish species along river networks. *Sci. Total Environ.* **2022**, *812*, 152437. [CrossRef]

73. Lyons, J.; Piette, R.R.; Niermeyer, K.W. Development, validation, and application of a fish-based index of biotic integrity for Wisconsin's large warmwater rivers. *Trans. Am. Fish. Soc.* **2001**, *130*, 1077–1094. [CrossRef]

74. Turgeon, K.; Turpin, C.; Gregory-Eaves, I. Dams have varying impacts on fish communities across latitudes: A quantitative synthesis. *Ecol. Lett.* **2019**, *22*, 1501–1516. [CrossRef] [PubMed]

75. Bernardo, J.M.; Ilhéu, M.; Matono, P.; Costa, A.M. Interannual variation of fish assemblage structure in a Mediterranean river: Implications of streamflow on the dominance of native or exotic species. *River Res. Appl.* **2003**, *19*, 521–532. [CrossRef]

76. Propst, D.L.; Gido, K.B. Responses of native and nonnative fishes to natural flow regime mimicry in the San Juan river. *Trans. Am. Fish. Soc.* **2004**, *133*, 922–931. [CrossRef]

77. Yoder, C.O.; Rankin, E.T. Biological response signatures and the area of degradation value: New tools for interpreting multimetric data. In *Biological Assessment and Criteria: Tools for Water Resources Planning and Decision Making*; Davis, W.S., Simon, T.P., Eds.; Lewis Publishers: Boca Raton, FL, USA, 1995.

78. Hughes, R.M.; Kaufmann, P.R.; Herlihy, A.T.; Kincaid, T.M.; Reynolds, L.; Larsen, D.P. A process for developing and evaluating indices of fish assemblage integrity. *Can. J. Fish. Aquat. Sci.* **1998**, *55*, 1618–1631. [CrossRef]
79. Sabater, S.; Elosegi, A.; Ludwig, R. *Multiple Stressors in River Ecosystems*; Elsevier: Amsterdam, The Netherlands, 2019.

Article

Contemporary Trends in the Spatial Extent of Common Riverine Fish Species in Australia's Murray–Darling Basin

Wayne Robinson [1,*], **John Koehn** [2] and **Mark Lintermans** [3]

[1] Gulbali Institute for Agriculture, Water and Environment, Charles Sturt University, P.O. Box 789, Albury, NSW 2640, Australia

[2] Department of Environment, Land, Water and Planning, Arthur Rylah Institute for Environmental Research, P.O. Box 137, Heidelberg, VIC 3084, Australia; john.koehn@delwp.vic.gov.au

[3] Centre for Applied Water Science, University of Canberra, Bruce, ACT 2617, Australia; mark.lintermans@canberra.edu.au

[*] Correspondence: wrobinson@csu.edu.au

Abstract: As one of the world's most regulated river basins, the semi-arid Murray–Darling Basin (MDB) in south-eastern Australia is considered at high ecological risk, with substantial declines in native fish populations already identified and climate change threats looming. This places great importance on the collection and use of data to document population trends over large spatial extents, inform management decisions, and provide baselines from which change can be measured. We used two medium-term data sets (10 MDB basin-wide fish surveys from 2004–2022) covering the 23 catchments and 68 sub-catchments of the MDB to investigate trends in the distribution of common riverine species at the entire basin scale. Fifteen native species were analysed for changes in their contemporary range, and whilst short-term changes were identified, all species showed no significant continuous trend over the study period. We further analysed the native species extent relative to their historic records, with bony herring and golden perch occurring in 78% and 68% of their historic river kilometres, respectively, whereas southern pygmy perch, northern river blackfish, silver perch, mountain galaxias, and freshwater catfish were all estimated to occur in less than 10% of their historic extent. Six established non-native species were also analysed and were very consistent in extent over the years, suggesting that they are near the available limits of expansion of their invasion. We provide effect sizes for the spatial extent index which can be used as baselines for future studies, especially those aiming to monitor changes in the spatial extent and population status of native species, or changes in the spatial extent of new or existing non-native species.

Keywords: spatial extent; native fish; non-native species; historical distributions; monitoring; baselines

Key Contribution: This monitoring and data analysis not only documents recent changes in population extent for 21 key freshwater species but also provides essential baseline data from which population recovery or the influence of climate change can be measured.

Citation: Robinson, W.; Koehn, J.; Lintermans, M. Contemporary Trends in the Spatial Extent of Common Riverine Fish Species in Australia's Murray–Darling Basin. *Fishes* **2024**, *9*, 221. https://doi.org/10.3390/fishes9060221

Academic Editors: Bror Jonsson, Robert L. Vadas Jr. and Robert M. Hughes

Received: 15 April 2024
Revised: 4 June 2024
Accepted: 7 June 2024
Published: 12 June 2024

1. Introduction

The Murray–Darling Basin (MDB) is one of the world's most regulated river basins [1,2], with the majority of its 23 catchments classified as having impaired connectivity [1] and considered in poor ecological condition [3–5]. Over-allocation of water, flow regulation, and environmental damage have been long-term concerns, e.g., [6–8], which are becoming more pressing as a result of climate change, e.g., [9–11]. Recent severe drought conditions, extensive fish kills, and extreme bushfires (2019–2020) have heightened concerns over the ecological health of the MDB [12–14]. Even prior to these recent concerns, MDB native fish were known to be in serious declines, with populations (viz distribution and abundance) of common riverine species estimated to be at <10% of pre-European settlement levels (early 1800s) [15,16].

Recognition of the need for higher confidence in native fish species status together with considerable investment in initiatives to improve river health, e.g., [17,18], require a significant monitoring program. Hence, the MDB Sustainable Rivers Audit (SRA) began in 2004 as an MDB-wide surveillance monitoring program to report on the status and long-term trends in the ecological health (condition) of rivers at the river valley (catchment/watershed) scale. One of the five ecological health themes included in the SRA was fish and this included the first ever attempt at an MDB-wide fishery-independent assessment program [4,5]. The SRA fish sampling methodology aimed to assess riverine species richness, only sampled in channel-permanent habitats (no floodplain or ephemeral habitats), and relied on a single sampling technique (electrofishing) [4,19]. Nine years after the SRA program was implemented, the fish component was transformed into the MDB Fish Survey (MDBFS) which revisited sites previously used in the SRA and maintained the same within-site sampling protocols but sampled fewer sites more often. Both data sets combined provide the first medium-term (10 basin-wide surveys over 19 years; 2004–2022) data set for fish in the MDB. The first 9 years of data have previously been used to report on river health using conceptual indicators of health, e.g., [5], and response to hydrological factors [19], but these data have not been widely analysed as a combined data set nor used to investigate the population trends for MDB fish species over time (see [20]).

Monitoring fish distributions is a useful measure of how populations are responding to management interventions (e.g., fish passage, habitat rehabilitation) or how climatic factors (drought, flood) are influencing populations. It is also a component of assessing conservation status and a key metric in assessing the recovery of threatened species where extent of occurrence and area of occurrence are widely used in classifications such as the IUCN red list [21]. The MDB also has a high proportion of non-native fish species, both from overseas and translocated native species [22]. Although several non-native species have their distribution in the MDB limited by environmental variables (e.g., Salmonids and water temperature), others (including those recently introduced) are habitat generalists (e.g., common carp *Cyprinus carpio*) with considerable potential for range expansions [22]. Non-native fish are significant hosts and vectors of novel pathogens [23,24], and tracking the extent of the distribution of non-native fish is a key concern for managers. Non-native species are an ongoing threat that may have high social capital but are rarely targeted in river health monitoring programs [25].

Aims of this Study

We used the SRA and MDBFS data sets to determine the trends in the contemporary spatial distribution (extent) of common, widespread, and abundant MDB fish species over the period 2004–2022. We further assessed spatial extent for each native species relative to their known historical distribution and the current spatial extent of non-native (Non-native) fish species in the MDB. The results provide trends and baseline assessments of the extent of common riverine species that may be used as reference for future comparisons.

2. Materials and Methods

2.1. Study Area

The Murray–Darling Basin in south-eastern Australia (Figure 1) covers more than a million km^2, or about 14% of Australia's total landmass, provides about 40% of Australia's total agricultural production, and accounts for 50% of the nation's irrigated agricultural water use [26,27]. Hence, there is conflict between water management for agriculture and that for environmental benefits and assets such as fish [28]. The MDB encompasses all or parts of five states and territories, with fish management remaining the responsibility of state/territory governments. Overlaying this structure is a Commonwealth responsibility for water management via the Basin Plan [27]. The MDB is subdivided into

23 valleys/catchments/watersheds and 68 sub-catchments for water use management and for SRA/MDFS fish sampling (Figure 1).

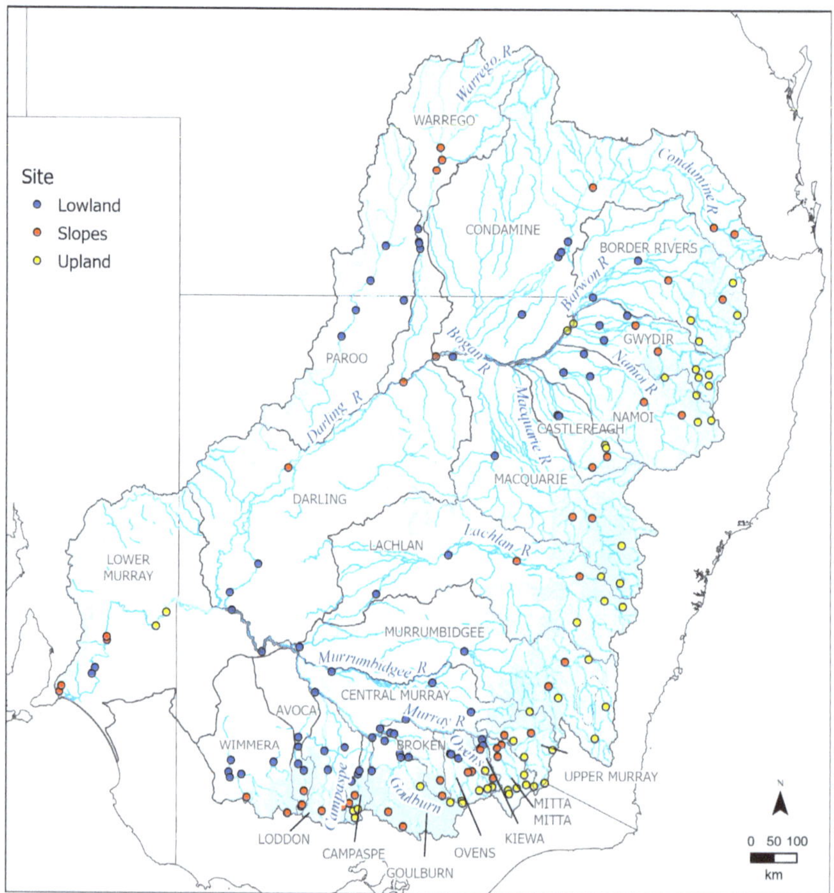

Figure 1. The Murray–Darling Basin, south-eastern Australia, is made up of 23 catchments. The SRA/MDBFS fish monitoring scheme sampled the basin triennially between 2004 and 2013 then annually or biennially up to 2022. The dots identify 169 sites that were sampled at least twice (min = 2, median = 9, max = 10 visits per site) through both programs that are analysed in the current paper. The sampling strategy included sites in lowland (0 < 200 m ASL), slope (200 < 600 m ASL), and upland (>600 m ALS) bioregions.

2.2. Data Sets

The MDBFS and SRA data sets offer medium-term data covering the whole of the MBD stream network with a standardised sampling effort for fish based on a probabilistic site selection strategy in a mixed longitudinal rotating panel design [29]. The sampling frame covers all streams in the MDB with at least 5 GL mean annual flow and does not include ephemeral or non-riverine (wetland and floodplain) habitats. Each valley is stratified into three sub-catchments that align with potential changes in fish assemblages. These sub-catchments include lowland (<200 m ASL), slope (200 < 600 m ASL), and upland (>600 m ASL) bioregions. Upland sites are sampled with relatively more spatial intensity than lowland or slope habitats [19]. Consequently, when aggregating to higher scale assessments (catchment or MDB), site parameters must be weighted by the selection probability of each site (weighting is easily interpreted as the kilometres of river represented by the site when it first enters the sampling program). Sites are 1 km in length and were selected from a stream length GIS layer; thus, after aggregating site data to larger extents, the assessments are best interpreted as river kilometres (e.g., fish/km) rather than as averages.

The SRA sampled more intensely, but less frequently, whilst the MBDFS sampled more frequently but less intensely:

- SRA pilot study (2003) = four MDB valleys and sub-catchments were sampled once with 21 to 26 sites per valley. A total of 92 sites sampled; the results are summarised in [19,30];
- SRA (2004 to 2013) = every MDB valley and its sub-catchment was sampled once every three years with 14 to 28 sites per valley. Approximately 450 sites sampled per year;
- MDBFS (2014 to 2022) = every MDB valley and its sub-catchment was sampled once every year except 2019 and 2020, during which half of the MDB was sampled. Approximately 145 sites were sampled per year (except 2019 and 2020), with 4–8 sites per valley, and all sites were previously sampled SRA sites.

Combined, the data sets include 1222 separate survey sites (each site represents 1 km of river) and 2368 sampling events. We treat sampling the whole MDB in single or consecutive years as a sampling round and, after 10 sampling rounds, the combined data set now offers the opportunity to analyse large spatial extent trends in the frequency of occurrence (extent) for common species that are ubiquitous and/or abundant in the data set.

2.3. Spatial Extent of Common Fish in the MDB (2004 to 2022)

We choose spatial extent as a practical measure for this data set because whilst abundance is also important, its estimation by any sampling method is difficult and comparing abundance through time is susceptible to many confounding factors. For example, fish have varying susceptibility between species (habits, habitats, and sizes) and within species (e.g., life phases) to different sampling seasons and methods, and detectability varies with environmental variables such as flow and turbidity. This makes the assessment of abundance best interpreted as a relative assessment, e.g., [31–33]. The extent of occurrence, on the other hand, the proportion of river sites or river kilometres where a fish occurs, offers a simpler assessment of fish population health over a large area as it only requires detecting the species presence as a standard effort. It relies simply on the assumption that the detection of the presence of a species within a site is more likely when the species is more abundant. This assumption is clearly the case for many fish species in the MDB when sampled via electrofishing (refer to Table 2 in [31]).

The within-site sampling methodology of the SRA was designed to target species presence rather than to measure abundance. It involved intensive electrofishing and returned the full list of species that are well-sampled by electrofishing in 174 of 180 (97%) of sites in the first year of implementation [19]. Thus, the data collected by the SRA/MDBFS programs are highly suitable for estimating the presence of common main-channel species that are susceptible to electrofishing within each site [19] and, consequently, their extent at the sub-catchment and basin scale. This includes most large-bodied species and riverine species that have a wide distribution in the MDB.

There are three aspects to the spatial extent analyses. Firstly, for each native species, we identified contemporary sub-catchments as sub-catchments where a species had been collected at least once since 2004 were also a historically (pre-1980) known sub-catchment for that species, based on [22]. As we use sites where the species had been collected at least once in the data sets, we treat this analysis as a prevalence assessment and refer to the results as an assessment of current or contemporary extent. We ignore sub-catchments where there was no history of occurrence to avoid misinterpretations from post-1980 translocations. In a second analysis, we also include an estimated absolute extent for each species relative to the species' known distribution [22]. In this calculation, sites that are in a zone where the species should be present are included, regardless of whether the species has ever been collected in that particular site. This is considered a conservative estimate as the sites are randomly selected, and a missing species may be a random effect (e.g., that site has an incorrect habitat for that species). Nevertheless, it is an empirical estimate that returns a relative assessment of the species' overall status that serves two purposes: (1) it allows us to identify species that have relatively better or worse spatial extent estimates than other species, and (2) it serves as a baseline for these species for this study period to allow managers to make future comparisons. The two assessments should be interpreted in conjunction. For example, a species may appear to have high contemporary prevalence—by consistently occurring in a number of sites in the current data set—yet those sites may only be from a fraction of its historic distribution.

On the other hand, we analyse non-native species relative to the entire sampling frame. In other words, the trend in extent of each non-native species is analysed relative to their distribution across the entire MDB, but for native species, trends in extent are relative to their historical distribution. In summary, the non-native species analyses include all 68 sub-catchments, whereas the number of sub-catchments used for native species are unique to each species.

To avoid potential confounding between the SRA and MDBFS site composition, only sites from the SRA program that were also sampled at least once in the MDBFS are included in the analyses. That is, many of the SRA sites were only sampled once, whilst almost all MDBFS sites were revisited annually or biennially, and all MDBFS sites were also sampled in the SRA. Single-visit SRA sites are not included in the analyses because they only occurred in the first three sampling rounds and could not contribute to a long-term trend analysis. For example, a single-visit site that that did (or did not) have a particular species collected can never contribute to the trend analysis because it is never sampled again. Meanwhile, a repeat-visit site that did (or did not) have the species collected can have the species collected (or not) in the repeat visits and can therefore contribute to the trend analysis of change in extent. There are 169 sites that were sampled in both the SRA and the MDBFS and these offer 1222 sampling events that are used in the current paper. The data do not include the pilot SRA as the sampling frame and site selection strategy changed after the pilot.

Only species that had been detected in at least 14 of the 68 (20%) MDB defined sub-catchments are included in the analyses. The 20% occurrence requirement returns enough data points to reasonably estimate the extent of each sampling round with 95% confidence intervals and constrains the analyses to species well collected by electrofishing. If such a species is abundant within a site, it is expected to be collected by the electrofishing survey. Accordingly, if a common species was not collected in a site or sub-catchment, it was either absent or in low abundance.

The data are assessed at the entire basin scale over 10 sampling rounds. The first 9 years (SRA) of data represent 3 rounds where the entire basin was sampled, whilst the remaining 8 years (MDBFS) represent 7 entire basin sampling rounds. That is, the first 3 rounds cover 3 years each and the remaining rounds are 1 or 2-year periods.

2.4. Statistical Analyses

The spatial extent index is the weighted average proportion of river kilometres in which each species was detected during each sampling round and is described in Equation (1).

$$I_{extent_{i,r}} = \frac{\sum d_{i_{[y=1]},r}}{\sum d_{E_{i_{[y=1]},r}}} \tag{1}$$

where i = species, r = sampling round, d_i = river kilometres sampled for species i, $y = 1$ is where the species was detected, and $dE_{i(y=1)}$ is the sampled river kilometres where species i was expected to be detected.

For native species, the denominator of Equation (1) is different for the prevalence analyses (expected river kilometres are constrained to known contemporary river kilometres) and the historical analyses (expected river kilometres include pre-1980 river kilometres).

For non-native species, the denominator is set as the entire basin river kilometres sampled each round.

To estimate trends, the index was estimated for all rounds using a generalised linear model fitted with a binomial response (species presence = 1, absence = 0) where sites are random replicates within sub-catchments which are repeated subjects with an independent correlation structure. Because there are differences in sampling efficiency between species, the index should be interpreted in a relative, not definitive, manner when comparing between species.

When graphing trends, the calculated index of extent for prevalence (contemporary distribution) for native species and overall extent for non-native species was plotted as a time series with a modelled three-period centred moving average for the index and its 95% confidence intervals. We assessed monotonic trends in the spatial index for each species over the 10 sampling rounds using the Kendall rank correlation test. Complex trends were not tested statistically as there were only 10 sampling rounds, and intervals are not strictly equidistant given each data points represent a 1, 2, or 3-year cycle. To investigate short or long-term changes in extent, we performed pairwise comparisons of spatial extent between every survey round for each species. To moderate the type-1 error rate (n = 45 comparisons per species), we only reported survey rounds where the comparison is significant at $p < 0.01$. To provide a baseline guide for monitoring, we estimated the effect size for the spatial extent index between survey rounds for each species. We plotted the magnitude of differences in extent between all sampling rounds with the probability of each difference being significant and fit a LOESS (locally estimated scatterplot smoothing) line to the plot. We estimated the effect size as the magnitude of difference between sampling rounds where the LOESS intersected with the 0.05 level of significance. We have provided a guide to the sensitivity of the spatial index for monitoring, calculated as the effect size expressed as a percentage relative to the long-term average. Trends for the historical analyses for native species are included on the same plots to aid in interpreting status comparison, and we have provided a summary table of all species extent relative to their historical distribution. All analyses were performed using SAS/STAT®14.1 [34].

3. Results

3.1. Common Fish Trends in Extent in the MDB (2004 to 2022)

Twenty-two fish species were detected in at least 20% of the MDB sub-catchments and 21 species were analysed for trends in spatial extent (Table 1). These included 15 native and 6 non-native species. Obscure galaxias were omitted from the analysis as they were undescribed at the start of the monitoring program—hence, no historical records. Non-native and native species are reported separately, and native species are grouped by life expectancy guild (Table 1) (long, intermediate, and short-lived species) to aid in interpretation. Short-lived species have life cycles of <3 years, intermediate-lived species have lifecycles from 3 to 6 years, and long-lived species have life cycles of 6 years or greater.

No species showed statistically significant monotonic trends throughout the 10 sampling rounds ($p > 0.05$). Golden perch (*Macquaria ambigua*) tended to show a consistent increase in spatial extent throughout the study period (Kendall's Tau = 0.47, $p < 0.07$), whereas two-spined blackfish (*Gadopsis bispinosa*) tended to decrease in extent (Tau = −0.42, $p < 0.09$). The effect size for assessing short-term changes in the spatial extent index varied considerably among species and tended to be smaller in species that were collected more often (Table 1). Ten of the twenty-one species had effect sizes less than 50% of the long-term mean, with bony herring (19%), golden perch (24%), and common carp (11%) being the most sensitive.

Table 1. Common fish species in the MDBFS/SRA data set between 2004 and 2022. Several *Hypseleotris* species were combined for the analyses because of taxonomic resolution differences in the early years of the monitoring programs, e.g., [35,36]. * Obscure galaxias: *Galaxias oliros* were not analysed for trends in spatial extent. The number of sub-catchments collected are cumulative over the 10 sampling rounds and used for the prevalence analysis. The number of historical sub-catchments expected are based on [22] and used to estimate overall status compared to historical status for native species. The effect size of the index is the magnitude of change in contemporary extent between any two sampling rounds that would be considered statistically significant ($p < 0.05$). The % change in the index is the effect size relative to the 20-year average from the current study. na = non-estimable.

Species	Common Name	Origin	Life Guild	Num. Sub-Catchments Collected	Num. Sub-Catchments Expected	Index Effect Size	Index % Change
Gambusia holbrooki	eastern gambusia	Non-native	Short-lived	60	0	0.17	36%
Cyprinus carpio	common carp	Non-native	Long-lived	54	0	0.09	11%
Retropinna semoni	Australian smelt	Native	Short-lived	51	66	0.23	48%
Carassius auratus	goldfish	Non-native	Long-lived	52	0	0.17	45%
Hypseleotris spp.	carp gudgeon complex	Native	Short-lived	47	56	0.17	31%
Macquaria ambigua	golden perch	Native	Long-lived	49	66	0.16	24%
Perca fluviatilis	redfin perch	Non-native	Long-lived	44	0	0.08	61%
Maccullochella peelii	Murray cod	Native	Long-lived	43	63	0.19	39%
Nematalosa erebi	bony herring	Native	Intermediate-lived	29	49	0.15	19%
Philypnodon grandiceps	flathead gudgeon	Native	Intermediate-lived	22	30	0.24	68%
Tandanus tandanus	freshwater catfish	Native	Long-lived	23	50	0.32	>100%
Galaxias olidus	mountain galaxias	Native	Intermediate-lived	26	51	0.4	89%
Salmo trutta	brown trout	Non-native	Long-lived	27	0	0.05	75%
Oncorhynchus mykiss	rainbow trout	Non-native	Intermediate-lived	26	0	0.05	94%
Gadopsis marmarata	northern river blackfish	Native	Intermediate-lived	20	53	0.18	28%
Melanotaenia fluviatilis	Murray–Darling rainbowfish	Native	Short-lived	23	46	0.26	46%
Leiopotherapon unicolor	spangled perch	Native	Intermediate-lived	23	33	0.28	53%
Bidyanus bidyanus	silver perch	Native	Long-lived	21	51	0.26	>100%
Craterocephalus stercusmuscarum fulvus	unspecked hardyhead	Native	Short-lived	20	40	na	na
Galaxias oliros *	obscure galaxias	Native	Intermediate-lived	21	41	na	na
Gadopsis bispinosa	two-spined blackfish	Native	Intermediate-lived	18	22	0.3	43%
Nannoperca australis	southern pygmy perch	Native	Short-lived	14	36	na	na

3.2. Short-Lived Native Fish Species

- Short-lived species all showed high inter-annual variability in their index of contemporary spatial extent (Figure 2).
- The carp gudgeon complex (*Hypseleotris* spp.) was estimated to occur in about 55% of its current distribution throughout the study, but the year-to-year variability in extent was high (range from 34% to 66%) (Figure 2a). The contemporary extent of carp gudgeons in 2010–2013 and 2020/21 was significantly lower compared to all sampling rounds from 2014 to 2020 ($p < 0.01$).
- Murray–Darling rainbowfish's (*Melanotaenia fluviatilis*) contemporary spatial extent was also highly variable between rounds (39% to 75%), and no persistent trends were found. The largest contemporary extent for the species was 75% in 2016/17, which

was significantly higher than the smallest extents of 2010–2013 and 2016/17 ($p < 0.01$) (Figure 2b).

- Southern pygmy perch (*Nannoperca australis*) decreased from a three-round average spatial occurrence of 68% at the start of the data set to an average of 53% over the final three rounds (Figure 2c), but the trend was not statistically significant ($p > 0.05$). The lack of significance and low effect size for this species can be partially attributed to wider confidence intervals because of the low number of sub-catchments that it was collected in (Table 1).
- Unspecked hardyhead (*Craterocephalus stercusmuscarum fulvus*) was generally less variable than the other short-lived species between years (46% to 66%) and averaged a contemporary extent of 55% for the study period. The effect size was non-estimable.
- Australian smelt's (*Retropinna semoni*) contemporary spatial extent was between 37% and 52% during the first nine sampling rounds but increased to 72% in 2021/22, which was significantly ($p < 0.01$) greater than in 2007–2013, 2017/18, or 2019–21 (Figure 2e).

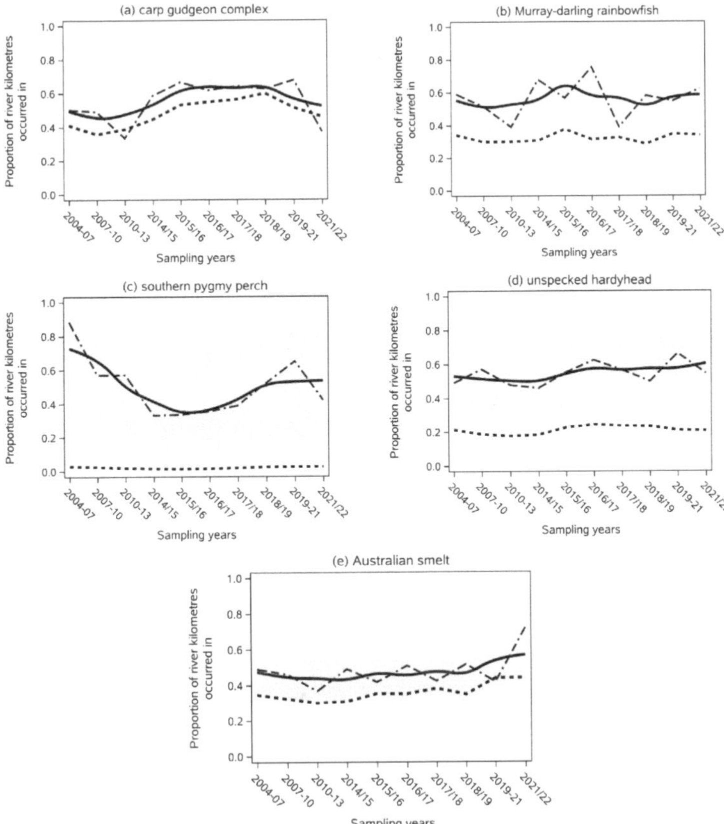

Figure 2. The proportion of contemporary river kilometres inhabited by five common and short-lived native fish species in the MDB from 2004 to 2022. The dashed line is the estimated proportion of contemporary river kilometres for each sampling round. The solid line is the 3-round moving average of occurrence within current river kilometres, and the grey shade indicates the 3-round moving average 95% confidence interval. The lower dotted line is an estimate of spatial extent relative to the species' historical distribution. The proportions are relative to each species' contemporary or historic distributions, not the entire MDB.

When considering the overall status estimates compared to historical records (lower dotted lines on Figure 2), the carp gudgeon complex (Figure 2a) had the highest spatial extent, 40–60%, of historical river kilometres compared to the other short-lived species throughout the study. Southern pygmy perch was regularly only collected in less than 5% of its historical river kilometres (Figure 2c).

3.3. Intermediate-Lived Native Fish Species

- Bony herring (*Nematalosa erebi*) had the highest contemporary spatial extent index of all native species and varied between 62 and 92% between sampling rounds (Figure 3a). The index achieved 92% in 2016/17 and 2021/22, which was significantly higher ($p < 0.01$) than the lower scores in 2017/18 to 2020/21 (Figure 3a).
- Flathead gudgeon (*Philypnodon grandiceps*) was very consistent in estimated contemporary spatial extent, averaging 35% throughout time (Figure 3b).
- Mountain galaxias averaged 45% and its peak extent of 78% in 2014/15 was significantly ($p < 0.01$) greater than that from 2010 to 2013 and in 2021/22 (Figure 3c).
- Northern river blackfish (*Gadopsis marmarata*) (average 65%) underwent slight but non-significant declines in their current extent in the last few years of the data (Figure 3d).
- Spangled perch (*Leiopotherapon unicolor*) was highly variable, occurring in between 21 and 80% of contemporary river kilometres in the study (Figure 3e). It had a significantly lower extent in 2004–2010 and 2018/19 than in 2019–2020 ($p < 0.01$) or 2010–2013 and 2021/2022 ($p < 0.02$).
- Two-spined blackfish (*Gadopsis bispinosa*) was also relatively consistent in contemporary extent, estimated to occur in ~70% of contemporary river kilometres throughout the study (Figure 3d). It also underwent slight but non-significant declines in its current extent in the last four rounds of the study (Figure 3d).

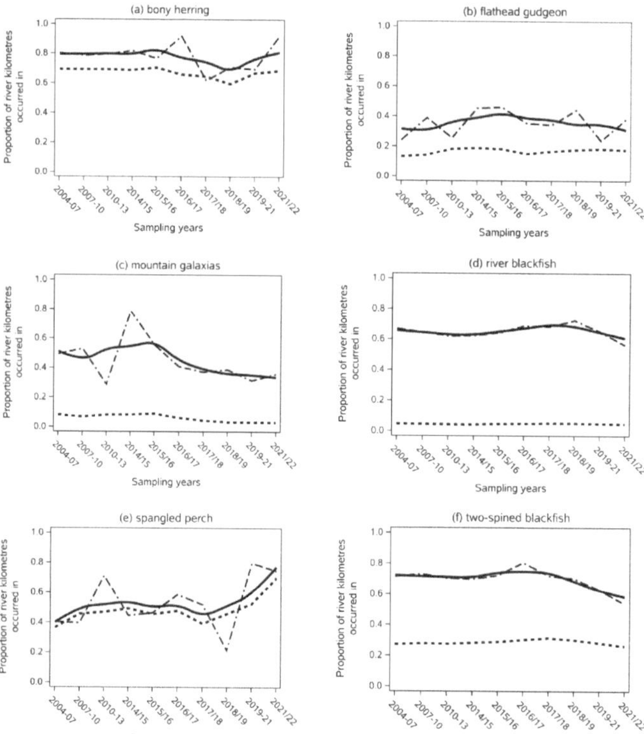

Figure 3. The proportion of contemporary river kilometres inhabited by six common and intermediate-lived native fish species in the MDB from 2004 to 2022. The dashed line is the estimated proportion

of contemporary river kilometres for each sampling round. The solid line is the 3-round moving average of occurrence within current river kilometres, and the grey shade indicates the 3-round moving average 95% confidence interval. The lower dotted line is an estimate of spatial extent relative to the species' historical distribution. The proportions are relative to each species' contemporary or historic distributions, not the entire MDB.

Compared to historic distributions, bony herring and spangled perch were the most widespread and were consistently collected in more than 60% and 40% of historical river kilometres, respectively (Figure 3). The other four intermediate-lived species were collected in less than 20% of their historical river kilometres, and notably, northern river blackfish and mountain galaxias were only collected in less than 5% of their historical river kilometres.

3.4. Long-Lived Native Fish Species

- Freshwater catfish (*Tandanus tandanus*) averaged 29% for the extent index throughout the study, with occasional dips to 17% and highs up to 51% (Figure 4a). The high variability between year-to-year estimates of spatial extent was high and the long-term average was low; consequently, the effect size for this species is more than 100% of the mean (Table 1).
- The contemporary spatial extent for golden perch averaged 68% and there was a visible but not significant overall increase in extent throughout the study (Figure 4b). The extent in 2016/17 (80%) and 2021/22 (84%) was significantly higher ($p < 0.01$) than in 2004–2007 (50%), 2007–2010 (61%), and 2017/18 (57%) (Figure 4b).
- Murray cod (*Maccullochella peelii*) averaged its contemporary distribution at 49% during the study (Figure 4c) and was significantly greater in extent ($p < 0.01$) from 2014 to 2017 than in 2010–2013 and 2018/19 (Figure 4c).
- Silver perch (*Bidyanus bidyanus*) was consistently low in river kilometres in which it was collected (average 19%) and had significantly low contemporary spatial extent from 2019 to 2021 (1%) compared to 2010–2013 (34%) and 2021/22 (32%) (Figure 4d).

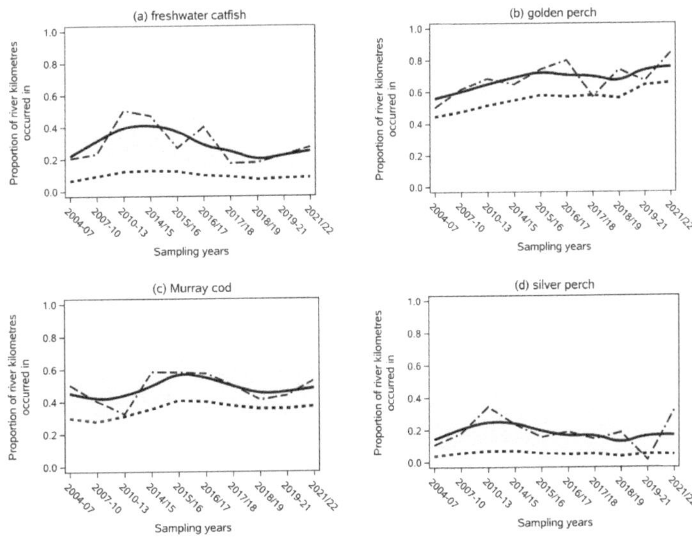

Figure 4. The proportion of contemporary river kilometres inhabited by four common and long-lived native fish species in the MDB from 2004 to 2022. The dashed line is the estimated proportion of contemporary river kilometres for each sampling round. The solid line is the 3-round moving average of occurrence within current river kilometres, and the grey shade indicates the 3-round moving average 95% confidence interval. The lower dotted line is an estimate of spatial extent relative to the species' historical distribution. The proportions are relative to each species' contemporary or historic distributions, not the entire MDB.

In comparison with known historical distributions, golden perch was collected in more than 50% of its historical river kilometres in every round after 2010, and Murray cod was consistently collected in 40–50% of its river kilometres throughout the 19-year study period (Figure 4). On the other hand, freshwater catfish and silver perch were only ever collected in less than 10% of their historic distribution (Figure 4).

3.5. Non-Native Species

- All of the non-native species had consistent extent distribution throughout the study, as indicated by narrow confidence intervals and smooth trend lines (Figure 5).
- Common carp was the most collected non-native species in the data set and was estimated to occur in between 74% and 90% of river kilometres throughout the study period (Figure 5b). Carp showed a consistent but non-significant ($p > 0.05$) increasing trend in extent throughout the monitoring period. Nevertheless, the final sampling round in 2021/2022 (90%) was significantly higher ($p < 0.01$) than the sampling rounds in 2004–2007, 2007–2010, and 2014/15 (Figure 5b).
- Eastern gambusia (*Gambusia holbrooki*) was detected in 60 of the 68 sub-catchments (Table 1), but it was rarely detected in more than 50% of its river kilometres in any sampling round and always between 35% and 62% of total river kilometres (Figure 5c). It was significantly higher ($p < 0.01$) in 2015/16 and 2016/17 than from 2004 to 2013 and in 2018/19 (Figure 5c).
- Goldfish's (*Carassius auratus*) spatial extent had several peaks and troughs between 25% and 51% and averaged at 38% of its river kilometres throughout all years (Figure 5d). Goldfish had a significant ($p < 0.01$) higher extent in 2007–2010 and 2016/17 than in 2014/15 and 2019–21 (Figure 5d).
- Redfin perch (*Perca fluviatilis*) (max 19%) and both brown and rainbow trout (*Salmo trutta* and *Oncorhynchus mykiss*) (<9%) generally occurred in low MDB river kilometres throughout the study period (Figure 5a,e,f). Redfin perch had a significantly lower spatial extent in 2014/15 than in the first and third sampling rounds (Figure 5f).

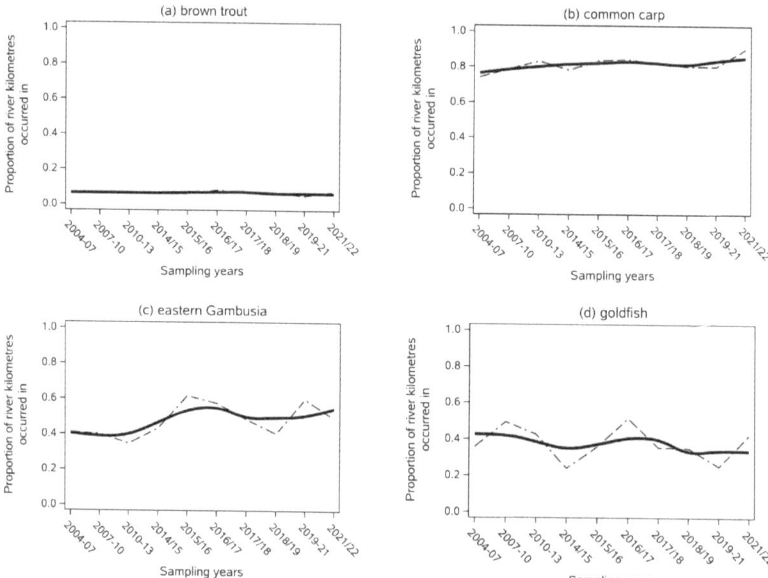

Figure 5. *Cont.*

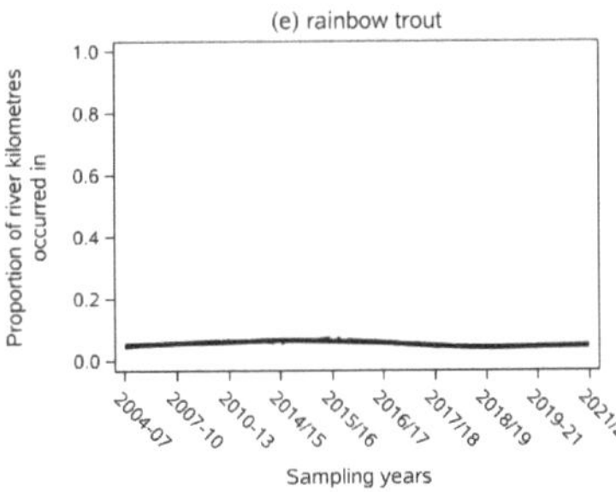
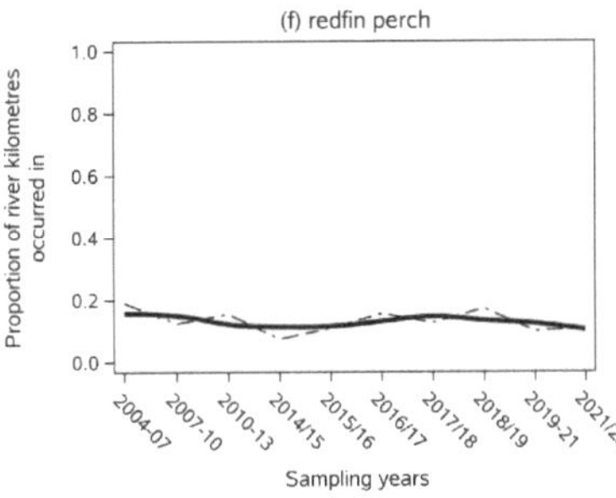

Figure 5. The proportion of total MDB riverine (>5 GL mean annual flow) river kilometres estimated to be inhabited by six common non-native fish species in the MDB from 2004 to 2022. The dashed line is the actual proportion for each sampling round. The solid line is the 3-round moving average and the grey shade indicates the 3-round moving average 95% confidence interval.

Overall, the non-native species appear stable in their current distributions, with only eastern gambusia and redfin perch showing some occasional inter-annual variability in spatial extent. Common carp is clearly the most widespread non-native species, followed by eastern gambusia and goldfish (Figure 5). Redfin perch and both trout species appear restricted in their ranges.

3.6. Species Summary of Baseline Assessments and Comparisons

As trends in all native species' spatial extents generally did not display consistent change throughout time, we compared their relative estimated average annual spatial extents during the monitoring program (Table 2). Spangled perch, carp gudgeon complex, bony herring, and golden perch all have better estimated contemporary distributions relative to their historical records than the other native species (Table 2). Native species that we assessed as having some reduction in spatial distribution, but with relatively stable current distributions, include Murray cod, Australian smelt, and Murray–Darling rainbowfish. The remaining native species have either (a) substantially reduced distributions, (b) are not a predominantly riverine species, or (c) are poorly sampled by electrofishing, or a combination of these factors.

Table 2. The average estimated spatial extent of 21 species in the Murray–Darling Basin between 2004 and 2022 using the SRA/MDBFS data set. The confidence in estimates reflects our perception of the sampling susceptibility of the species to electrofishing and whether it is predominantly a riverine species. The average extent for native species is the average percent of the historical (pre-1980) river kilometres that the species was collected in. For non-native species, the average extent is relative to the entire 64,000 km of stream kilometres in the sampling frame.

Species	Confidence in Estimate	Average Extent (%)	Interpretation of Extent
	Native Species—relative to pre-1980 distribution		
Short-lived			
unspecked hardyhead	Medium	21	Substantially declined in riverine habitats.
carp gudgeon complex	Medium	48	Widespread and abundant.
Murray–Darling rainbowfish	High	33	Has declined and is now patchily distributed.
southern pygmy perch	Low	2	Rare in main channel riverine habitats when using electrofishing.
Australian smelt	Medium	36	Widespread and abundant in lowland habitats.

Table 2. *Cont.*

Species	Confidence in Estimate	Average Extent (%)	Interpretation of Extent
	Native Species—relative to pre-1980 distribution		
Intermediate-lived			
two-spined blackfish	High	28	Declined and fragmented distribution.
northern river blackfish	Low	5	Declined significantly in larger streams but assessment confounded by historic taxonomy.
mountain galaxias	Medium	6	Greatly reduced, especially in lowland streams or where trout are present.
spangled perch	Medium	48	Widespread in Northern Basin but penetrate Southern Basin rarely.
bony herring	High	68	Widespread and abundant.
flathead gudgeon	Low	17	Poorly sampled in riverine habitats by using electrofishing.
Long-lived			
Murray cod	High	35	Has declined but widely distributed. Stocked
golden perch	High	56	Remain widespread in lowlands. Declined in uplands. Stocked.
freshwater catfish	Medium	10	Substantially declined in riverine habitats.
silver perch	Medium	5	Substantially declined in riverine habitats. Stocked
	Non-Native Species—relative to all MDB riverine habitats		
Short-lived			
eastern gambusia	Medium	48	Successful invader. Widely distributed.
Intermediate-lived			
goldfish	High	38	Successful invader in lowland rivers and some slope regions.
rainbow trout	High	5	Distributed widely in cool upland streams. Stocked.
Long-lived			
common carp	High	81	Highly successful and widely distributed.
redfin perch	High	13	Absent from warmer waters in Northern MDB.
brown trout	High	7	Widely distributed but restricted to cool upland streams. Stocked.

Of the non-native species, common carp are clearly the most successful invader and have a wider distribution than any native fish (Table 2). Eastern gambusia is also widespread, but not as well sampled by electrofishing and therefore does not demonstrate as wide a distribution as carp. Goldfish are widespread but less so than carp. The other three non-native species have restricted distributions in comparison (Table 2).

4. Discussion

4.1. Summary

The recognition of severely degraded native fish populations in Australia's Murray–Darling Basin in the early 2000's led to the setting up of a monitoring program to assess long-term cumulative changes in riverine fish populations at the entire MDB scale [4]. We used these large-scale data to investigate early 21st century trends in the spatial extent of 15 common native and 6 non-native riverine fish species. We found several common riverine native species displayed short-term fluctuations in extent of occurrence at the entire basin scale, but most species remained relatively stable in their contemporary distribution. All native species were collected in fewer sub-catchments compared to their pre-1980 distribution. Three non-native species—common carp, eastern gambusia, and goldfish—occurred in more sub-catchments than any single native species, but all also showed no consistent change in spatial distribution during the study period.

4.2. Species Trends

Short-lived native species generally had wider confidence intervals in their trend estimates, reflecting high year-to-year variability in their distributions, which likely resulted from their short life cycles, electrofishing sampling relative inefficiency, and a faster response to varying environmental conditions. Some smaller-bodied fish can be poorly sampled through electrofishing. Relevant to this study, carp gudgeons (CG) and flathead gudgeons (FHG) were poorly sampled in the pilot SRA (2004) where they were collected through electrofishing in 22 of 49 and 3 of 10 sites that they were known to occur in, respectively [19]. Southern pygmy perch (SPP) has an unknown susceptibility to electrofishing and is a cryptic species that favours non-riverine and/or densely vegetated habitats, and the results are treated cautiously in this riverine study. Even if there is higher sampling variability for these short-lived species, trends in estimated extent are assumed to reflect changes in actual extent, even if the extent estimates are conservative (less than actual). That is, after 10 sampling rounds, we believe that changes in estimated extent throughout time reflect true changes in extent and note that the assessments will be better understood as more years are sampled.

Carp gudgeons (CGs) are one of the most widespread and abundant taxa in the MDB and were detected in 47 of their 56 known historical sub-catchments. Only bony herring had more individual fish collected in the entire data set. Overall, we estimated that CG occurs in an average of 48% of their historic distribution in any sampling round, and because they are probably poorly sampled by electrofishing, this is quite a conservative estimate.

At the entire MDB extent, we estimated flathead gudgeon's (FHG) current extent to be at less than 20% of its historic distribution. Although our estimate is likely to be quite conservative, it was also collected in low numbers using the multiple gear types of the NSW Rivers Survey in the 1990s [37]. FHG may be recovering in some catchments [22], but confirmation of improved population status requires more targeted or flexible sampling [38] than the current data set.

Small-bodied species that were well-detected through electrofishing as compared to other methods in riverine habitats in the pilot SRA in 2004 include eastern gambusia (collected through electrofishing in 26 of 29 sites), Australian smelt (43 of 47), mountain galaxias (12 of 15), unspecked hardyhead (16 of 21), and Murray–Darling rainbowfish (28 of 28) [19]. The pilot was conducted in dry antecedent conditions and these species may be less well-sampled under higher antecedent flow conditions. Even so, we maintain confidence in comparing *relative* extent between these small-bodied species.

Mountain galaxias was deemed to have a very low overall occurrence (<7% of river kilometres) relative to its known historical distribution, even though this species is known to be widespread [22]. The very low historical extent estimate may reflect lower detectability of this small species using electrofishing in lowland rivers, which have the most river kilometres in the sampling frame. In upland sites, mountain galaxias, along with other small species, is known to be particularly susceptible to predation by species such as trout and redfin [39,40]. Other factors affecting its local abundance include drought and climate change [41].

Estimates of spatial extent for spangled perch were the most variable of all common species, but the species was found in up to 80% and 75% of its contemporary distribution in the final two rounds. This increase could be associated with the species' ability to rapidly colonise new areas following rainfall [42]. The species is common in the warmer, northern MDB and individuals have occasionally been recorded in the cooler Murray River system (Southern MDB) after extensive flooding in the northern Darling catchment [42], but they have not persisted in the Murray River.

Northern river blackfish's contemporary extent appears stable; however, the species is extremely sparsely distributed relative to its historical distribution. There is a possibility that some of the southern historic records refer to two-spined blackfish prior to its recognition and description in 1984 [43], but the two taxa prefer significantly different habitats [22] and potential confusion is minimal. Nevertheless, anecdotally, the species has disappeared from many the larger streams that are more impaired by river regulation, barriers, and coldwater pollution. There is real concern for MDB blackfish population persistence [44,45].

Golden perch and Murray cod are subject to both hatchery stocking and recreational harvest [46]. Golden perch is a highly mobile species, operating across large riverscapes over its life cycle, showing fast responses to extensive flooding events [47], and being collected more frequently when the sampling site had above average flows in the 3 months to 3 years prior to sampling [19]. Murray cod showed a decline in the estimated spatial extent in the third sampling round, from 2010 to 2013, but was very steady from 2014 onwards. The patterns observed in these two species using our extent index are consistent with recent population abundance estimates [20] that assessed data from multiple research programs within NSW only (including sites outside the MDB). These combined results point to a recovery in distribution for Murray cod after the 1990s [20,46]. The general trends for Murray cod, golden perch, and Murray–Darling rainbowfish were also similar to those recently predicted using population models in the southern portion of the MDB [48].

Freshwater catfish's overall spatial distribution remained fairly constant compared to its historic levels. In contrast, this species was found to be decreasing in average fish size and increasing in abundance from the 1990s through to the present by [20].

Silver perch was widespread historically but had declined over most of its range prior to the 1990s [22]. This highly mobile species has been badly affected by river regulation [49] as it relies on long stretches of river uninterrupted by weirs to maintain successful recruitment [50]. Only 9 silver perch were recorded using multiple sampling gear types in a two-year survey of 40 randomly selected sites in the NSW MDB in the mid-1990s [37]. In the SRA/MDBFS data sets, it was detected in just 21 sub-catchments across the entire 19 years, but typically only in about 20% of these in any sampling round. Compared to its historical distribution of 51 sub-catchments, it was collected in less than 10% of its historic spatial extent and in very low numbers in the northern MDB. We consider it the large-bodied native species in poorest condition relative to historical extent and it is listed as critically endangered nationally [51].

Common carp was the most frequent non-native species and estimated to occur in ~80% of MDB river kilometres. It was consistently widely distributed and was slightly more widespread in the most recent three sampling rounds compared to the first three rounds. Carp is more frequently collected when the preceding 3 months' or 3 years' flow levels are higher than average [19] and may be particularly more successful than the other non-native species because its populations respond to overbank flows and are favoured by some current water management practices [52,53]. The trend lines for the spatial extent indices for carp and golden perch were very similar, with both species generally increasing in extent throughout the study period but suffering slight decreases in 2014 and 2019. We suggest that this is because these two species respond in a similar fashion to flood events and flow in general, which can both operate at a basin-wide scale.

Eastern gambusia's extent fluctuated throughout, but it is still considered a very successful invader, occurring in ~48% of river kilometres on average in our data sets. It does not migrate and is a weak swimmer, relying on flooding, drift, and rapid breeding for range expansion. This small-bodied species is more likely to be collected when the site has low flow [19], presumably as it is more concentrated within the site and the water has lower turbidity. When present, it can be locally abundant and may be reasonably well collected through electrofishing for a small-bodied species, especially in small streams. However, anecdotally, we feel that this current spatial estimate is low because it was collected in more sub-catchments than any other species. Non-detections in some large lowland streams using large boat electrofishing gear may be spurious because of its habit of occupying shallow, shoreline vegetated areas. This species is a successful invader as it is tolerant to a wide range of water temperatures, oxygen levels, salinities, and turbidities (e.g., [54–56]). Its local abundance declines in the winter seasons in cooler parts of the basin.

Goldfish and redfin perch were less widely distributed and their spatial extent in the MDB have remained relatively constant throughout the first 20 years of this century. Both trout species are also stable in extent and are likely limited by their low tolerance to high water temperatures [22], occurring mostly in upland and montane streams. Hence, they are unlikely to expand their distribution from their current extent naturally because of increased temperatures and reduced water availability in the MDB through climate change [10]. Nevertheless, both species are stocked and could be translocated to new catchments by angler groups.

4.3. Surveillance Monitoring Returns Coarse Assessments

Not all MDB fish species were included in the analyses because small-order streams (<5 GL average annual flow), floodplains, and wetlands were not sampled, and rare, cryptic, and/or threatened species are not well assessed by generic monitoring methods. There are 64,000 km of riverine streams in the sampling frame, and when assessing species status and trends in fish species across the whole MDB, by necessity, generalisations must be made. The restrictions on stream size and the coarse sampling method lend themselves to sampling common riverine species, which may select many species likely to be more adaptable to environmental change and habitat degradation and hence show smaller changes in trends throughout time. They are, however, also likely to be impaired by spatially extensive events, such as the drying of waterways under the expected higher frequency of extreme events predicted from climate heating, and therefore provide important data on broader extents. Furthermore, multiple common riverine species included here, i.e., Murray cod, golden perch, common carp, silver perch, and river blackfish, have been the target of multiple interventions during the past 20 years [18,28] and their spatially extensive change is of interest. Given the widespread degradation of the MDB (and many other river basins worldwide), together with investments in rehabilitation activities, data sets and analyses such as the ones used in this study provide important baselines from which to measure improvements or further declines.

On the other hand, many native fish species in Australia have evolved to respond to 'boom and bust' climate fluctuations (e.g., [57]) and are known responders to rapid changes in habitat. In this study, we include several species known to expand in distribution via dispersal and spawning following flood events (e.g., [3,58]), but these expansions typically occur at finer scales such as in reaches or sub-catchments. We suggest that common carp and golden perch trends in spatial extent are similar in our study because they respond in a similar manner to flow events at any spatial extent. Inevitably, when looking at spatial patterns and temporal trends both within and among species at the basin scale, there are complex and intricate influencing factors that may be considered, but most are beyond the scope of this paper. The data collected here are not aimed toward identifying causes of change but merely to identify short-term and long-term trends and overall patterns. The lack of detection of sustained long-term change in extent for common riverine species at the MDB scale only means that the cumulative effect of multiple small-scale management interventions or local impacts are not yet discernible at this extent for those species. Assessment of finer-extent objectives requires targeted follow-up surveys for specific species at finer spatial scales.

4.4. Factors That Can Influence Population Extent at a Larger Scale

Apart from the cumulative effect of local-scale interventions, factors that can influence fish populations at the entire basin scale are typically climate- and flow-related. The MDB is a highly variable, semi-arid environment that can exhibit extremes in environmental conditions, especially flow rates and water quality. For example, the Millenium Drought of 1997–2010 [59] had significant impacts on freshwater fish populations and their habitats [60], including major fish kills and deteriorated wetland extent/quality affecting many species [12]. The sampling frame for the current data sets attempts to lessen drought influences on the collection of data by restricting sampling to permanent streams. But fish assemblages in these streams remain affected by climate change, reduced connectivity, fish kills, and changes in water quality well after such events. Alternately, high-flow years provide reproductive and relocation opportunities for many flow-responsive species such as golden and spangled perches [42] and common carp. Climate change is likely to exacerbate flood and drought frequencies and magnitudes [9,41,61], and there is need for future management adaptation and ongoing monitoring [11]. This highlights the need to consider scale, environment, and sampling conditions, as well as the ecology of individual species when interpreting survey results.

Given the size of the MDB, spatially extensive scale assessments are subject to spatial–environmental variations, and in some cases further, finer-scale (e.g., river reach) or targeted, individual species analyses over smaller distributions (e.g., for trout cod *Maccullochella maquariensis*) [62] may be needed. This may apply, in particular, in relation to assessing rehabilitation projects that may have been conducted, or for range-limited and fragmented rare or threatened taxa. For example, such studies may be needed for assessing the distribution of migratory or highly mobile species following the installation of fishways (see [63]).

We recognise that the methods used for general, spatially extensive condition monitoring may not be equally applicable to all species or habitats [31,33,64]. The sampling methods used here were designed to be consistent over large spatial extents (in this case, 1 M km^2) and to assess river kilometres, not individual fish species or fish populations. These monitoring programs are early-warning, surveillance-type programs, where there is no attempt to associate assessments to individual stressors or management actions [4]. Consequently, the methodology allows for the assessments to be representative at the MDB scale, but simultaneously, the interpretation of assessments is complex and subject to many confounding factors. Species extent estimations may vary throughout time for two reasons: (1) changes in each species true extent throughout time and (2) changes in factors that affect the calculations. Factors that could affect the true extent of each species include climatic (wet and dry years affecting flow levels), extreme event (bushfires), fishery management (stocking, harvest seasons, slot sizes, etc.), riverscape management (e.g., connectivity, habitat enhancement, or degradation) (e.g., [15,65–68]), and competition with non-native species [69]. Factors that could affect the extent estimates for each species include differences in susceptibility to electrofishing among species or different size classes (e.g., [31,64,70,71]), or antecedent conditions that influence sampling efficiency [31]. These sampling efficiency factors are of little concern with a long-term data set, however, because they do not create a bias in the trend assessments within a species or size class; (1) susceptibility to electrofishing between sizes and species remains constant—the assessments throughout time remain relative to each species, and (2) conditions also do not create a bias as they affect all species or surveys randomly across the 20 years and at the same time. In brief, a widespread, impaired sampling effect would potentially produce similar patterns in trends or changes in extent among multiple species, and this did not occur.

We acknowledge that finer-extent sampling is required to interpret finer-extent responses such as individual management actions or local climatic events. Conversely, it is clear that no common riverine species increased or decreased in its basin-wide spatial extent for a prolonged period during the study period.

5. Conclusions

We found no significant consistent trend or sustained long-term change in the spatial extent of any common riverine fish species in the entire MDB between 2004 and 2022. We believe this is a clear indication that any changes in common riverine fish species distribution are not occurring at the basin scale or for sustained periods of time.

It is important to note that these data come only from the past 20 years. The decline in many popular and well-known MDB fish species occurred well before this time period, 1950s for golden perch [72] and circa 1900 for Murray cod [73], providing an important example of the trap of 'shifting baselines' [74]. In that context, it should be recognised that populations and distributions of MDB native fishes were once much greater and more widespread than they are now, but we provide valuable, contemporary data over two decades for comparison with future monitoring of riverine native and non-native species via electrofishing. Across the 21 species, there were considerable differences in the variability of assessments. Species that we identified as having effect sizes > 50% of the mean should be subjected to increased sampling efforts if monitoring for changes in their extent becomes a priority. This should be expected for any monitoring program across multiple species and reinforces the need for long-term data sets to allow for a better understanding of inherent variability, in our case, in short-lived species especially.

Our results also highlight the challenges of variable population responses in highly variable environments across a large spatial scale. The detection of significant trends in highly variable ecosystems requires long-term data sets, and additional years of sampling and interpretation will add value to the existing data sets. This first 20 years of surveillance monitoring data have provided a valuable contribution to the assessment of the contemporary spatial extent of native and non-native fish populations in the MDB and have provided a baseline from which changes in the status of fishes, their protection, and population recovery or decline can occur. This is pertinent to river basins worldwide in the face of changes in hydrological regimes from climate change. The data sets use a well-designed probabilistic sample and, as such, can readily be supplemented with data collected as part of targeted monitoring programs (e.g., [75]). For example, targeted sampling (e.g., from wetland habitats, small streams, cryptic or rare species) that is informed by the ecology of each fish species may also improve the data sets. Similarly, the inclusion of other parameters that may be useful for management, such as indicators of population dynamics (reproduction, recruitment, disease prevalence, intensity, etc.) and being more predictive using population modelling (e.g., [76]), will add value.

These data and their analyses provide an important step forward to improving the management of native fishes of the MDB in Australia. The importance of long-term monitoring to guide and evaluate the benefits from the implementation of major water reforms under the Murray–Darling Basin Plan is essential. As many river basins throughout the world are under threat and have similarly reduced fish populations to the MDB, this approach is applicable to many river basins globally. Ongoing spatially extensive monitoring is also important to identify the occurrence and potential expansion of new non-native species, (e.g., oriental weatherloach *Misgurnus anguillicaudatus*) or to detect new incursions, especially by *Tilapia* species that currently occur in catchments close to the northern MDB [77].

Author Contributions: Conceptualisation, J.K. and M.L.; methodology, W.R., J.K. and M.L.; formal analysis, W.R.; investigation, W.R.; resources, M.L. and W.R.; writing—original draft preparation, W.R, J.K. and M.L. All authors have read and agreed to the published version of the manuscript.

Funding: This monitoring programs were funded by the Murray-Darling Basin Authority. This paper is an independent use of the data.

Institutional Review Board Statement: Not applicable. This research only performs analyses and interpretation of previously collected data.

Informed Consent Statement: Not applicable.

Data Availability Statement: The authors were not involved in the data collection process. The data used in this study are attributed to the Murray Darling Basin Authority, Commonwealth of Australia, Murray-Darling Basin Fish and Macroinvertebrate Survey, sourced 19 January 2023. The data are provided under a Creative Commons Attribution 3.0 Australia license. https://data.gov.au/data/dataset/murray-darling-basin-fish-and-macroinvertebrate-survey.

Acknowledgments: The data used in this study were generated by hundreds of fishery scientists and support staff from five state and territory jurisdictions. The monitoring programs came about through the vision of Peter Cullen AO and Don Blackmore more than two decades ago. The coordination of data collation and management was provided by Rhonda Butcher and Peter Cottingham. Shane Brooks collated and prepared the data prior to analyses. Improvements to early versions of the manuscript were made after review by Alison King. Deanna Duffy provided Figure 1.

Conflicts of Interest: The authors declare no conflicts of interest. The funders had no role in the design of the study; in the collection, analyses, or interpretation of the data; in the writing of the manuscript; or in the decision to publish the results.

References

1. Grill, G.; Lehner, B.; Thieme, M.; Geenen, B.; Tickner, D.; Antonelli, F.; Babu, S.; Borrelli, P.; Cheng, L.; Crochetiere, H.; et al. Mapping the world's free-flowing rivers. *Nature* **2019**, *569*, 215–221. [CrossRef] [PubMed]
2. Nilsson, C.; Reidy, C.A.; Dynesius, M.; Revenga, C. Fragmentation and flow regulation of the world's large river systems. *Science* **2005**, *308*, 405–408. [CrossRef] [PubMed]
3. Gehrke, P.C.; Brown, P.; Schiller, C.B.; Moffatt, D.B.; Bruce, A.M. River regulation and fish communities in the Murray-Darling River system, Australia. *Regul. Rivers Res. Manag.* **1995**, *11*, 363–375. [CrossRef]
4. Davies, P.E.; Harris, J.H.; Hillman, T.; Walker, K.F. The Sustainable Rivers Audit: Assessing river ecosystem health in the Murray-Darling Basin, Australia. *Mar. Freshw. Res.* **2010**, *61*, 764–777. [CrossRef]
5. Davies, P.E.; Harris, J.H.; Hillman, T.J.; Walker, K.F. *SRA Report 1: A Report on the Ecological Health of Rivers in the Murray–Darling Basin, 2004–2007*; Prepared by the Independent Sustainable Rivers Audit Group for the Murray–Darling Basin Ministerial Council; Murray–Darling Basin Commission: Canberra, Australian, 2008.
6. Walker, K.F.; Sheldon, F.; Puckridge, J.T. A perspective on dryland river ecosystems. *Regul. Rivers Res. Manag.* **1995**, *11*, 85–104. [CrossRef]
7. Kingsford, R.T. Ecological impacts of dams, water diversions and river management on floodplain wetlands in Australia. *Austral Ecol.* **2000**, *25*, 109–127. [CrossRef]
8. Lester, R.E.; Webster, I.T.; Fairweather, P.G.; Young, W.J. Linking water-resource models to ecosystem-response models to guide water-resource planning—An example from the Murray–Darling Basin, Australia. *Mar. Freshw. Res.* **2011**, *62*, 279–289. [CrossRef]
9. CSIRO. *Water Availability in the Murray-Darling Basin: A Report from CSIRO to the Australian Government*; CSIRO: Canberra, Australian, 2008.
10. Whetton, P.; Chiew, F. Chapter 12—Climate change in the Murray–Darling Basin. In *Ecohydrology from Catchment to Coast, Murray-Darling Basin, Australia*; Hart, B.T., Bond, N.R., Byron, N., Pollino, C.A., Stewardson, M.J., Eds.; Elsevier: Amsterdam, The Netherlands, 2021; Volume 1, pp. 253–274.
11. Prosser, I.P.; Chiew, F.H.S.; Stafford Smith, M. Adapting water management to climate change in the Murray–Darling Basin, Australia. *Water* **2021**, *13*, 2504. [CrossRef]
12. Sheldon, F.; Barma, D.; Baumgartner, L.J.; Bond, N.; Mitrovic, S.M.; Vertessy, R. Assessment of the causes and solutions to the significant 2018–19 fish deaths in the Lower Darling River, New South Wales, Australia. *Mar. Freshw. Res.* **2022**, *73*, 147–158. [CrossRef]
13. Legge, S.; Woinarski, J.C.Z.; Scheele, B.C.; Garnett, S.T.; Lintermans, M.; Nimmo, D.G.; Whiterod, N.S.; Southwell, D.M.; Ehmke, G.; Buchan, A.; et al. Rapid assessment of the biodiversity impacts of the 2019–2020 Australian megafires to guide urgent management intervention and recovery and lessons for other regions. *Divers. Distrib.* **2022**, *28*, 571–591. [CrossRef]
14. Rumpff, L.; Legge, S.M.; van Leeuwen, S.; Wintle, B.A.; Woinarski, J.C.Z. *Australia's Megafires: Biodiversity Impacts and Lessons from 2019–2020*; CSIRO Publishing: Clayton, Australia, 2023; p. 512.
15. Koehn, J.D.; Lintermans, M.; Copeland, C. Laying the foundations for fish recovery: The first 10 years of the Native Fish Strategy for the Murray-Darling Basin, Australia. *Ecol. Manag. Restor.* **2014**, *15*, 3–12. [CrossRef]
16. Koehn, J.D.; Lintermans, M. A strategy to rehabilitate fishes of the Murray-Darling Basin, south-eastern Australia. *Endanger. Species Res.* **2012**, *16*, 165–181. [CrossRef]
17. Murray-Darling Basin Authority. *The Water Act 2007—Basin Plan 2012*; Murray-Darling Basin Authority: Canberra, Australia, 2012.
18. Barrett, J.; Lintermans, M.; Broadhurst, B. *Key Outcomes of Native Fish Strategy Research: Technical Report*; Institute for Applied Ecology, University of Canberra: Canberra, Australia, 2013.
19. Robinson, W.A.; Lintermans, M.; Harris, J.H.; Guarino, F. *A Landscape-Scale Electrofishing Monitoring Program Can Evaluate Fish Responses to Climatic Conditions in the Murray–Darling River System, Australia*; American Fisheries Society Symposium; American Fisheries Society: Bethesda, MD, USA, 2019; Volume 90, pp. 179–201.
20. Crook, D.A.; Schilling, H.T.; Gilligan, D.M.; Asmus, M.; Boys, C.A.; Butler, G.L.; Cameron, L.M.; Hohnberg, D.; Michie, L.E.; Miles, N.G.; et al. Multi-decadal trends in large-bodied fish populations in the New South Wales Murray–Darling Basin, Australia. *Mar. Freshw. Res.* **2023**, *74*, 899–916. [CrossRef]
21. IUCN. *IUCN Red List Categories and Criteria*, 2nd ed.; IUCN: Cambridge, UK, 2012; p. 32.
22. Lintermans, M. *Fishes of the Murray–Darling Basin: An Introductory Guide*, 2nd ed.; Australian River Restoration Centre, Canberra: Canberra, Australia, 2023.
23. Kaminskas, S. Alien pathogens and parasites impacting native freshwater fish of southern Australia: A scientific and historical review. *Aust. Zool.* **2021**, *41*, 696–730. [CrossRef]
24. Britton, J.R. Contemporary perspectives on the ecological impacts of invasive freshwater fishes. *J. Fish Biol.* **2023**, *103*, 752–764. [CrossRef] [PubMed]
25. Feio, M.J.; Silva, J.P.; Hughes, R.M.; Aguiar, F.C.; Alves, C.B.M.; Birk, S.; Callisto, M.; Linares, M.; Macedo, D.R.; Pompeu, P.S.; et al. The impacts of non-native species on river bioassessment. In Proceedings of the XXII Congress of the Iberian Association of Limnology, Vigo, Spain, 23–28 June 2024.
26. MDBA. *Delivering a Healthy Working Basin*; About the Draft Basin Plan; Murray–Darling Basin Authority: Canberra, Australia, 2011.
27. MDBA. *Guide to the Proposed Basin Plan: Overview*; Murray-Darling Basin Authority: Canberra, Australia, 2010; p. 223.

28. Koehn, J.D. Managing people, water, food and fish in the Murray–Darling Basin, south-eastern Australia. *Fish. Manag. Ecol.* **2015**, *22*, 25–32. [CrossRef]
29. Lesser, V.M.; Kalsbreek, W.D. A comparison of periodic survey designs employing multi-stage sampling. *Environ. Ecol. Stat.* **1997**, *4*, 117–130. [CrossRef]
30. Lintermans, M.; Robinson, W.; Harris, J.H. Identifying the health of fish communities in the Murray-Darling Basin: The Sustainable Rivers Audit. In Proceedings of the 4th Australian Stream Management Conference: Linking Rivers to Landscapes, Launceston, Australia, 19–22 October 2004; p. 6.
31. Lyon, J.P.; Bird, T.; Nicol, S.; Kearns, J.; O'Mahony, J.; Todd, C.R.; Cowx, I.G.; Bradshaw, C.J.A. Efficiency of electrofishing in turbid lowland rivers: Implications for measuring temporal change in fish populations. *Can. J. Fish. Aquat. Sci.* **2014**, *71*, 878–886. [CrossRef]
32. Lyon, J.; Stuart, I.; Ramsey, D.; O'Mahony, J. The effect of water level on lateral movements of fish between river and off-channel habitats and implications for management. *Mar. Freshw. Res.* **2010**, *61*, 271–278. [CrossRef]
33. Lintermans, M. Finding the needle in the haystack: Comparing sampling methods for detecting an endangered freshwater fish. *Mar. Freshw. Res.* **2016**, *67*, 1740–1749. [CrossRef]
34. SAS Institute. *SAS/STAT®14.1, v.14.1*; SAS Institute: Cary, NC, USA, 2015.
35. Schmidt, D.J.; Bond, N.R.; Adams, M.; Hughes, J.M. Cytonuclear evidence for hybridogenetic reproduction in natural populations of the Australian carp gudgeon (*Hypseleotris*: Eleotridae). *Mol. Ecol.* **2011**, *20*, 3367–3380. [CrossRef] [PubMed]
36. Thacker, C.E.; Geiger, D.L.; Unmack, P.J. Species delineation and systematics of a hemiclonal hybrid complex in Australian freshwaters (Gobiiformes: Gobioidei: Eleotridae: *Hypseleotris*). *R. Soc. Open Sci.* **2022**, *9*, 220201. [CrossRef] [PubMed]
37. Harris, J.H.; Gehrke, P.C. (Eds.) *Fish and Rivers in Stress: The NSW Rivers Survey*; NSW Fisheries Office of Conservation and CRC for freshwater Ecology: Sydney, Australia, 1997; p. 298.
38. Dunn, C.G.; Paukert, C.P. A flexible survey design for monitoring spatiotemporal fish richness in nonwadeable rivers: Optimizing efficiency by integrating gears. *Can. J. Fish. Aquat. Sci.* **2020**, *77*, 978–990. [CrossRef]
39. Lintermans, M.; Geyle, H.M.; Beatty, S.; Brown, C.; Ebner, B.C.; Freeman, R.; Hammer, M.P.; Humphreys, W.F.; Kennard, M.J.; Kern, P.; et al. Big trouble for little fish: Identifying Australian freshwater fishes in imminent risk of extinction. *Pac. Conserv. Biol.* **2020**, *26*, 365–377. [CrossRef]
40. Lintermans, M. Recolonization by the mountain galaxias *Galaxias olidus* of a montane stream after the eradication of rainbow trout *Oncorhynchus mykiss*. *Mar. Freshw. Res.* **2000**, *51*, 799–804. [CrossRef]
41. Balcombe, S.R.; Sheldon, F.; Capon, S.J.; Bond, N.R.; Hadwen, W.L.; Marsh, N.; Bernays, S.J. Climate-change threats to native fish in degraded rivers and floodplains of the Murray–Darling Basin, Australia. *Mar. Freshw. Res.* **2011**, *62*, 1099–1114. [CrossRef]
42. Ellis, I.; Whiterod, N.; Linklater, D.; Bogenhuber, D.; Brown, P.; Gilligan, D. Spangled perch (*Leiopotherapon unicolor*) in the southern Murray-Darling Basin: Flood dispersal and short-term persistence outside its core range. *Austral Ecol.* **2015**, *40*, 591–600. [CrossRef]
43. Sanger, A. Description of a new species of *Gadopsis* (Pisces: Gadopsidae) from Victoria. *Proc. R. Soc. Vic.* **1984**, *96*, 93–97.
44. Unmack, P.J.; Sandoval-Castillo, J.; Hammer, M.P.; Adams, M.; Raadik, T.A.; Beheregaray, L.B. Genome-wide SNPs resolve a key conflict between sequence and allozyme data to confirm another threatened candidate species of river blackfishes (Teleostei: Percichthyidae: *Gadopsis*). *Mol. Phylogenetics Evol.* **2017**, *109*, 415–420. [CrossRef]
45. Hammer, M.P.; Unmack, P.J.; Adams, M.; Raadik, T.A.; Johnson, J.B. A multigene molecular assessment of cryptic biodiversity in the iconic freshwater blackfishes (Teleostei: Percichthyidae: *Gadopsis*) of south-eastern Australia. *Biol. J. Linn. Soc.* **2014**, *111*, 521–540. [CrossRef]
46. Forbes, J.; Watts, R.J.; Robinson, W.; Baumgartner, L.J.; McGuffie, P.; Cameron, L.M.; Crook, D. Assessment of stocking effectiveness for Murray cod (*Maccullochella peelii*) and golden perch (*Macquaria ambigua*) in rivers and impounds of south eastern Australia. *Mar. Freshw. Res.* **2015**, *67*, 1410–1419. [CrossRef]
47. Koehn, J.D.; Raymond, S.M.; Stuart, I.; Todd, C.R.; Balcombe, S.R.; Zampatti, B.P.; Bamford, H.; Ingram, B.A.; Bice, C.M.; Burndred, K.; et al. A compendium of ecological knowledge for restoration of freshwater fishes in Australia's Murray–Darling Basin. *Mar. Freshw. Res.* **2020**, *71*, 1391–1463. [CrossRef]
48. Yen, J.D.L.; Thomson, J.R.; Lyon, J.P.; Koster, W.M.; Kitchingman, A.; Raymond, S.; Stamation, K.; Tonkin, Z. Underlying trends confound estimates of fish population responses to river discharge. *Freshw. Biol.* **2021**, *99*, 1799–1812. [CrossRef]
49. Poomchaivej, T.; Robinson, W.; Ning, N.; Baumgartner, L.J.; Huang, X. Suitability of tropical river fishes for PIT tagging: Results for four Lower Mekong species. *Fish. Res.* **2024**, *272*, 106930. [CrossRef]
50. Tonkin, Z.; King, A.; Mahoney, J.; Morrongiello, J. Diel and spatial drifting patterns of silver perch Bidyanus bidyanus eggs in an Australian lowland river. *J. Fish Biol.* **2007**, *70*, 313–317. [CrossRef]
51. TSSC. Conservation Advice Bidyanus bidyanus (silver perch). Threatened Species Scientific Committee, Environment Australia. 2013. Available online: https://www.environment.gov.au/biodiversity/threatened/species/pubs/76155-conservation-advice.pdf (accessed on 6 May 2024).
52. Koehn, J.D. Carp (*Cyprinus carpio*) as a powerful invader in Australian waterways. *Freshw. Biol.* **2004**, *49*, 882–894. [CrossRef]
53. Todd, C.R.; Koehn, J.D.; Stuart, I.G.; Wootton, H.F.; Zampatti, B.P.; Thwaites, L.; Conallin, A.; Ye, Q.; Stamation, K.; Bice, C. Balancing beneficial environmental flows in rivers with controlling invasive common carp: A modelling approach to assessing flow-release strategies. *Biol. Invasions* **2024**, *26*, 1437–1456. [CrossRef]

54. Chervinski, J. Salinity tolerance of the mosquito fish, *Gambusia affinis* (Baird and Girard). *J. Fish Biol.* **2006**, *22*, 9–11. [CrossRef]

55. El-Boray, K. Tolerance of mosquitofish *Gambusia holbrooki* (Girard, 1859) to temperature, salinity and pH. *Aljouf Sci. Eng. J.* **2014**, *1*, 10–14. [CrossRef]

56. Pyke, G.H. Plague minnow or mosquito fish? A review of the biology and impacts of introduced *Gambusia* species. *Annu. Rev. Ecol. Evol. Syst.* **2008**, *39*, 171–191. [CrossRef]

57. Arthington, A.H.; Balcombe, S.R. Extreme flow variability and the 'boom and bust' ecology of fish in arid-zone floodplain rivers: A case history with implications for environmental flows, conservation and management. *Ecohydrology* **2011**, *4*, 708–720. [CrossRef]

58. O'Connor, J.P.; O'Mahony, D.J.; O'Mahony, J.M. Movements of *Macquaria ambigua*, in the Murray River, south-eastern Australia. *J. Fish Biol.* **2005**, *66*, 392–403. [CrossRef]

59. van Dijk, A.I.J.M.; Beck, H.E.; Crosbie, R.S.; de Jeu, R.A.M.; Liu, Y.Y.; Podger, G.M.; Timbal, B.; Viney, N.R. The millennium drought in southeast Australia (2001–2009): Natural and human causes and implications for water resources, ecosystems, economy, and society. *Water Resour. Res.* **2013**, *49*, 1040–1057. [CrossRef]

60. Lennox, R.J.; Paukert, C.P.; Aarestrup, K.; Auger-Méthé, M.; Baumgartner, L.; Birnie-Gauvin, K.; Bøe, K.; Brink, K.; Brownscombe, J.W.; Chen, Y.; et al. One hundred pressing questions on the future of global fish migration science, conservation, and policy. *Front. Ecol. Evol.* **2019**, *7*, 286. [CrossRef]

61. Morrongiello, J.R.; Beatty, S.J.; Bennett, J.C.; Crook, D.A.; Ikedife, D.N.E.N.; Kennard, M.J.; Kerezsy, A.; Lintermans, M.; McNeil, D.G.; Pusey, B.J.; et al. Climate change and its implications for Australia's freshwater fish. *Mar. Freshw. Res.* **2011**, *62*, 1082–1098. [CrossRef]

62. Koehn, J.D.; Lintermans, M.; Lyon, J.P.; Ingram, B.A.; Gilligan, D.M.; Todd, C.R.; Douglas, J.W. Recovery of the endangered trout cod, *Maccullochella macquariensis*: What have we achieved in more than 25 years? *Mar. Freshw. Res.* **2013**, *64*, 822–837. [CrossRef]

63. Baumgartner, L.; Zampatti, B.; Jones, M.; Stuart, I.; Mallen-Cooper, M. Fish passage in the Murray-Darling Basin, Australia: Not just an upstream battle. *Ecol. Manag. Restor.* **2014**, *15*, 28–39. [CrossRef]

64. Ebner, B.C.; Thiem, J.D.; Gilligan, D.M.; Lintermans, M.; Wooden, I.J.; Linke, S. Estimating species richness and catch per unit effort from boat electro-fishing in a lowland river in temperate Australia. *Austral Ecol.* **2008**, *33*, 891–901. [CrossRef]

65. Crook, D.A.; O'Mahony, D.J.; Gillanders, B.M.; Munro, A.R.; Sanger, A.C.; Thurstan, S.; Baumgartner, L.J. Contribution of stocked fish to riverine populations of golden perch (*Macquaria ambigua*) in the Murray–Darling Basin, Australia. *Mar. Freshw. Res.* **2016**, *67*, 1401–1409. [CrossRef]

66. Zampatti, B.; Leigh, S. Effects of flooding on recruitment and abundance of golden perch (*Macquaria ambigua ambigua*) in the lower River Murray. *Ecol. Manag. Restor.* **2013**, *14*, 135–143. [CrossRef]

67. MDBC. *Native Fish Strategy. 2006–2007 Annual Implementation Report*; Murray-Darling Basin Commission: Canberra, Australia, 2008; p. 52.

68. Barrett, J. Introducing the Murray-Darling Basin Native Fish Strategy and initial steps towards demonstration reaches. *Ecol. Manag. Restor.* **2004**, *5*, 15–23. [CrossRef]

69. Manchester, S.J.; Bullock, J.M. The impacts of non-native species on UK biodiversity and the effectiveness of control. *J. Appl. Ecol.* **2000**, *37*, 845–864. [CrossRef]

70. Zalewski, M.; Cowx, I.G. Factors affecting the efficiency of electrofishing. In *Fishing with Electricity*; Cowx, I.G., Lamarque, P., Eds.; Blackwell Science: London, UK, 1990; pp. 89–111.

71. Penczak, T.; Głowacki, Ł. Evaluation of electrofishing efficiency in a stream under natural and regulated conditions. *Aquat. Living Resour.* **2008**, *21*, 329–337. [CrossRef]

72. Reid, D.; Harris, J.H. *New South Wales Inland Fisheries Data Analysis*; Fisheries Research and Development Corporation: Canberra, Australia, 1997; p. 64.

73. Dakin, W.J.; Kesteven, G.L. *The Murray Cod (Maccullochella Macquariensis (Cuv. et Val.)) Bulletin of New South Wales State Fisheries*; Department of Fisheries: Nelson Bay, Australia, 1938; pp. 1–18.

74. Humphries, P.; Winemiller, K.O. Historical impacts on river fauna, shifting baselines, and challenges for restoration. *BioScience* **2009**, *59*, 673–684. [CrossRef]

75. Stein, E.D.; Bernstein, B. Integrating probabalistic and targeted compliance monitoring for comprehensive watershed assessment. *Environ. Monit. Assess.* **2008**, *144*, 117–129. [CrossRef]

76. Todd, C.; Koehn, J.; Pearce, L.; Dodd, L.; Humphries, P.; Morrongiello, J. Forgotten fishes: What is the future for small threatened freshwater fish? Population risk assessment for southern pygmy perch, *Nannoperca australis*. *Aquat. Conserv. Mar. Freshw. Ecosyst.* **2017**, *27*, 1290–1300. [CrossRef]

77. Arthington, A.; McKay, R.; Russell, D.; Milton, D. Occurence of the introduced cichlid *Oreochromis mossambicus* (Peters) in Queensland. *Mar. Freshw. Res.* **1984**, *35*, 267–272. [CrossRef]

 *fishes*

Article

Seasonal and Daily Movement Patterns by Wels Catfish (*Silurus glanis*) at the Northern Fringe of Its Distribution Range

Kristofer Bergström *, Hanna Berggren, Oscar Nordahl, Per Koch-Schmidt, Petter Tibblin and Per Larsson

Ecology and Evolution in Microbial Model Systems, EEMiS, Department of Biology and Environmental Science, Linnaeus University, SE-391 82 Kalmar, Sweden; hanna.e.berggren@gmail.com (H.B.); oscar.nordahl@lnu.se (O.N.); info@ghosttackle.se (P.K.-S.); petter.tibblin@lnu.se (P.T.); per.larsson@lnu.se (P.L.)
* Correspondence: kristofer.bergstrom@lnu.se

Abstract: Fish behavior often varies across a species' distribution range. Documenting how behaviors vary at fringes in comparison to core habitats is key to understanding the impact of environmental variation and the evolution of local adaptations. Here, we studied the behavior of Wels catfish (*Silurus glanis*) in Lake Möckeln, Sweden, which represent a European northern fringe population. Adult individuals (101–195 cm, N = 55) were caught and externally marked with data storage tags (DSTs). Fifteen DSTs were recovered a year after tagging, of which 11 tags contained long-term high-resolution behavioral data on the use of vertical (depth) and thermal habitats. This showed that the catfish already became active in late winter (<5 °C) and displayed nocturnal activity primarily during summer and late autumn. The latter included a transition from the bottom to the surface layer at dusk, continuous and high activity close to the surface during the night, and then descent back to deeper water at dawn. During the daytime, the catfish were mainly inactive in the bottom layer. These behaviors contrast with what is documented in conspecifics from the core distribution area, perhaps reflecting adaptive strategies to cope with lower temperatures and shorter summers.

Keywords: peripheral population; *Siluriformes*; behavior; data storage tags; biologgers; apex predator; freshwater

Key Contribution: Adult Wels catfish in a northern fringe population exhibit distinct behavior compared to those in core habitats in Europe. Unexpectedly, the catfish did not seek the warmest habitats in the lake, which may reflect adaptations to the colder climate.

Citation: Bergström, K.; Berggren, H.; Nordahl, O.; Koch-Schmidt, P.; Tibblin, P.; Larsson, P. Seasonal and Daily Movement Patterns by Wels Catfish (*Silurus glanis*) at the Northern Fringe of Its Distribution Range. *Fishes* **2024**, *9*, 280. https://doi.org/10.3390/fishes9070280

Academic Editors: Robert L. Vadas, Jr. and Robert M. Hughes

Received: 11 June 2024
Revised: 5 July 2024
Accepted: 10 July 2024
Published: 14 July 2024

1. Introduction

Fish in lakes and rivers at high latitudes must cope not only with short summers but also with low winter temperatures, low light levels, and the formation of ice. Several species have adapted to these conditions, although some are regarded as warm-water fishes and are limited in foraging activity, growth, and reproduction by the low temperatures [1,2]. One such species is the Wels catfish (*Silurus glanis*, hereafter "catfish"), which is a relic from warm postglacial times in temperate ecosystems and naturally inhabits only three water systems in northern Europe, namely those in southern Sweden. Because the core distribution area for catfish is found in eastern, central, and southern Europe, these Swedish populations have been isolated for a long time [3]. Whether these isolated populations have adapted to the low temperatures or how their activity is affected by these cold conditions remain unknown.

Movement behavior allows fish to adjust to environmental changes and disturbances, as well as intra- and interspecific competition [4,5]. This includes fundamental abilities such as feeding, breeding, and predator avoidance [6], which govern fitness [7]. During climate change, movement can also be of vital importance for coping with unfavorable temperatures. The movement behavior of fish is consequently adapted to patterns of environmental variation at different temporal scales (i.e., seasonal or daily) that optimize growth, reproductive output, and survival [7–10]. The most prominent temporal pattern is the earth's 24 h rotation around its own axis making up day (light) and night (dark). Many biological and environmental patterns are linked to this diel rhythm, such as temperature, light, food availability, and migration. To be able to anticipate and prepare for rhythmic events, organisms are controlled by a biological clock to "be on time".

Biological clocks are controlled and adjusted by different factors. The most important factor is considered to be photoperiod [8,11], but temperatures [12] and access to food also help synchronize biological rhythms in both fishes and mammals [13–15]. The diel activity rhythm (diurnal, crepuscular, or nocturnal) in animals is species-specific but may also differ within species due to age, size, or social status [9,10]. Further, activity rhythms are not fixed but plastic, and individuals may change their rhythms over seasons, with food availability, habitats, or other environmental parameters [8,13,14,16].

In fish species like catfish, particularly for native Swedish populations, there are knowledge gaps regarding rhythms in movement and behavioral patterns on yearly and daily bases. Catfish are the largest freshwater fish in Europe. The species is well-documented to reach lengths of 2.7 m and weights of 130 kg in Europe [17], and rumored to reach very large sizes (length of 5 m). Catfish are apex predators with a varied diet and ability to adapt to new food resources [2,18–22]. There are several studies investigating various spatiotemporal behavioral patterns of catfish in Europe. Juvenile catfish have been shown to prefer nocturnal feeding in laboratory experiments, especially when in groups [13,23]. In the wild, results are varied, but a nocturnal predilection has been shown that varies over seasons and locations. For instance, juvenile catfish exhibited strict nocturnal patterns during spring, summer, and autumn in a Czech lake [24]. Another study of both juvenile and adult catfish found diurnal activity during winter and spring, all-hours activity during summer, and then nocturnal activity during autumn [25]. A recent radiotelemetry study in River Po, Italy also indicated increased nocturnal and movement activity during spring and summer compared to the rest of the year [26–28]. However, tracking energy usage patterns showed no consistent, detectable relation to the diurnal cycle [29]. In the Řimov Reservoir, catfish were observed shifting from deeper waters during periods with low temperatures to shallower waters in warmer seasons. Additionally, they displayed activity peaks during both cold and warm seasons [28]. Disregarding the time of day, several studies agree that catfish movements and activity increase with increasing temperature [24,25,30].

Few previous studies of catfish ecology have examined movement patterns in native wild populations using a high frequency of measurements throughout the year. The aim of this study was to investigate yearly and daily activity patterns in catfish in Lake Möckeln, which is a northern fringe habitat in Sweden where catfish have been isolated from the main distribution area since the warmer postglacial era 9500–8000 years ago. This isolation has led to distinct genetic differences for catfish in Lake Möckeln compared to populations in southern and central Europe [31,32]. Here, we aimed to increase our understanding of whether this northern fringe population of catfish displays different seasonal and daily activity patterns compared to those documented in the core distribution area, thus reflecting latitudinal variation in temperature and light conditions. We hypothesized that we would see strict nocturnal activity and minimal diurnal activity over the seasons. Secondly, we expected to see long periods of low activity during the months with low temperatures (November–April), and activity peaking during summer (June–July) when the temperature was highest. Particular interest centered around the onset of activity in spring and the activity patterns during catfish spawning season.

2. Materials and Methods

2.1. Study Area

This study was conducted in Lake Möckeln (56°39'48.9″ N 14°8'57.7″ E), situated in the upper part of the Helge River system in southern Sweden. Helge River harbors one of three native populations of catfish in Sweden, with a population size of 720 ± 80 mature individuals [33]. Lake Möckeln has an area of 46 km^2 and a mean depth of 3 m with a maximal depth of 12 m. The lake is a mesotrophic (Tot-P 23 ug/L [34]), brown water (130 mg Pt/L [34]) lake with an annual mean temperature of 9.4 °C [34] (1982–2020). Other fish predators present in the lake are pike (*Esox lucius*), perch (*Perca fluviatilis*), and pikeperch (*Sander lucioperca*), whereas the main prey species are roach (*Rutilus rutilus*), bream (*Abramis brama*), zope (*Abramis ballerus*), and silverbream (*Blicca bjoerkna*).

2.2. Fishing and Tagging

Fishing was conducted with longlines in the deepest part of the lake (12 m depth), with the objective of capturing and equipping adult catfish (>100 cm) with data storage tags (DST G5, and G5 PDST, Cefas, Suffolk, UK) to record the depth (every minute) and temperature (every 5 min). Longlines consisted of a floating mainline (400 m) with monofilament (1.2 mm) leaders (1–1.5 m) attached every 10–20 m [35]. Leaders were fitted with single treble hooks and baited with native cyprinids. After capture, a DST was attached to the catfish using a braided fishing line (0.35 mm) tied around the base of the first fin ray of the pectoral fin. During August 2018, catfish (N = 20, size range 105–195 cm) were captured and fitted with G5 PDSTs (pop-ups) that were programmed to release after one year. Additionally, another 35 catfish were tagged in August 2019 (size range 101–167 cm) with a combination of both a G5 DST and a radio transmitter (transmitter F1580, ATS, Isanti, MN, USA) to facilitate recovery of the tags. In addition, all individuals were measured to the nearest cm (total length) and weighed to the closest 0.1 kg (Berkley 50lbs/22 kg, Columbia, SC, USA or Steinberg Systems SBS-KW-300/100-O, Berlin, Germany). A passive integrated transponder (PIT, Biomark, Boise, Idaho, USA, 23 mm HDX) was also injected into the pelvic girdle, or the abdominal cavity, to allow identification of recaptures regardless of whether the DST stayed attached.

Several searches for detached DSTs were conducted in Lake Möckeln and adjacent streams (2019, 2020), with the aid of a radio receiver (ATS, Isanti, MN, USA, R410) and antenna (ATS, Isanti, MN, USA, 5 element Yagi). Data from all recovered DSTs were initially checked by plotting both depth and temperature against time to determine when the DST had detached from the catfish. The DST was determined to be free-floating when the depth was constant (at the water surface) and showed no further amplitude changes, and the temperature showed indications of being affected by direct sunlight through sharp peaks during the daytime.

2.3. Estimates of Daylength

Seasonal patterns of daylength in the study area were based on photo-active radiation (PAR) data downloaded from the STRÅNG archive (Swedish Meteorological and Hydrological Institute, SMHI, Norrköping, Sweden), covering the period 1 August 2018–1 September 2020 (https://strang.smhi.se, accessed on 5 October 2022). STRÅNG provides hourly estimates of several radiation parameters each day of the year. Here, we defined PAR > 0 as day and PAR = 0 as night, resulting in a daylength varying between 7 and 17 h throughout the year.

2.4. Data Handling and Statistical Analysis

To estimate catfish activity, we utilized information about vertical movement (i.e., vertical activity) by calculating the delta values of depth (m) between each timestamp (every minute). To assess seasonal and circadian patterns in activity, we calculated the mean delta values per month (from m/min, Figure 1A) and hour of each day (i.e., 00-23, henceforth TOD), respectively. The calculations were carried out for each catfish.

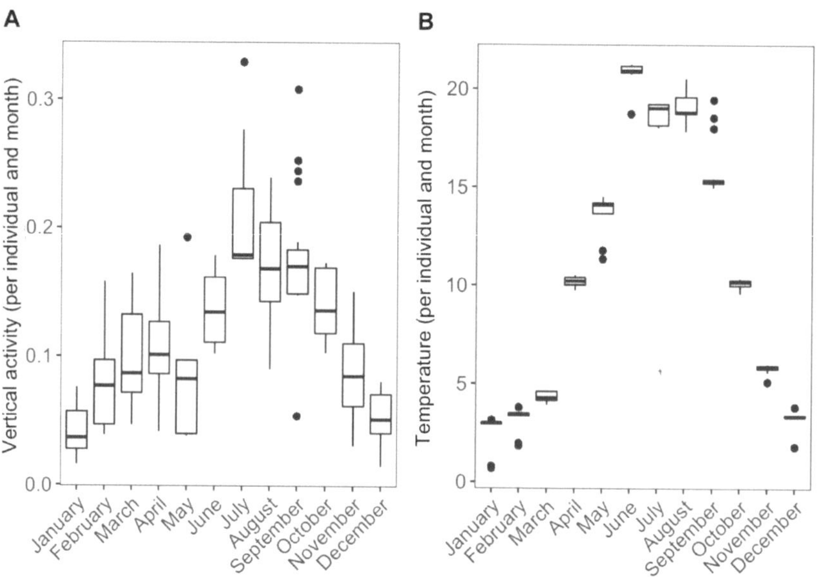

Figure 1. (**A**) The monthly vertical activity of catfish measured as delta values of depth change per time unit (m/min, recorded by DST). The boxes show the average activity for the 15 catfish. (**B**) The monthly mean water temperature recorded by DSTs in °C for the catfish (n = 15). Box-plot elements: center line: median; box limits: upper and lower quartiles; whiskers: 1.5× interquartile range; dots: outliers.

To visually explore seasonal patterns in daily activity (i.e., circadian rhythm), we plotted mean activity per hour of the day by calculating a monthly mean activity per TOD (Figure 2). Visual inspections of plots indicated that activity generally peaked at night (Figure 2). Consequently, we hypothesized that the circadian rhythm could be described with a cosine function peaking in activity at night. The cosine function to model circadian activity patterns was specified as follows: circadian rhythm = $\cos(2\pi t/\tau)$, where t is the time variable (TOD) and τ is the period of a cycle (i.e., 24 h). To investigate whether this variable could describe the hourly activity patterns (calculated for each individual and day), we performed a generalized linear mixed model (GLMM) with gamma distribution and a log-link function. Circadian rhythm and month were included as fixed explanatory variables with an interaction between them, and an individual was included as a random factor. This was performed with glmer in the lme4 package (v1.1-30) [36]. For this analysis, we added a constant (1) to the response variable because observations of zero activity cannot be used with gamma distribution [37].

Due to a significant interaction effect between the circadian rhythm and month, we analyzed each month separately using paired *t*-tests to evaluate if there were any significant differences in activity between hours categorized as night or day. Due to multiple testing, p-values were adjusted with the Bonferroni method in the p.adjust function in R. All data handling, statistical analysis, and graphics were performed with R (v. 4.2) [38] and RStudio (v2022.07.2) [39] using packages like dplyr (v1.0.10), ggplot2 (v3.3.6), ggpubr (v0.4.0), and lubridate (v1.8.0) [40–43]. Data and R-scripts are available in the Supplementary Materials (https://doi.org/10.5061/dryad.nk98sf82f).

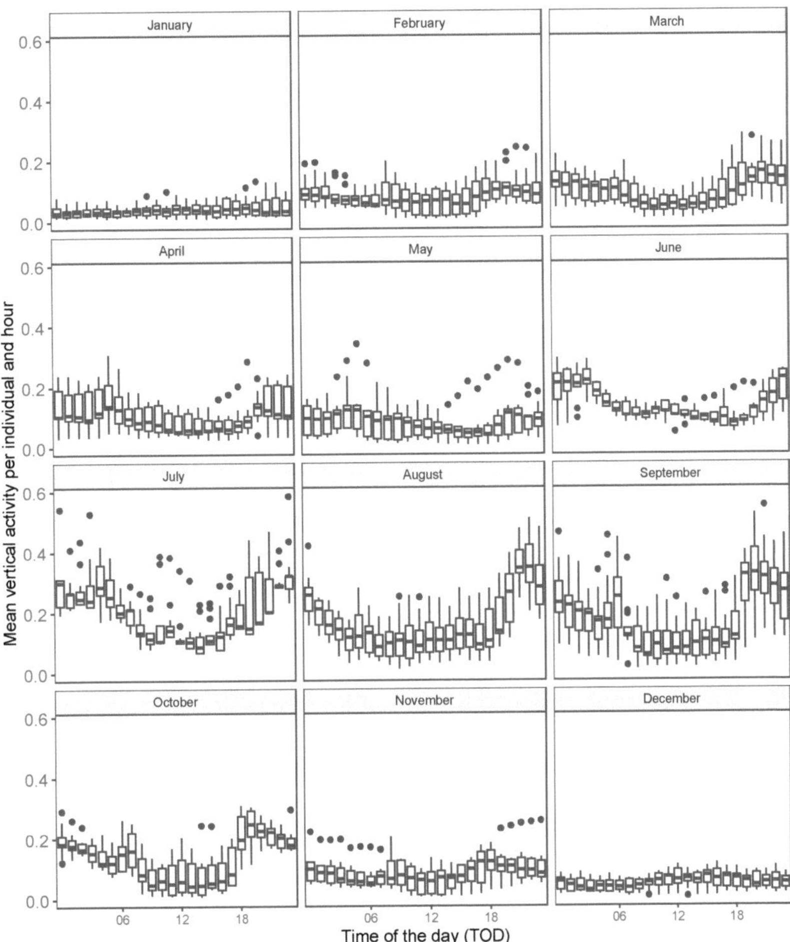

Figure 2. Vertical activity for catfish (delta values of depth m/min) per hour of the day for each month. Values plotted are a monthly average in activity for every individual and TOD (n ≤ 15 values per box). Box-plot elements: center line: median; box limits: upper and lower quartiles; whiskers: 1.5× interquartile range; dots: outliers.

3. Results

During 2018, catfish were tagged (N = 20) with a G5 PDST, and two of those loggers were recovered in 2019, which contained data for 260 days/individual. During the second tagging event in 2019, G5 DSTs with an attached radio transmitter were attached to catfish (n = 35), and 13 were recovered in 2020. In total, tags from 15 individuals (length 101 to 155 cm; seven males, six females, and two unidentified) were recovered containing temperature and depth data from 22 to 364 days. The recovery of pop-up DSTs was 10% after the first year, and the recovery of tags increased to 37% the second year after adding a radio transmitter. Of the 15 tags that were reclaimed, 11 tags represented data from a time span longer than 300 days while four DSTs contained less than one month of data.

The DST data showed marked variations in activity of the catfish during a full year. The mean vertical activity (delta values for depth changes) per individual and month, together with the ambient temperature experienced by the catfish, brought forth yearly activity patterns (Figure 1). High activity (0.198 to 0.162 m/min averaged per month) was recorded during the summer from July to August when water temperatures were about

20 °C. High activity (0.167 m/min) was maintained throughout September although the water temperatures decreased to 15 °C. The water temperatures (falling from 10 to 5 °C) and activity then decreased from October to November, from 0.140 to 0.087 m/min per month. During midwinter, in December and January, the catfish were inactive (0.050–0.041 m/min). Activity then continuously increased from February to April, starting at a water temperature <5 °C, from 0.083 to 0.108 m/min per month. In May, activity decreased to 0.085 m/min with increasing water temperatures but increased again to 0.138 m/min during June.

The GLMM revealed that catfish daily activity patterns varied across the year (the effect of interaction between the circadian rhythm and month: $\chi^2 = 2236.2$, $df = 11$, $p < 0.0001$; Figure 2). Pairwise comparisons between day and night, repeated for every month, showed that the catfish were significantly more active during the night compared to the day (when the length of day varied between 7 and 17 h) during the months of February–March and July–October (paired *t*-test, $p < 0.05$; Figure 3). From April to June, however, no differences in activity between night and day were recorded. In December and January, when the catfish were passive, no diurnal activity occurred. However, as overall activity increased in the following months, a pattern of higher activity during the night and lower during the day emerged. This pattern of high night activity then became more pronounced during July–October ($p < 0.01$), but the pattern disappeared in November (not significant).

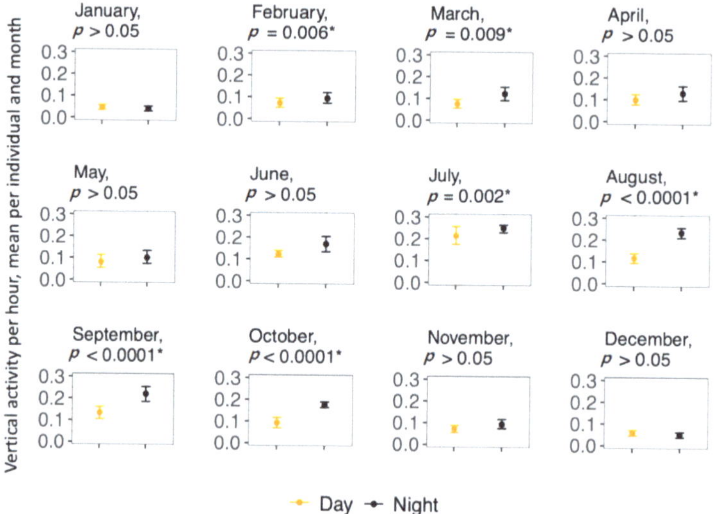

Figure 3. Monthly vertical activity (means) for catfish during the day versus night for the year. The length of day varied between 7 and 17 h for the full year. *p*-values from paired *t*-tests were adjusted with the Bonferroni method. Error bars denote 95% confidence intervals. * denotes statistically significant values.

The high nocturnal activity from summer to autumn coincided with a diurnal activity pattern exemplified in Figure 4. During the day, the catfish were passive in the bottom habitat of the deeper part of the lake (Figure 5). At dusk, the catfish started to move and ascended quickly from the bottom (within minutes) to the surface water. During the night, activity was high with movements up and down at a scale of a few meters, whereas at dawn, the catfish returned to the bottom water. There were no indications of long horizontal movements, e.g., to the shallow littoral habitats (a few hundred meters from the deep part of the lake), which would have been recorded on a DST as a continuous decrease in depth over a longer time span.

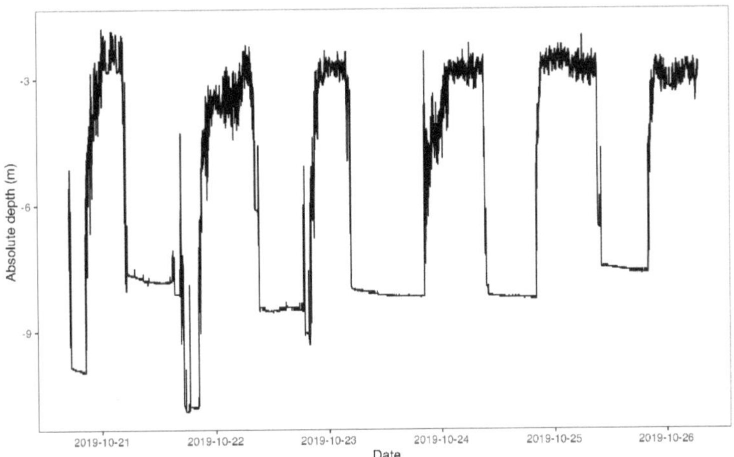

Figure 4. A representative example of the diurnal behavior of one adult catfish (male, length 155 cm, weight 19.9 kg) during the warmer parts of the year (here, in October). Grey areas indicate night hours, whereas white areas indicate hours of daylight. For this specific period, the daylength was 10 h.

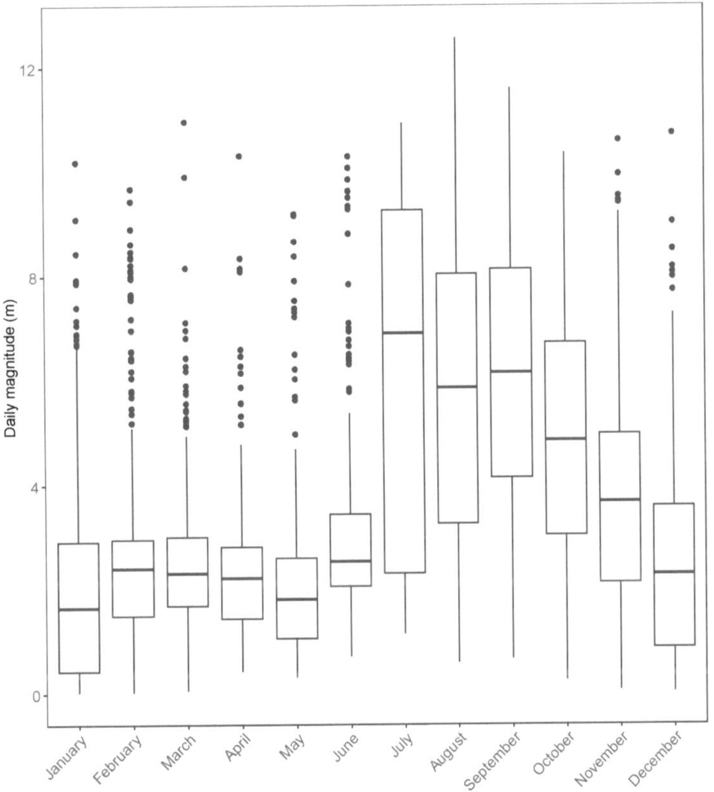

Figure 5. Boxplot showing daily depth magnitude of catfish plotted for each month (summarized for N = 15 individuals). Box-plot elements: center line: median; box limits: upper and lower quartiles; whiskers: 1.5× interquartile range; dots: outliers.

4. Discussion

4.1. Daily Activity

Catfish in the core distribution area of central and southern Europe are known to display a circadian rhythm, with night activity and day resting being the predominant pattern, although this may vary across seasons and locations [24–26,44]. We found a similar behavior in this northern population for parts of the year. In late fall and early winter (November to January), the overall activity of catfish was low with no differences registered between night and day. In February, a significant circadian rhythm could already be distinguished, and the pattern continued throughout March. From April to June, however, no differences in activity between night and day were recorded. The spring period was characterized by increasing water temperatures in the lake and increasing day length (from 13 to 17 h).

A general and daily behavior of adult catfish was evident from July until the end of October: an ascension from the bottom to surface water at dusk, continuous and high activity during night, and descension back to bottom water layers at dawn. Those activity periods were often initiated with an extensive vertical migration of 6–8 m followed by high activity of minor vertical movements in surface waters. During the daytime, the catfish were mainly inactive in the bottom layer. This behavior was repeated over longer time periods, and occasionally interrupted by inactive periods during the night. High activity thus occurred during the night (darkness) for the period July–October. It has been shown experimentally that Wels catfish use the lateral line to detect and pursue prey [45,46]. Swimming by prey creates water movements that remain briefly in the water. These "wakes" are detected by catfish; the prey is then followed and attacked [45,46]. This predation behavior is disturbed by obstacles in the water, e.g., bottom vegetation, stones, or sunken trees. Consequently, catfish most likely hunt and forage in the open water column where prey, such as zooplanktivorous zopes and other cyprinids, can be detected by the lateral line. We, therefore, suggest that the observed movement of adult catfish from the bottom up toward the surface at dusk reflects their main foraging behavior during the warmer season. The fast ascension that was followed by high activity at night in the presumed open water column for several hours, along with the continued high frequency of vertical movement of a lower magnitude, is a behavior that we interpreted as prey search and/or hunting. At dawn, the catfish returned to deeper water near the bottom and remained inactive until the next foraging cycle was initiated the following night. This detailed activity pattern has not been shown for catfish in their central distribution area. During no part of the year were catfish in this northern fringe population more active by day than night; the latter behavior being reported for winter and spring in River Berounka in the Czech Republic [25]. Such differences in behavior may stem from differences in foraging strategies, which could be linked to variations in prey type and behavior. Alternatively, these discrepancies might represent different adaptations to competing needs, like behavioral thermoregulation, foraging, and predator avoidance. The isolated Swedish population under study has long been separated from the core distribution area [3] and is genetically differentiated from other European populations [31,32]. This genetic divergence raises the possibility of adaptations to the colder climate, potentially influencing behaviors and physiological functions.

4.2. Seasonal Activity

The highest activity during the year, and thus the most pronounced foraging, occurred in warm summer months and early autumn (July to October). The warmest period, however, was June when the catfish showed considerably lower activity than in July. Because the lower activity in June was preceded by even lower activity in May, this may indicate a pre-spawning and/or spawning behavior of the catfish [29], which we discuss further below.

The lowest activity during the year was registered in December and January, which were the cold winter months when ice generally covered the lake, with water temperatures at or below 4 °C. In February, their activity increased despite temperatures still being low

(<5 °C), and this pattern continued during March and April at temperatures increasing up to 10 °C. The catfish were active in water temperatures below 5 °C, unlike conspecifics in the central distribution area of Europe, where catfish are reported to be "dormant" and inactive below 8–10 °C [2,47]. The demonstrated low-temperature activity in Swedish catfish may be adaptive for catfish in a northern population. Indeed, the ability to be active in lower temperatures increases the annual time window for foraging and energy gain for spawning and growth. Any trait that contributes to an ability to cope with long winters, with frequent ice cover events during the long lifetime of the catfish [1], will enhance survival and reproductive success. The main environmental variable affecting catfish in these areas—compared to the central distribution region—is the colder climate and thus a considerably lower annual water temperature. It is thus plausible to suggest that catfish in these peripheral, northern regions show adaptation to these harsh conditions, even though the catfish is defined as a warm-water species [2].

In the high water temperatures of July, the catfish once again displayed circadian rhythms with high night activity. This behavior included swimming from bottom to surface waters at dusk, foraging in free water during the night, and returning to bottom waters in the morning. This pattern of diel vertical migration is a known phenomenon among both freshwater and marine species, and is frequently demonstrated by planktivorous fish species [48,49]. It is plausible that the pattern of depth utilization observed here is indicative of catfish tracking the movements of their planktivorous prey. The circadian behavior with high nocturnal activity continued until the end of October and indicates intense foraging, as well as experiencing water temperatures decreasing from 20 to 10 °C. We suggest that the time span from July to October is the main foraging and growth period for catfish during the year. Catfish then became inactive for the coldest midwinter months when water temperatures were below 4 °C.

4.3. Do Early Summer Movement Patterns Coincide with Pre-Spawning Behavior?

The main population of adult catfish spawn in a small, shallow (<2 m), creek (Agunnarydsån, with a mean yearly water discharge of <2 m^3/s in the northern part of the lake [33], where the outlet opens into a shallow bay. Pre-spawning behavior includes a migration from central parts of the lake (where tagging took place) to the spawning creek, which is a distance of at least 5 km [33]. This is also supported by tagged-catfish presence in the spawning stream during parts of May–June, where four individuals marked with both a DST and PIT were detected at a PIT station (in operation for another study). When catfish aggregate in these areas before and during spawning, vertical movements are restricted by the shallow water, and their movement activity (as defined) is decreased. Moreover, catfish utilizing this shallow habitat for reproduction is corroborated by the absence of deeper descensions in the DST pressure data and the day–night fluctuation in temperature, which is characteristic of shallow, running water being cooled at night and warmed during the day. The decrease in activity during May–June compared to the following summer months indicate that this time of the year is devoted to pre-spawning and spawning behavior in shallower areas and in the creek. A similar reduction of activity from mid-May to the end of June was also observed by Říha et al., and hypothesized to be linked to spawning [28]. The behavior was not dominated by night activity. The time spent at the spawning grounds varied for the adult catfish, from a few days to months.

5. Conclusions

Our findings suggest that the behavior of adult Wels catfish in this northern fringe population differs from those in the core habitats of eastern, central, and southern Europe. This adds to a previous study finding that these fringe populations differ by growth rate and longevity [1]. These behavioral discrepancies likely arise from distinct thermal conditions and seasonal patterns. Consequently, those may shift as conditions change with global warming. The management of these peripheral populations should prioritize enabling their ability to adapt to changing environmental conditions, which includes ensuring connectivity within and among habitats to facilitate behavioral flexibility for successful feeding and reproduction. Our results suggest that the catfish do not, as would be expected from a warm-water species in a northern peripheral population, seek the warmest habitats of the lake, which most likely would be shallow, vegetated, and sheltered bays. Instead, the catfish displayed distinct behaviors and rhythms, depending on the season, which may reflect adaptations to the colder climate. The activity of the catfish had already started in late winter when the lake is generally frozen, and was so at the time of the study. This was followed by the main foraging period in the summer and early autumn when the catfish utilized the deeper bottom habitat diurnally and the free water toward the surface for nocturnal hunting and foraging. These findings were made possible through the rare use of in situ mark–recapture using data storage tags, which provided high-resolution data on focal catfish behavior and opened up new research to understand its underlying mechanisms and adaptive value.

Supplementary Materials: The following supporting information can be downloaded at: https://doi.org/10.5061/dryad.nk98sf82f (accessed on 2 July 2024).

Author Contributions: Conceptualization, K.B. and P.L.; methodology, K.B., P.K.-S., O.N. and P.L.; formal analysis, H.B. and O.N.; investigation, K.B., P.K.-S., O.N. and P.L.; resources, P.L.; data curation, H.B. and K.B.; writing—original draft preparation, K.B. and P.L.; writing—review and editing, K.B., H.B., P.T., O.N. and P.L.; visualization, H.B.; supervision, P.L. and P.T.; project administration, K.B. and P.L.; funding acquisition, P.L. All authors have read and agreed to the published version of the manuscript.

Funding: This research was funded by the County Administration Board of Kronoberg, and the strategic research program ECOCHANGE.

Institutional Review Board Statement: The study was conducted in accordance with the Declaration of Helsinki, and approval (Dnr 168677-2018) for this study was granted by the Ethical Committee on Animal Experiments in Linköping, Swedish Board of Agriculture, Sweden.

Data Availability Statement: Data are contained within the article and Supplementary Materials.

Acknowledgments: We thank Henrik Flink for assistance in the field, whereas Anders Johnson (Linnaeus University) provided valuable input on language.

Conflicts of Interest: The authors declare no conflicts of interest.

References

1. Bergström, K.; Nordahl, O.; Söderling, P.; Koch-Schmidt, P.; Borger, T.; Tibblin, P.; Larsson, P. Exceptional longevity in northern peripheral populations of Wels catfish (*Silurus glanis*). *Sci. Rep.* **2022**, *12*, 8070. [CrossRef] [PubMed]
2. Copp, G.H.; Britton, J.R.; Cucherousset, J.; García-Berthou, E.; Kirk, R.; Peeler, E.; Stakėnas, S. Voracious invader or benign feline? A review of the environmental biology of European catfish *Silurus glanis* in its native and introduced ranges. *Fish Fish.* **2009**, *10*, 252–282. [CrossRef]
3. Vinterstare, J.; Nathanson, J.E. Åtgardsprogram för mal. *Hav. Och. Vattenmyndigheten* **2017**, *33*, 2017.
4. Flink, H.; Tibblin, P.; Hall, M.; Hellström, G.; Nordahl, O. Variation among bays in spatiotemporal aggregation of Baltic Sea pike highlights management complexity. *Fish. Res.* **2023**, *259*, 106579. [CrossRef]
5. Nordahl, O.; Tibblin, P.; Koch-Schmidt, P.; Berggren, H.; Larsson, P.; Forsman, A. Sun-basking fish benefit from body temperatures that are higher than ambient water. *Proc. R. Soc. B Biol. Sci.* **2018**, *285*, 20180639. [CrossRef] [PubMed]
6. Nathan, R.; Monk, C.T.; Arlinghaus, R.; Adam, T.; Alós, J.; Assaf, M.; Baktoft, H.; Beardsworth, C.E.; Bertram, M.G.; Bijleveld, A.I.; et al. Big-data approaches lead to an increased understanding of the ecology of animal movement. *Science* **2022**, *375*, eabg1780. [CrossRef] [PubMed]
7. Benito, J.; Benejam, L.; Zamora, L.; García-Berthou, E. Diel cycle and effects of water flow on activity and use of depth by Common carp. *Trans. Am. Fish. Soc.* **2015**, *144*, 491–501. [CrossRef]
8. Reebs, S.G. Plasticity of diel and circadian activity rhythms in fishes. *Rev. Fish Biol. Fish.* **2002**, *12*, 349–371. [CrossRef]
9. Alanärä, A.; Burns, M.D.; Metcalfe, N.B. Intraspecific resource partitioning in brown trout: The temporal distribution of foraging is determined by social rank. *J. Anim. Ecol.* **2001**, *70*, 980–986. [CrossRef]
10. Brännäs, E. Temporal resource partitioning varies with individual competitive ability: A test with Arctic charr *Salvelinus alpinus* visiting a feeding site from a refuge. *J. Fish Biol.* **2008**, *73*, 524–535. [CrossRef]

11. Froland Steindal, I.A.; Whitmore, D. Circadian clocks in fish-what have we learned so far? *Biology* **2019**, *8*, 17. [CrossRef] [PubMed]
12. Volpato, G.L.; Trajano, E. Biological rhythms. *Fish Physiol.* **2005**, *21*, 101–153.
13. Bolliet, V.; Aranda, A.; Boujard, T. Demand-feeding rhythm in rainbow trout and European catfish. Synchronisation by photoperiod and food availability. *Physiol. Behav.* **2001**, *73*, 625–633. [CrossRef] [PubMed]
14. Boujard, T.; Leatherland, J. Circadian rhythms and feeding time in fishes. *Environ. Biol. Fishes* **1992**, *35*, 109–131. [CrossRef]
15. Mistlberger, R.E. Circadian food-anticipatory activity: Formal models and physiological mechanisms. *Neurosci. Biobehav. Rev.* **1994**, *18*, 171–195. [CrossRef]
16. Valdimarsson, S.K.; Metcalfe, N.B.; Thorpe, J.E.; Huntingford, F.A. Seasonal changes in sheltering: Effect of light and temperature on diel activity in juvenile salmon. *Anim. Behav.* **1997**, *54*, 1405–1412. [CrossRef]
17. Boulêtreau, S.; Santoul, F. The end of the mythical giant catfish. *Ecosphere* **2016**, *7*, e01606. [CrossRef]
18. Bouletreau, S.; Gaillagot, A.; Carry, L.; Tetard, S.; De Oliveira, E.; Santoul, F. Adult Atlantic salmon have a new freshwater predator. *PLoS ONE* **2018**, *13*, e0196046. [CrossRef]
19. Carol, J.; Benejam, L.B.; García-Berthou, E. Growth and diet of European catfish (*Silurus glanis*) in early and late invasion stages. *Fundam. Appl. Limnol.* **2009**, *174*, 317–328. [CrossRef]
20. Cucherousset, J.; Bouletreau, S.; Azemar, F.; Compin, A.; Guillaume, M.; Santoul, F. "Freshwater killer whales": Beaching behavior of an alien fish to hunt land birds. *PLoS ONE* **2012**, *7*, e50840. [CrossRef]
21. Syväranta, J.; Cucherousset, J.; Kopp, D.; Martino, A.; Cereghino, R.; Santoul, F. Contribution of anadromous fish to the diet of European catfish in a large river system. *Naturwissenschaften* **2009**, *96*, 631–635. [CrossRef] [PubMed]
22. Syväranta, J.; Cucherousset, J.; Kopp, D.; Crivelli, A.; Céréghino, R.; Santoul, F. Dietary breadth and trophic position of introduced European catfish *Silurus glanis* in the River Tarn (Garonne River basin), southwest France. *Aquat. Biol.* **2010**, *8*, 137–144. [CrossRef]
23. Boujard, T. Diel rhythms of feeding activity in the European catfish, *Silurus glanis*. *Physiol. Behav.* **1995**, *58*, 641–645. [CrossRef] [PubMed]
24. Daněk, T.; Horký, P.; Kalous, L.; Filinger, K.; Břicháček, V.; Slavík, O. Seasonal changes in diel activity of juvenile European catfish *Silurus glanis* (Linnaeus, 1758) in Byšická Lake, Central Bohemia. *J. Appl. Ichthyol.* **2016**, *32*, 1093–1098. [CrossRef]
25. Slavík, O.; Horký, P.; Bartoš, L.; Kolářová, J.; Randák, T. Diurnal and seasonal behaviour of adult and juvenile European catfish as determined by radio-telemetry in the River Berounka, Czech Republic. *J. Fish Biol.* **2007**, *71*, 101–114. [CrossRef]
26. Nyqvist, D.; Calles, O.; Forneris, G.; Comoglio, C. Movement and activity patterns of non-native Wels catfish (*Silurus glanis* Linnaeus, 1758) at the confluence of a large river and its colder tributary. *Fishes* **2022**, *7*, 325. [CrossRef]
27. Naughton, G.P.; Hogan, Z.S.; Campbell, T.; Graf, P.J.; Farwell, C.; Sukumasavin, N. Acoustic telemetry monitors movements of wild adult catfishes in the Mekong River, Thailand and Laos. *Water* **2021**, *13*, 641. [CrossRef]
28. Říha, M.; Rabaneda-Bueno, R.; Jarić, I.; Souza, A.T.; Vejřík, L.; Draštík, V.; Blabolil, P.; Holubová, M.; Jůza, T.; Gjelland, K.; et al. Seasonal habitat use of three predatory fishes in a freshwater ecosystem. *Hydrobiologia* **2022**, *849*, 3351–3371. [CrossRef]
29. Slavik, O.; Horky, P. Diel dualism in the energy consumption of the European catfish *Silurus glanis*. *J. Fish Biol.* **2012**, *81*, 2223–2234. [CrossRef]
30. Capra, H.; Pella, H.; Ovidio, M. Movements of endemic and exotic fish in a large river ecosystem (Rhône, France). In Proceedings of the 10th International Conference on Ecohydraulics, Trondheim, Norway, 22–26 June 2014.
31. Palm, S.; Vinterstare, J.; Nathanson, J.E.; Triantafyllidis, A.; Petersson, E. Reduced genetic diversity and low effective size in peripheral northern European catfish *Silurus glanis* populations. *J. Fish Biol.* **2019**, *95*, 1407–1421. [CrossRef]
32. Jensen, A.; Lillie, M.; Bergstrom, K.; Larsson, P.; Hoglund, J. Whole genome sequencing reveals high differentiation, low levels of genetic diversity and short runs of homozygosity among Swedish wels catfish. *Heredity* **2021**, *127*, 79–91. [CrossRef] [PubMed]
33. Nordahl, O.; Bergström, K.; Koch-Schmidt, P.; Söderling, P.; Sunde, J.; Skov, C.; Tibblin, P.; Larsson, P. Wels catfish (*Silurus glanis*)—An apex predator in northern peripheral habitats. 2023, Unpublished material intended for publication.
34. Miljödata, M.V.M. Swedish University of Agricultural Sciences (SLU). National Data Host Lakes and Watercourses, and National Data Host Agricultural Land. 2021. Available online: https://miljodata.slu.se/mvm/ (accessed on 16 May 2021).
35. Vejřík, L.; Vejříková, I.; Kočvara, L.; Blabolil, P.; Peterka, J.; Sajdlová, Z.; Jůza, T.; Šmejkal, M.; Kolařík, T.; Bartoň, D.; et al. The pros and cons of the invasive freshwater apex predator, European catfish *Silurus glanis*, and powerful angling technique for its population control. *J. Environ. Manag.* **2019**, *241*, 374–382. [CrossRef] [PubMed]
36. Bates, D.; Maechler, M.; Bolker, B.M.; Walker, S.C. Fitting Linear mixed-effects models using lme4. *J. Stat. Softw.* **2015**, *67*, 1–48. [CrossRef]
37. Lecomte, J.; Benoît, H.P.; Ancelet, S.; Etienne, M.; Bel, L.; Parent, E. Compound Poisson-gamma vs. delta-gamma to handle zero-inflated continuous data under a variable sampling volume. *Methods Ecol. Evol.* **2013**, *4*, 1159–1166. [CrossRef]
38. *R: A Language and Environment for Statistical Computing*; R Foundation for Statistical Computing: Vienna, Austria, 2013.
39. *RStudio: Integrated Development for R*; RStudio, Inc.: Boston, MA, USA, 2019.
40. Kassambara, A. ggpubr: 'ggplot2' Based Publication Ready Plots. *R Package Version* **2018**, 2.
41. Wickham, H. *Elegant Graphics for Data Analysis*; Springer: New York, NY, USA, 2016.
42. dplyr: A Grammar of Data Manipulation. Available online: https://dplyr.tidyverse.org/ (accessed on 14 October 2022).
43. Grolemund, G.; Wickham, H. Dates and times made easy with lubridate. *J. Stat. Softw.* **2011**, *40*, 1–25. [CrossRef]

44. Carol, J.; Zamora, L.; García-Berthou, E. Preliminary telemetry data on the movement patterns and habitat use of European catfish (*Silurus glanis*) in a reservoir of the River Ebro, Spain. *Ecol. Freshw. Fish* **2007**, *16*, 450–456. [CrossRef]
45. Pohlmann, K.; Atema, J.; Breithaupt, T. The importance of the lateral line in nocturnal predation of piscivorous catfish. *J. Exp. Biol.* **2004**, *207*, 2971–2978. [CrossRef]
46. Pohlmann, K.; Grasso, F.W.; Breithaupt, T. Tracking wakes: The nocturnal predatory strategy of piscivorous catfish. *Proc. Natl. Acad. Sci. USA* **2001**, *98*, 7371–7374. [CrossRef]
47. Ciolac, A. Migration of fishes in Romanian Danube River (N° 1). *Appl. Ecol. Environ. Res.* **2004**, *2*, 143–163. [CrossRef]
48. Mehner, T. Diel vertical migration of freshwater fishes—Proximate triggers, ultimate causes and research perspectives. *Freshw. Biol.* **2012**, *57*, 1342–1359. [CrossRef]
49. Zaret, T.M. *Predation and Freshwater Communities*; Yale University Press: New Haven, CT, USA, 1980; 187p.

MDPI AG
Grosspeteranlage 5
4052 Basel
Switzerland
Tel.: +41 61 683 77 34

Fishes Editorial Office
E-mail: fishes@mdpi.com
www.mdpi.com/journal/fishes